Selected Titles in This Series

190 **Alexander N. Starkov,** Dynamical systems on homogeneous spaces, 2000

189 **Mitsuru Ikawa,** Hyperbolic partial differential equations and wave phenomena, 2000

188 **V. V. Buldygin and Yu. V. Kozachenko,** Metric characterization of random variables and random processes, 2000

187 **A. V. Fursikov,** Optimal control of distributed systems. Theory and applications, 2000

186 **Kazuya Kato, Nobushige Kurokawa, and Takeshi Saito,** Number theory 1: Fermat's dream, 2000

185 **Kenji Ueno,** Algebraic Geometry 1: From algebraic varieties to schemes, 1999

184 **A. V. Mel′nikov,** Financial markets, 1999

183 **Hajime Sato,** Algebraic topology: an intuitive approach, 1999

182 **I. S. Krasil′shchik and A. M. Vinogradov, Editors,** Symmetries and conservation laws for differential equations of mathematical physics, 1999

181 **Ya. G. Berkovich and E. M. Zhmud′,** Characters of finite groups. Part 2, 1999

180 **A. A. Milyutin and N. P. Osmolovskii,** Calculus of variations and optimal control, 1998

179 **V. E. Voskresenskiĭ,** Algebraic groups and their birational invariants, 1998

178 **Mitsuo Morimoto,** Analytic functionals on the sphere, 1998

177 **Satoru Igari,** Real analysis—with an introduction to wavelet theory, 1998

176 **L. M. Lerman and Ya. L. Umanskiy,** Four-dimensional integrable Hamiltonian systems with simple singular points (topological aspects), 1998

175 **S. K. Godunov,** Modern aspects of linear algebra, 1998

174 **Ya-Zhe Chen and Lan-Cheng Wu,** Second order elliptic equations and elliptic systems, 1998

173 **Yu. A. Davydov, M. A. Lifshits, and N. V. Smorodina,** Local properties of distributions of stochastic functionals, 1998

172 **Ya. G. Berkovich and E. M. Zhmud′,** Characters of finite groups. Part 1, 1998

171 **E. M. Landis,** Second order equations of elliptic and parabolic type, 1998

170 **Viktor Prasolov and Yuri Solovyev,** Elliptic functions and elliptic integrals, 1997

169 **S. K. Godunov,** Ordinary differential equations with constant coefficient, 1997

168 **Junjiro Noguchi,** Introduction to complex analysis, 1998

167 **Masaya Yamaguti, Masayoshi Hata, and Jun Kigami,** Mathematics of fractals, 1997

166 **Kenji Ueno,** An introduction to algebraic geometry, 1997

165 **V. V. Ishkhanov, B. B. Lur′e, and D. K. Faddeev,** The embedding problem in Galois theory, 1997

164 **E. I. Gordon,** Nonstandard methods in commutative harmonic analysis, 1997

163 **A. Ya. Dorogovtsev, D. S. Silvestrov, A. V. Skorokhod, and M. I. Yadrenko,** Probability theory: Collection of problems, 1997

162 **M. V. Boldin, G. I. Simonova, and Yu. N. Tyurin,** Sign-based methods in linear statistical models, 1997

161 **Michael Blank,** Discreteness and continuity in problems of chaotic dynamics, 1997

160 **V. G. Osmolovskiĭ,** Linear and nonlinear perturbations of the operator div, 1997

159 **S. Ya. Khavinson,** Best approximation by linear superpositions (approximate nomography), 1997

158 **Hideki Omori,** Infinite-dimensional Lie groups, 1997

157 **V. B. Kolmanovskiĭ and L. E. Shaĭkhet,** Control of systems with aftereffect, 1996

156 **V. N. Shevchenko,** Qualitative topics in integer linear programming, 1997

155 **Yu. Safarov and D. Vassiliev,** The asymptotic distribution of eigenvalues of partial differential operators, 1997

(*Continued in the back of this publication*)

Dynamical Systems on Homogeneous Spaces

Translations of
MATHEMATICAL
MONOGRAPHS

Volume 190

Dynamical Systems on Homogeneous Spaces

Alexander N. Starkov

American Mathematical Society
Providence, Rhode Island

А. Н. Старков

ДИНАМИЧЕСКИЕ СИСТЕМЫ НА ОДНОРОДНЫХ ПРОСТРАНСТВАХ

Издательство ФАЗИС, Москва, 1999

Translated from the Russian by the author.

2000 *Mathematics Subject Classification.* Primary 37A17;
Secondary 11H55, 22E40, 37–02, 37A45.

ABSTRACT. The book provides an extensive survey of the theory of dynamical systems on homogeneous spaces (the so-called homogeneous flows). In particular, the author discusses ergodicity and mixing of homogeneous flows, the theory of unipotent flows (including recent proofs of the Raghunathan and Dani conjectures), and the dynamics of nonunipotent flows. The final chapter covers applications of homogeneous flows to number theory, mainly to the theory of Diophantine approximations. In particular, the author describes in detail the famous proof of the Oppenheim-Davenport conjecture using ergodic properties of homogeneous flows.

The book will be useful to graduate students and research mathematicians working in dynamical systems, ergodic theory, and number theory.

Library of Congress Cataloging-in-Publication Data

Starkov, Alexander N., 1960–
[Dinamicheskie sistemy na odnorodnykh prostranstvakh. English]
Dynamical systems on homogeneous spaces / Alexander N. Starkov.
p. cm. — (Translations of mathematical monographs, ISSN 0065-9282 ; v. 190)
Includes bibliographical references and index.
ISBN 0-8218-1389-7 (hardcover : alk. paper)
1. Flows (Differentiable dynamical systems) 2. Ergodic theory. 3. Number theory. I. Title. II. Series.
QA614.82.S7313 2000
515′.352—dc21 00-022001

∞ The paper used in this book is acid-free and falls within the guidelines
established to ensure permanence and durability.
Visit the AMS home page at URL: http://www.ams.org/

10 9 8 7 6 5 4 3 2 1 05 04 03 02 01 00

Contents

Introduction

In 1980, while a third year student at Moscow State University, the author was given[1] a rather simply formulated problem to solve. Take a connected Lie group G and a discrete subgroup Γ with the manifold G/Γ compact (such a subgroup Γ is called a *uniform lattice* in G). Now multiply Γ on the left by a one-parameter subgroup $g_{\mathbb{R}} \subset G$ and take the closure $\overline{g_{\mathbb{R}}\Gamma} \subset G/\Gamma$. The question: Is this closure a submanifold in G/Γ? If one lifts everything to the group G, the problem seems purely algebraic (and easy to solve if G is commutative or compact).

One can somewhat extend the problem and consider all *orbits* $g_{\mathbb{R}}x$, $x \in G/\Gamma$, of the action $(t, x) \to g_t x$, called a *homogeneous flow*. A more cautious formulation is the following: Is it true that the closure $\overline{g_{\mathbb{R}}x} \subset G/\Gamma$ is a manifold for a generic orbit (that is, for almost all $x \in G/\Gamma$)? This question is related to ergodic theory (an area about which the author then had only a rather vague idea). Experts advised him in this connection to read the book of Auslander, Green and Hahn [**AGH**] "Flows on homogeneous spaces", written in the early sixties. This is how the author found himself in the area that combines ergodic theory and Lie groups. As became clear later, the problem of orbit closures was already at the focus of attention of experts due to applications in number theory, but the main events were yet to follow.

Having no idea about the story, the author first decided to study the case of nilpotent Lie groups G, the next in complexity after the commutative case. Fortunately, the book [**AGH**] contained all the information necessary to solve the problem: Green's ergodicity criterion and a result due to Auslander saying that any ergodic nilflow $(G/\Gamma, g_{\mathbb{R}})$ is *minimal*, that is, all orbits are dense in G/Γ. An easy induction shows that in the nonergodic case, the closures $\overline{g_{\mathbb{R}}x}$, as for commutative groups G, are not only submanifolds but even *homogeneous subspaces* (that is, for every point $x \in G/\Gamma$ there exists a subgroup $F \subset G$ such that $\overline{g_{\mathbb{R}}x} = Fx$). The next class – compact extensions of nilpotent groups – revealed that all orbit closures are still smooth but not always algebraic.

To report the results obtained, the author came to (and became a constant participant in) the seminar on dynamical systems headed by D. V. Anosov and A. M. Stepin. It became clear at once that in the general case one cannot hope that all orbits are "good". As an example one takes either the geodesic flow on a compact surface of constant curvature -1 (here $G = \mathrm{SL}(2, \mathbb{R})$), or the suspension over the automorphism $\left(\begin{smallmatrix} 2 & 1 \\ 1 & 1 \end{smallmatrix}\right)$ of the torus $\mathbb{T}^2$ (this gives rise to a three-dimensional solvable group of exponential type). In both cases one obtains *Anosov flows*, the theory of which was developed by Anosov, Sinai, Smale, et al. Orbits of such flows were studied in detail by means of symbolic dynamics (see, for instance, [**Al**] or

[1] In connection with integration of Hamiltonian systems [**N, KK, MF**].

[**Bo79**]). Moreover, long ago, back in the twenties, Morse had constructed an orbit of the geodesic flow whose closure is nowhere locally connected (and locally looks like the product of a Cantor set and an interval)! (Note that the algebraic origin of the geodesic flow was first explored only in the fifties by Gelfand and Fomin [**GF**], and this led to the modern theory of dynamical systems on homogeneous spaces.)

Still, one could hope that the closure of a generic orbit is a manifold. This is supported by the easy observation that almost all orbits of an ergodic flow are everywhere dense. Moreover, it is natural to extend the class of homogeneous spaces under question by considering all spaces G/D of *finite volume*, that is, those carrying a smooth finite G-invariant measure (such spaces need not be compact, as is seen from the example $\mathrm{SL}(2,\mathbb{R})/\mathrm{SL}(2,\mathbb{Z})$). Thus, one has to study *nonergodic* flows on spaces of finite volume. It turned out that the ergodicity criterion for homogeneous flows was essentially known by that time. More precisely, in the sixties Moore [**Mo66**] examined the semisimple case, and Auslander [**Au**] the solvable case. Spectral invariants of ergodic flows in these cases were found by Stepin [**Ste69,73**] and Safonov [**Saf**]. The key step in reducing the general case to the semisimple and solvable cases was made by Dani [**Da77**]. This stage of the theory of homogeneous flows was summarized in the paper of Brezin and Moore [**BM**], where they found the ergodicity criterion and calculated the spectrum of homogeneous flows on the so-called admissible spaces of finite volume.[2] It remained to construct explicitly the *ergodic decomposition* of homogeneous flows. The fact that the space G/Γ breaks into closed invariant (not necessarily homogeneous) submanifolds such that the flow on each of them is ergodic relative to a smooth invariant measure was not difficult to prove [**St83**]. But even this implies that the closure of a generic orbit is smooth (more specifically, is equal to the corresponding ergodic submanifold). More surprising was that every ergodic manifold can be finitely covered (together with the flow thereon) by a homogeneous space of finite volume (see [**St86b,89**]). Therefore the theory of homogeneous flows was reduced to the ergodic case.

These results are expounded in a fairly detailed way in Chapter 1 of the present book. It is worth noting that the ergodicity criterion is based on the Mautner phenomenon [**Mo80**] for unitary representations; to establish the latter we apply a nice argument suggested by Margulis [**Ma91b**] which does not involve spectral theory. While studying homogeneous spaces we apply the methods of algebraic groups.

As for *individual orbits*, it became clear after the example with the geodesic flow that everything depends on the subgroup $g_{\mathbb{R}} \subset G$. On one hand, it may be the case that the action of $g_{\mathbb{R}}$ by left translations on the group G has at most a polynomial rate of divergence of close orbits relative to the right-invariant metric (then $g_{\mathbb{R}}$ is called a *quasi-unipotent* subgroup). Considering the adjoint action of g_t on the Lie algebra of the group G shows that such a subgroup must have all eigenvalues of absolute value 1 (if all eigenvalues are equal to 1, the subgroup $g_{\mathbb{R}}$ is called *unipotent*). On the other hand, the rate of divergence may be exponential (then $g_{\mathbb{R}}$ is said to be *partially-hyperbolic*). In this case the flow on G/Γ induced

[2] Still, the author was there in time to make "one or two lucky misimprovements" [**St86a,87a**]. In particular, he proved that every space of finite volume is admissible (see also [**Wu, Wi87, Z87**]). Thus, this subject was closed.

by $g_{\mathbb{R}}$ is *uniformly partially-hyperbolic*.[3] Such a flow possesses the *contracting and expanding* smooth foliations formed by orbits of the associated *horospherical* subgroups. For homogeneous actions, the techniques of partially-hyperbolic flows were already applied in the sixties by Auslander, Moore and Stepin; the general theory of smooth uniformly partially-hyperbolic flows was built in the late seventies by Pesin and Brin. An impression emerged that all orbit closures of a quasi-unipotent flow are always smooth (this was supported by examining the class of solvable groups G), whereas a partially-hyperbolic flow must have a "bad" orbit (see [**St87b**]).

In the spring of 1985, G. A. Margulis gave a talk at the seminar about a certain application of homogeneous flows to number theory, and clarified the situation with the orbit closures conjecture. First, any (not necessarily homogeneous) uniformly partially-hyperbolic flow on a compact manifold has an orbit which does not come back to a neighborhood of the original point (and hence its closure is not a manifold). Second, as followed from the talk, the situation with (quasi)unipotent flows is much more delicate. More precisely, in the seventies Raghunathan noticed a connection between unipotent orbits on $\mathrm{SL}(3,\mathbb{R})/\mathrm{SL}(3,\mathbb{Z})$ and the long-standing Oppenheim–Davenport conjecture on values of quadratic forms at integral points: the latter would be settled if one managed to prove that closures of all bounded orbits of the unipotent subgroup

$$u_t = \begin{pmatrix} 1 & t & t^2/2 \\ 0 & 1 & t \\ 0 & 0 & 1 \end{pmatrix}$$

are algebraic. The solution of this problem was precisely the subject of the talk (see [**Ma87**]). There is a more general conjecture (let us call it the *Raghunathan topological conjecture*, though it was formulated by Dani in [**Da81**]) stating that all orbits of a unipotent flow must have algebraic closures (that is, they must be homogeneous subspaces of finite volume).[4] Moreover, all ergodic measures of such a flow must have an algebraic origin (the *Dani measure conjecture*).

By that time, the case of a solvable group G was completely understood (see [**St84a**]). However, the most important affirmations of these conjectures come from the study of the *horocycle flow* induced by the subgroup $h_t = \left(\begin{smallmatrix} 1 & t \\ 0 & 1 \end{smallmatrix}\right) \subset G = \mathrm{SL}(2,\mathbb{R})$. Back in the thirties, Hedlund (by purely geometric methods) had proved that h_t has only periodic and everywhere dense orbits provided that the space G/Γ is of finite volume; moreover, the flow h_t is minimal if G/Γ is a compact space! The corresponding results on ergodic measures were obtained by Furstenberg [**Fu72**] and by Dani and Smillie [**DS**]. The conjectures were also supported by results of [**Bo76, V75, V77, EP**] on actions of horospherical subgroups on compact homogeneous spaces. Every horospherical subgroup is unipotent, but not vice versa: for instance, the main difficulty in the study of the flow $(\mathrm{SL}(3,\mathbb{R})/\mathrm{SL}(3,\mathbb{Z}), u_{\mathbb{R}})$ is that the subgroup $u_{\mathbb{R}}$ is not horospherical. In the general case, as was mentioned by Margulis, the Raghunathan–Dani conjectures seem to be very difficult to prove (especially the measure part).

[3] Experts in smooth dynamical systems will notice at once that the quasi-unipotent and partially-hyperbolic cases correspond to whether or not the entropy of the flow is trivial.

[4] Thanks to the above remark of Margulis, the Raghunathan conjecture easily implies the criterion for all orbit closures to be smooth [**St90**].

The difficulties in studying unipotent actions were already clear from an early result of Margulis [**Ma71**] saying that no unipotent trajectory in $\mathrm{SL}(n,\mathbb{R})/\mathrm{SL}(n,\mathbb{Z})$ exits to infinity (this was utilized in one of the proofs of his celebrated arithmeticity theorem). The proof was highly nontrivial (now it is more accessible – see [**DM90b, KM98**]), but the result did not even ensure that such a trajectory is recurrent (that is, comes back to any neighborhood of the original point)! Nevertheless, it proved to be one of the most important tools for studying dynamics of unipotent flows – especially after Dani [**Da84,86b**] established a general result of this kind stating that any unipotent trajectory on a space of finite volume visits a compact subset with positive density of times.

In the eighties and nineties the study of unipotent actions took the central position in the theory. Partial results (again for horospherical subgroups) were obtained in [**Da86a**] and [**St91**], and for unipotent subgroups of $\mathrm{SL}(3,\mathbb{R})$ in [**DM89,90a,b**]. On the other hand, in the early eighties Ratner discovered surprising rigidity properties of the horocycle flow [**R82a,b,83**]. For instance, a measure-theoretic isomorphism of two horocycle flows turned out to have an algebraic origin.[5] The proof utilized an H-property of the horocycle flow which made the polynomial nature of divergence more precise and, like the Margulis homecoming theorem, was based on certain properties of polynomials (subsequently Witte [**Wi85**], by exploring Ratner's idea, established the measure-theoretic rigidity of all ergodic unipotent flows). It was Ratner who by applying the so-called R-property of arbitrary unipotent flows (a modification of the H-property of the horocycle flow) managed to prove the measure conjecture in a series of papers [**R90a,b,c,91a**] whose length totals more than 100 pages. It is important to emphasize that the algebraicity of finite ergodic measures was proved in a much more general situation than it was originally conjectured: for *arbitrary* discrete subgroups $\Gamma \subset G$. Quite soon, by applying the measure conjecture and the Dani–Margulis results on homecoming, she managed to settle the topological conjecture as well [**R91b**] (simultaneously, Shah [**Sh91**], making use of [**R91a**], proved the conjecture for regular unipotent subgroups of semisimple groups of rank 1). In a sense, these two theorems of Ratner topped off the present stage of the theory of homogeneous flows. Margulis and Tomanov [**MT94,96**] (by using some ideas of Ratner as well as some from earlier papers of Dani and Margulis) suggested a more accessible proof of Ratner's measure theorem. Earlier Dani and Margulis [**DM93**] gave an alternative proof of the topological theorem.

In Chapter 2 we try to give an idea of what is involved in studying unipotent flows. The exposition starts with a detailed treatment of the geodesic and horocycle flows. Unfortunately, in the general case the proof of the measure conjecture is too complicated to reproduce within the framework of this book. To give its flavour, we examine the proof on the examples $G = \mathrm{SL}(2,\mathbb{R})$ and $G = \mathrm{SL}(2,\mathbb{C})$, following [**R92**] and [**MT96**] respectively. The proof of Margulis' homecoming theorem for unipotent trajectories is also demonstrated on the simplest example $G = \mathrm{SL}(2,\mathbb{R})$. The topological conjecture is proved in full scope along the lines of [**DM93**].

For one-parameter unipotent flows there is a stronger result: every trajectory not only has its closure homogeneous, but is uniformly distributed therein [**R91b**]. We expound various generalizations of the assertion which are due to Dani and Margulis [**DM93**], Ratner [**R94**], Eskin, Mozes and Shah [**MS**], [**Sh94**], [**EMS96,97**].

[5] For nilflows, the corresponding results were earlier proved by Parry [**P71,73**].

Essentially, all of them are concerned with the convergence in the weak* topology on the space of probability measures on G/Γ.

Of course, the proof of the Raghunathan–Dani conjectures gave a new stimulus to research in this area and served as a base for numerous applications. For instance, in the class of arbitrary one-parameter homogeneous flows one managed to settle Rokhlin's problem on multiple mixing and to better understand their minimal sets (Starkov [**St93,95a**]). There has been great progress in studying orbits and ergodic measures for actions of arbitrary subgroups $F \subset G$ (Ratner [**R94**], Mozes [**Moz95a**], Shah and Weiss [**Sh96, W, SW**], Katok and Spatzier [**KS96**], Margulis and Tomanov [**MT96**]).

Also, in Chapter 3 we discuss the following topics. After Dani [**Da85,86c**] found the Hausdorff dimension of the family of bounded orbits in a special case, the full solution of the problem was given by Kleinbock and Margulis [**KM96**]. Nice results on topological rigidity are due to Markus [**Mar83**], Benardete [**Ben**] and Witte [**Wi90**]. The structure of time changes for homogeneous flows is not well understood: we only mention results of Livshitz [**Liv**] and Katok and Spatzier [**KS94**] for Anosov actions, a classical theorem of Kolmogorov [**Ko**] for rectilinear flows on the 2-torus (subsequently developed by many authors) and a result of Ratner [**R78,79**] for horocycle flows. In recent times, the phenomenon of exponential mixing of certain homogeneous K-flows became one of the most powerful tools (Moore [**Mo87**], Ratner [**R87a**], Katok and Spatzier [**KS94**], Kleinbock and Margulis [**KM96,99**]).

It is worth emphasizing that, due to newly found *applications in number theory*, the interest in homogeneous flows rose considerably. For instance, at two recent International Congresses of Mathematicians three lectures were devoted to the theory of homogeneous flows, and two of them (those of Margulis at ICM-90 and of Ratner at ICM-94) were plenary ones. One of the first applications of homogeneous flows in number theory was given in [**AGH**], where nilflows were utilized to prove Weyl's theorem on uniform distribution of the fractional parts of values of polynomials. A classical result due to Kolmogorov [**Ko**] on time changes for rectilinear flows on the 2-torus reveals a connection with Diophantine properties of the rotation number. Dani [**Da85**] showed how Diophantine properties of reals affect the behavior of corresponding orbits of the geodesic flow. There exist numerous refinements of this connection; in particular, a well-known conjecture of Littlewood reduces to a certain (not yet established) statement on the structure of orbits of the Cartan diagonal subgroup on $\mathrm{SL}(3,\mathbb{R})/\mathrm{SL}(3,\mathbb{Z})$. After Margulis settled the Oppenheim–Davenport conjecture, Dani and Margulis [**DM93**], and also Eskin, Mozes and Margulis [**EMM**], found the exact asymptotics of the number of Diophantine solutions of the inequality $a < Q(x) < b$ for indefinite quadratic forms.

Kleinbock and Margulis [**KM98**] obtained remarkable results proving a general conjecture of Sprindzhuk [**Sp80**] on Diophantine approximation on manifolds. Skriganov [**Skr98**] used homogeneous flows to obtain asymptotic estimates of the number of lattice points inside polyhedra (the first results in the two-dimensional case go back to the twenties and are due to Hardy and Littlewood [**HL**] and Khintchine [**Kh23**]). Eskin, Mozes and Shah [**EMS96**] sharpened results of [**DRS**] and [**EM**] on the asymptotics of the number of lattice points on homogeneous manifolds in $\mathbb{R}^n$. All of these questions are discussed in Chapter 4. Before that, we give preliminaries from the theory of Diophantine approximation.

A considerable number of recently found applications show that the theory of homogeneous flows is far from being completed. The reader especially interested in number theory is referred to surveys of Borel [**B**] and Margulis [**Ma97**].

For various reasons, we do not touch upon certain subjects. These are Bernoulli properties of homogeneous flows, Markov partitions and symbolic dynamics (see [**Si68**], [**Kat**], [**Bo79**], [**OW**]); the Bernoullicity criterion was studied by Dani [**Da76a**] but in the general case it is not yet proved. All Lie groups throughout the book are real; hence we are not concerned with dynamical systems on homogeneous spaces over other local fields; we only note that the Raghunathan–Dani conjectures are settled in this setup as well (see [**R93,95a,98**] and [**MT94,96**]). Entropy theory will also play a minor role. We do not touch upon the theory of flows on spaces of infinite volume. Whereas the solvable case is completely understood [**St87b**], the case $G = \mathrm{SO}(1, n)$ presents an independent and rather delicate object of research in the theory of Fuchsian and Kleinian groups. Here the results are not definitive; see the book [**Ni**] and the survey [**St95b**].

Other subjects we try to present in as much detail as possible. In addition to [**AGH**], we intensively used the surveys [**Da96,99**], [**Gh**], [**Ma91b,97**], [**R84,95b**], [**SSS**], [**St97**]. The reader may consult the list of open problems given in [**Ma99**] to find directions for future research.

The author is very grateful to A. S. Mishchenko, whose question introduced him into the topic. Since then, the author had a happy opportunity to interact with many remarkable experts in ergodic theory and Lie groups: D. V. Anosov, S. Dani, A. B. Katok, G. A. Margulis, M. E. Ratner, A. M. Stepin, E. B. Vinberg. The author is also happy to acknowledge the great impact of conversations with A. Eskin, D. Kleinbock, S. Mozes, V. V. Ryzhikov, N. Shah, G. Tomanov, D. Witte, and many others.

The author was supported by the Russian Foundation of Basic Research (grant No 95-01-02804).

List of Notations

$\mathbb{Z}$ — the ring of integers
$\mathbb{N}$ — the semigroup of positive integers
$\mathbb{Q}$, $\mathbb{R}$, $\mathbb{C}$ — the fields of rational, real and complex numbers
G, H, F, etc. — Lie groups
$\mathfrak{g}$, $\mathfrak{h}$, $\mathfrak{f}$, etc. — Lie algebras
$\Gamma \subset G$ — discrete subgroup of a Lie group G
G_s — the semisimple splitting of a Lie group G
Ad — the adjoint representation
$\mathrm{Aut}(G)$ — the automorphism group of a Lie group G
$D_0 \subset G$ — the identity component of a closed subgroup $D \subset G$
$\overline{D} \subset G$ — the closure of a subset $D \subset G$
$g_{\mathbb{R}} \subset G$ — one-parameter subgroup of a Lie group G
$g^{\mathbb{Z}} \subset G$ — the cyclic group generated by an element $g \in G$
G/D — the homogeneous space of a Lie group G by a closed subgroup $D \subset G$
$(G/D, g_{\mathbb{R}})$ — one-parameter homogeneous flow
$E(g_{\mathbb{R}})$, $\widetilde{E}(g_{\mathbb{R}})$ — the ergodic partition and decomposition of a homogeneous flow
$\mathrm{SL}(n, \mathbb{R})$ — the group of unimodular real $n \times n$ matrices
A^{T} — the transpose matrix
$\mathrm{diag}(t_1, \ldots, t_n)$ — the diagonal matrix with elements $t_1, \ldots, t_n$
det — the determinant of a matrix
card — the cardinality of a set
(X, μ) — measure space
$\mathcal{P}(X)$ — the space of Borel probability measures on a topological space
$C_c(X)$ — the space of continuous functions with compact support on a topological space
$\mathbb{H}$ — Hilbert space
$\mathrm{Im}(\phi)$ — the image of a homomoprhism ϕ
$\mathrm{Ker}(\phi)$ — the kernel of a homomorphism ϕ
$\mathbb{H}^n$ — the n-dimensional Lobachevskii space
$\mathcal{H}^n$ — the n-dimensional Hausdorff measure

Preliminaries

We recall some definitions from topology which are relevant to our purposes. All topological spaces X are assumed to be *metrizable*, *locally compact* (that is, every point $x \in X$ has a neighborhood with compact closure) and separable (that is, X contains a countable dense subset). In particular, X always possesses a *countable base* of open sets (that is, satisfies the *second countability axiom*).

Borel sets in X are sets obtained from open sets by at most countably many operations of union, intersection and complementation. The family of Borel sets forms a σ-algebra (in particular, a countable intersection and a countable union of Borel sets are Borel sets).

A subset $A \subset X$ is said to be *locally closed* if for every point $x \in A$ there exists a neighborhood $O(x) \subset X$ such that the intersection $A \cap O(x)$ is closed in $O(x)$. A subset $A \subset X$ is *locally connected* if for any point $x \in A$ and any neighborhood $O(x) \subset X$ there exists a neighborhood $O'(x) \subset O(x) \subset X$ such that the intersection $A \cap O'(X)$ is connected.

A. Ergodic theory

The information to follow is taken from [**KSF**].

1. Measure theory. A *measure* μ on a set X is a nonnegative σ-additive function $\mu : \Omega(X) \to \mathbb{R}^+ \cup \infty$ defined on a σ-algebra $\Omega(X)$ of subsets in X. Subsets in $\Omega(X)$ are called *measurable sets*. The measure μ is said to be *complete* if from $\mu(A) = 0$ and $A \in \Omega(X)$ it follows that $B \in \Omega(X)$ for every $B \subset A$.

A measure μ on X is called *absolutely continuous* relative to a measure μ' (we denote this by $\mu \leq \mu'$) if all μ'-measurable sets are μ-measurable and from $\mu'(A) = 0$ it follows that $\mu(A) = 0$. If $\mu \leq \mu'$ and $\mu' \leq \mu$, then the measures μ and μ' are called *equivalent* ($\mu \sim \mu'$). Measures μ and μ' are *mutually singular* if there exists a subset $A \subset X$ such that $\mu(A) = 0$ and $\mu'(X - A) = 0$.

A measure μ is said to be *finite* if $\mu(X) < \infty$, and *σ-finite* if $X = \bigcup_{i \in \mathbb{N}} X_i$, where $\mu(X_i) < \infty$. A finite measure μ normalized by the condition $\mu(X) = 1$ is called a *probability measure*. A pair (X, μ) (we omit $\Omega(X)$ to simplify notation) will be called a *measure space*. In what follows, we will always work with sufficiently good measures. To be precise, under rather natural conditions ($\Omega(X)$ has a countable basis (mod 0), the measure μ is complete and σ-finite, etc.; see [**KSF**] for more), the space (X, μ) is measure-theoretically isomorphic to the real line $\mathbb{R}$ equipped with a *Lebesgue–Stieltjes* measure (finite or infinite). If this is the case, the space (X, μ) will be called a *Lebesgue space*.[1] If in addition $\mu(X) = 1$, the pair (X, μ) is called a *probability space*.

[1]We emphasize that in contrast to the common definition we do not require the measure μ to be finite.

We recall that a Lebesgue–Stieltjes measure m on $\mathbb{R}$ is given by a nondecreasing left continuous function $F : \mathbb{R} \to \mathbb{R}$. Here

$$m(a,b) = F(b) - F(a+0), \qquad m[a,b] = F(b+0) - F(a).$$

By a standard construction one obtains a σ-additive measure defined on the σ-algebra of Borel subsets in $\mathbb{R}$. The completion of this measure is called a Lebesgue–Stieltjes measure (note that the family of measurable sets depends on the function F). The usual Lebesgue measure l on $\mathbb{R}$ is given by the function $F(x) = x$.

If $m \leq l$, the measure m is called *absolutely continuous.* For such a measure there exists a nonnegative measurable (relative to the Lebesgue measure) function $s : \mathbb{R} \to \mathbb{R}$ such that

$$\int_{\mathbb{R}} f dm = \int_{\mathbb{R}} f(x)s(x)dx$$

for any Lebesgue measurable function f. If the measure m is supported on a finite or countable[2] subset in $\mathbb{R}$, it is called a *discrete* measure. If every countable set is of zero m-measure and at the same time $m(\mathbb{R} - A) = 0$ for some set $A \subset \mathbb{R}$ of zero Lebesgue measure, then m is said to be a *singular* measure. It is known that every Lebesgue–Stieltjes measure is the sum of an absolutely continuous measure, a discrete measure, and a singular measure.

Let a map $f : X \to X'$ be given and let the space X be equipped with a measure μ. Then X' can be equipped with a measure $\mu' = f_*\mu$ by the rule

$$\mu'(A) = \mu(f^{-1}(A)) \text{ for any } A \subset X' \text{ with measurable inverse image } f^{-1}(A)$$

(note that $(X', f_*\mu)$ can fail to be a Lebesgue space).

Let (X, μ) and (X', μ') be two measure spaces. A map $f : X \to X'$ is called *measurable* if the inverse image $f^{-1}(A)$ of each measurable set $A \subset X'$ is measurable. A measurable map $f : X \to X'$ is called *measure-preserving* if $f_*\mu = \mu'$, that is, $\mu(f^{-1}(A)) = \mu'(A)$ for each measurable $A \subset X'$.

2. Actions on a measure space. An *automorphism* of a measure space (X, μ) is a measure-preserving invertible (mod 0) map $f : X \to X$. By $\mathrm{Aut}(X, \mu)$ we denote the group of all automorphisms of the space (X, μ) equipped with the following topology:

$$f_n \to f \text{ if } \mu(f_n(A) \cap B) \to \mu(f(A) \cap B) \text{ for all measurable sets } A, B \subset X.$$

Let F be a topological group (for instance, the group $\mathbb{R}$ of reals or the group $\mathbb{Z}$ of integers). In what follows, we assume that *F is locally compact and second countable.*[3] By a *continuous action* of F on a space (X, μ) we understand a continuous homomorphism $\phi : F \to \mathrm{Aut}(X, \mu)$. Throughout the book we consider only continuous actions, and instead of $\phi_g(x)$ we usually write gx, where $g \in F$, $x \in X$. In the case $F = \mathbb{R}$, a continuous action is called a *flow*, and in the case $F = \mathbb{Z}$ a *one-generator action.*

A measurable set $A \subset X$ is called *F-invariant* if $gA = A \pmod 0$ for all $g \in F$. Under our assumptions (the space (X, μ) is Lebesgue and F is second countable and locally compact) this is equivalent to the property $FA = A \pmod 0$ [**Ver**]. An action of F on (X, μ) is *ergodic* if for every F-invariant set $A \subset X$ either $\mu(A) = 0$ or $\mu(X - A) = 0$ (in this case μ is called an *F-ergodic measure*).

[2]Not necessarily discrete.

[3]Actually, all our groups will be finite-dimensional Lie groups.

3. Unitary representations. Let $L^2(X,\mu)$ be the Hilbert space of (equivalence classes of) complex-valued functions on X square integrable relative to μ. We equip the space with the scalar product

$$(f, f') = \int_X f(x)\overline{f'(x)}d\mu_x.$$

A norm in the space $L^2(X,\mu)$ is given as usual: $||f||^2 = (f,f)$. Since (X,μ) is always assumed to be a Lebesgue space, it follows that $L^2(X,\mu)$ is separable.

Let a probability measure μ be invariant for an action (X,F). Then one defines a continuous unitary representation $\rho : F \to \mathbb{U}(L^2(X,\mu))$ by the rule

$$(\rho_g f)(x) = f(g^{-1}x), \qquad g \in F, \quad x \in X, \quad f \in L^2(X,\mu)$$

(one says that F is *unitary* if

$$(\rho_g f, \rho_g f') = (f, f'), \qquad g \in F, \quad f, f' \in L^2(X,\mu),$$

and this always holds whenever μ is F-invariant).

Since the action is continuous, it follows, in particular, that the *stabilizer*

$$\mathrm{Stab}(f) = \{g \in F : \ \rho_g f = f\}$$

for each element $f \in L^2(X,\mu)$ is a closed subgroup of F.

In the case $F = \mathbb{R}$, a continuous flow on (X,μ) will be denoted by $(X, \phi_{\mathbb{R}})$. An element $f \in L^2(X,\mu)$ is called an *eigenfunction* if there exists $\lambda \in \mathbb{C}$ such that $\rho_t f = \lambda^t f$, $t \in \mathbb{R}$ (here the equality means that the functions coincide almost everywhere on X). From the unitary property it follows that $|\lambda| = 1$; the case $\lambda = 1$ corresponds to an *invariant function.* In this setting the flow $(X, \phi_{\mathbb{R}})$ is ergodic if and only if apart from constants there are no $\phi_{\mathbb{R}}$-invariant functions in $L^2(X,\mu)$.

4. Ergodicity of rectilinear flows on the torus. Let X be the torus $\mathbb{T}^n = \mathbb{R}^n/\mathbb{Z}^n$ equipped with the usual $\mathbb{R}^n$-invariant Lebesque measure μ, and let $v \in \mathbb{R}^n$. The *rectilinear flow* on the torus $\mathbb{T}^n$ induced by the vector v is a one-parameter group of translations:

$$\phi_t(x + \mathbb{Z}^n) = (tv + x + \mathbb{Z}^n) \in \mathbb{R}^n/\mathbb{Z}^n, \qquad x \in \mathbb{R}^n.$$

One says that the coordinates $v_1, \ldots, v_n$ are *linearly independent over the field* $\mathbb{Q}$ of rationals if

$$(*) \qquad (\alpha_1 v_1 + \cdots + \alpha_n v_n = 0, \ \alpha_i \in \mathbb{Q}) \Rightarrow (\alpha_i = 0, \ i = 1, \ldots, n).$$

The equation $\alpha_1 x_1 + \cdots + \alpha_n x_n = 0$, $x \in \mathbb{R}^n$, defines a subspace $L \subset \mathbb{R}^n$. If all α_i are rational, then the subspace is called *rational* and, as is known, intersects the lattice $\mathbb{Z}^n \subset \mathbb{R}^n$ in a lattice (that is, $L/\mathbb{Z}^n \cap L$ is a torus). Therefore if the coordinates of v are not independent over $\mathbb{Q}$, the torus $\mathbb{T}^n$ fibers into a family of invariant subtori of equal dimensions. Clearly, in this case the rectilinear flow is not ergodic. On the contrary, if the coordinates of v are independent over $\mathbb{Q}$, then the flow is ergodic. In fact, $\mathbb{T}^n$ carries a transitive action of the group $\mathbb{R}^n$, and one can consider a continuous unitary representation $\rho : \mathbb{R}^n \to \mathbb{U}(L^2(\mathbb{T}^n,\mu))$. Assume that $f \in L^2(\mathbb{T}^n,\mu)$ and $\rho(tv)f = f$, $t \in \mathbb{R}$. The stabilizer $H = \mathrm{Stab}(f)$ is a closed subgroup of $\mathbb{R}^n$, and $\mathbb{R}v$, $\mathbb{Z}^n \subset H$. It suffices to prove that $H = (\overline{\mathbb{R}v\mathbb{Z}^n})_0 = \mathbb{R}^n$. Since the group $\mathbb{R}^n$ is commutative, it follows that the subgroup H is closed, contains $\mathbb{R}v$ and intersects $\mathbb{Z}^n$ in a lattice. If $\dim H < n$, we arrive at a contradiction because the coordinates of v are rationally independent. Hence $H = \mathbb{R}^n$ and $f \equiv \text{const}$. We have proved *Kronecker's ergodicity criterion* for rectilinear flows on the torus.

5. Mixing. A flow $(X, \phi_{\mathbb{R}})$ on a space (X, μ) with probability measure μ is called *weak mixing* if apart from constants there are no eigenfunctions in $L^2(X, \mu)$. Clearly, weak mixing implies ergodicity. Note that rotations of the circle form an ergodic flow relative to the Lebesgue measure that fails to be weak mixing. A criterion for a flow to be weak mixing is that the diagonal flow $(X \times X, \phi_t \times \phi_t)$ be ergodic relative to $\mu \times \mu$.

A flow on a space (X, μ) with probability measure is called *mixing* if for any measurable $A, B \subset X$ one has

$$\lim_{t \to \infty} \mu(\phi_t A \cap B) = \mu(A)\mu(B).$$

This is equivalent to the property that

$$\int_X (\rho_t f_1)\overline{f}_2 d\mu \to \int_X f_1 d\mu \int_X \overline{f}_2 d\mu, \qquad t \to \infty, \text{ for any } f_1, f_2 \in L^2(X, \mu).$$

It is easy to see that a mixing flow is always weak mixing. The converse is not true in general.

More generally, a flow is called *k-mixing* if for any $k+1$ measurable sets $A_0, \dots, A_k \subset X$ and any $k+1$ sequences $\{t_i(0)\}, \dots, \{t_i(k)\}$ such that

$$|t_i(l) - t_i(m)| \to \infty, \qquad i \to \infty,\ l \neq m,$$

one has

$$\lim_{i \to \infty} \mu\Big(\bigcap_{l=0}^{k} \phi_{t_i(l)} A_l\Big) = \prod_{l=0}^{k} \mu(A_l).$$

Clearly, k-mixing implies mixing of smaller orders. In general it is not known (*Rokhlin's problem*) whether mixing of all orders follows from the usual mixing.

6. The main ergodic theorem. Assume that μ is an invariant probability measure for a flow $(X, \phi_{\mathbb{R}})$.

THEOREM A.1 (The Birkhoff–Khintchine ergodic theorem). *For every $f \in L^1(X, \mu)$ and for μ-almost every point $x \in X$ the following convergence holds:*

$$(**) \qquad s_f(T, x) = \frac{1}{T} \int_0^T f(\phi_t x) dt \to f^*(x), \qquad T \to \infty,$$

where $f^* \in L^1(X, \mu)$.

Here the function f^* is $\phi_{\mathbb{R}}$-invariant. Clearly, a flow is ergodic $\iff$ for each $f \in L^1(X, \mu)$ one has $s_f(T, x) \to \mu(f) = \int_X f d\mu$ for μ-almost each $x \in X$.

Actually, to verify whether a flow is ergodic, it suffices to check the property on a dense subset of $L^1(X, \mu)$ (since any convergent sequence in $L^1(X, \mu)$ contains a subsequence converging pointwise almost everywhere). In addition, it follows from the Birkhoff–Khintchine theorem that any two distinct ergodic probability measures are mutually singular.

7. Entropy. In ergodic theory there is an important concept of *metric entropy* $0 \leq e(\phi) \leq \infty$, which measures "the degree of instability" of trajectories of an automorphism $\phi \in \mathrm{Aut}(X, \mu)$. A precise definition can be found in [**KSF**]. For us only the criterion for the entropy of a homogeneous flow to be trivial will be important; therefore we do not reproduce the full definition. Entropies of translations involved in a flow are related by the formula $e(\phi_t) = |t|e(\phi_1)$. According to Pinsker [**Pi**], every flow admits a unique (up to isomorphism) maximal quotient flow of zero entropy. Fibers of the corresponding cover form the *Pinsker partition* of the space X. If the maximal quotient flow of zero entropy is k-mixing, the original flow is also k-mixing (Thouvenot; see [**Ka**]). If every quotient flow of zero entropy is trivial, then the original flow is called a *K-flow* or a *flow with* the *K-property*[4]. It is known that any K-flow is mixing of all orders. The converse statement is not true in general.

8. Equivalence relations. Let $(X, \phi_{\mathbb{R}})$ and $(X', \phi'_{\mathbb{R}})$ be two flows on measure spaces. A measurable map $\alpha : X \to X'$ is called a *conjugacy* of flows (or *equivariant* relative to the flows) if

$$\alpha(\phi_t x) = \phi'_t \alpha(x) \quad \text{for } \mu\text{-almost all } x \in X \text{ and all } t \in \mathbb{R}.$$

If α is equivariant and measure-preserving, it is called a *measure-theoretic conjugacy*. Here the flow $(X, \phi_{\mathbb{R}})$ is said to be a *covering*, and $(X', \phi'_{\mathbb{R}})$ is a *quotient flow*. If in addition α is invertible, the flows are called *measure-theoretically isomorphic*. Ergodicity, k-mixing, the K-property, and entropy are measure-theoretic invariants; that is, they are preserved by isomorphisms. Ergodicity, k-mixing, and the K-property are stable under taking quotient flows, whereas entropy does not increase.

Flows $(X, \mu, \phi_{\mathbb{R}})$ and $(X', \mu', \phi'_{\mathbb{R}})$ are called *Kakutani equivalent* if there exists a map $\alpha : X \to X'$ such that $\alpha_* \mu \sim \mu'$ and α sends (almost all) $\phi_{\mathbb{R}}$-orbits to $\phi'_{\mathbb{R}}$-orbits preserving the direction (that is, positive semiorbits are sent to positive semiorbits, and negative to negative). Kakutani equivalence preserves the ergodicity property, but does not preserve the mixing property. Entropy, generally speaking, is not preserved either, but the properties of triviality and finiteness of entropy are preserved under Kakutani equivalence.

9. Joinings. Let $(X, \mu, \phi_{\mathbb{R}})$ and $(X', \mu', \phi'_{\mathbb{R}})$ be measure-preserving flows. By

$$p : X \times X' \to X, \qquad p' : X \times X' \to X'$$

we denote the natural projections. A *joining* of these flows is a measure m on $X \times X'$ which is invariant under the diagonal flow $\phi_t \times \phi'_t$ and such that p and p' are measure-preserving (that is, $m(A \times X') = \mu(A)$ for every measurable $A \subset X$ and $m(X \times A') = \mu'(A')$ for every measurable $A' \subset X'$).

If there are no other joinings for $(X, \mu, \phi_{\mathbb{R}})$ and $(X', \mu', \phi'_{\mathbb{R}})$ except the trivial joining $\mu \times \mu'$, then the measures μ and μ' are called *disjoint*. In the case $X = X'$, $\mu = \mu'$, a joining m on $X \times X$ is called a *self-joining* of the measure μ. Apart from $\mu \times \mu$, there are other examples of self-joinings on $X \times X$: the diagonal measure $\widehat{\mu}(A \times B) = \mu(A \cap B)$ and its translates $\widehat{\mu}_t(A \times B) = \mu(\phi_t A \cap B)$. These self-joinings will be called *diagonal*. Joinings are especially useful in the study of multiple mixing; a detailed account can be found in the survey of Thouvenot [**Th**].

[4]One usually uses an equivalent original definition due to Kolmogorov [**KSF**].

10. Measurable partitions. A *partition* of a measure space (X, μ) is a family $\xi = \{C_\alpha\}$ of disjoint nonempty subsets in X such that $\bigcup_\xi C_\alpha = X \pmod 0$. One says that a partition $\xi = \{C_\alpha\}$ is *coarser* than a partition $\eta = \{D_\beta\}$ (we denote this by $\xi \leq \eta$) if μ-almost every[5] element $C_\alpha \in \xi$ is a union of some elements $D_\beta \in \eta$. A set $A \subset X$ which is a union of some elements $C_\alpha \in \xi$ is called *ξ-measurable.*

A partition ξ is called *measurable* if there exists a countable family of ξ-measurable sets $\{B_i\}$ such that for every $C, C' \in \xi$ there exists i such that either $C \subset B_i$, $C' \not\subset B_i$, or $C' \subset B_i$, $C \not\subset B_i$. Let X/ξ be the quotient space of X by the partition ξ (that is, the set of partition elements), and let $\pi : X \to X/\xi$ be the covering map. The basic property is that the quotient space $(X/\xi, \pi_*\mu)$ of a Lebesgue space (X, μ) by a measurable partition ξ is still a Lebesgue space.

Not every partition is measurable (example: the partition of the 2-torus into orbits of an ergodic rectilinear flow).

A *canonical system of conditional measures* for a measurable partition ξ of a Lebesgue space is a system of measures μ_α defined on elements $C_\alpha \in \xi$ such that

a) almost all (C_α, μ_a) are Lebesgue spaces, and

b) for every μ-measurable set $A \subset X$, the intersection $A \cap C_\alpha$ is almost always μ_α-measurable, the function $\mu_\alpha(A \cap C_\alpha)$ on X/ξ is $\pi_*\mu$-measurable, and

$$\mu(A) = \int_{X/\xi} \mu_\alpha(A \cap C_\alpha) d(\pi_*\mu).$$

The basic result in the theory of measurable partitions is

THEOREM A.2 (Rokhlin [**Ro**]). *Every measurable partition of a Lebesgue space has a canonical system of conditional measures, and it is unique* (mod 0).

B. Topological dynamics

Here we use [**Z**], [**KSF**] and [**AGH**].

1. Basic definitions. Let X be a topological space which as usual is assumed to be *metrizable, locally compact and separable*[6]. An *action* of a topological group F on X is a continuous homomorphism $\alpha : F \to \text{Homeo}(X)$ to the group of all homeomorphisms of the space X equipped with the topology of uniform convergence on compact sets.

An action is called *topologically transitive* if there exists a point $x \in X$ with dense F-orbit[7] $Fx \subset X$, and *minimal* if all F-orbits are dense. In essence, we have already proved that an ergodic rectilinear flow on the torus is minimal. An F-invariant closed subset $A \subset X$ is called *minimal* if every F-orbit inside A is dense in A. By Zorn's lemma, any action on a compact space has a minimal set.

2. Behavior of trajectories. Let $(X, \phi_\mathbb{R})$ be a one-parameter flow on a topological space X. Every point $x \in X$ has two semiorbits: the positive semiorbit $\phi_{\mathbb{R}^+}x$ and the negative semiorbity $\phi_{\mathbb{R}^-}x$. The positive semiorbit of a point $x \in X$ is

---

[5] That is, the union of remaining elements is of zero measure.

[6] Actually, we will consider only closed subsets in homogeneous spaces of Lie groups.

[7] One usually excludes the degenerate case when the dense trajectory is locally closed and noncompact (this can be the case when $\dim X = \dim F$).

called *recurrent*[8] if it comes back to an arbitrarily small neighborhood of the point x, that is, there exists a sequence $t_k \to +\infty$ such that $\phi_{t_k} x \to x$ (in a similar way one defines recurrence in the negative direction).

A nonrecurrent semiorbit is locally closed. The positive semiorbit is called *divergent* or *exiting to infinity* if for any compact set $K \subset X$ there exists a number $T > 0$ such that $\phi_t x \notin K$ for all $t \geq T$. If the closure of the semiorbit is compact, it is called a *bounded* semiorbit.

A point $x \in X$ is said to be *wandering* if there exist a neighborhood $O(x) \subset X$ and a number $T > 0$ such that

$$\phi_t(O(x)) \cap O(x) = \varnothing, \qquad \text{for all } t \geq T.$$

Clearly, both semiorbits of a wandering point are nonrecurrent. The *nonwandering set* $\Omega \subset X$, that is, the set of all nonwandering points, is closed and invariant.

The *limit set* $\Lambda(x) \subset X$ of a semiorbit $\phi_{\mathbb{R}^+} x$ consists of all points $y \in X$ such that $\phi_{t_n} x \to y$ for a sequence $t_n \to +\infty$. It is easily seen that the limit set is closed and invariant. Besides, $\Lambda(x) = \varnothing \iff$ the positive semiorbit $\phi_{\mathbb{R}^+} x$ is divergent.

3. Mixing. A flow $(X, \phi_{\mathbb{R}})$ is *topologically mixing* if for any open sets $U, V \subset X$ there exists $T > 0$ such that

$$\phi_t(U) \cap V \neq \varnothing \qquad \text{for all } t \geq T.$$

A flow on a metric space (X, d) is called *uniformly continuous* if for every $\varepsilon > 0$ there exists $\delta > 0$ such that $d(\phi_t x, \phi_t y) < \varepsilon$ for all $t \in \mathbb{R}$ whenever $d(x, y) < \delta$. The closure of any orbit of a uniformly continuous flow is a minimal set. Hence orbit closures of a uniformly continuous flow form a *topological partition* of the space X; that is, any two of them either coincide or are disjoint.

A flow is called *distal* if for each $x, y \in X$, $x \neq y$, there exists $\delta > 0$ such that $d(\phi_t x, \phi_t y) > \delta$ for all $t \in \mathbb{R}$. A uniformly continuous flow is always distal.

THEOREM B.1 (Ellis, see [**AGH**]). *The closure of a bounded orbit of a distal flow is a minimal set.*

It is not hard to show that a distal flow cannot be topologically mixing.

4. Borel measures. Assume that μ is a *Borel measure* on a topological space X. This means that the σ-algebra of measurable sets is the completion of the σ-algebra of Borel sets in X. In the sequel Borel measures are assumed to be *locally finite*, that is, finite on compact subsets in X, and hence the space (X, μ) will always be Lebesgue. Since X is metrizable, it follows that every Borel measure μ is *regular*, that is, for any measurable set $A \subset X$ of finite measure and any $\varepsilon > 0$ there exist a compact set $K \subset A$ and an open set $U \supset A$ such that $\mu(U - K) < \varepsilon$.

Given a topological space X, let $C_c(X)$ be the space of continuous functions with compact support, equipped with the following topology[9] : $f_i \to f$ if there exists a compact set $K \subset X$ such that $f_i, f \equiv 0$ outside K and the convergence is uniform on K.

---

[8] We use the terminology of Gottschalk and Hedlund [**GH**]. According to Birkhoff, the property of recurrence is much stronger, and in the compact case it means the minimality of the closure.

[9] If the space X is compact, this topology is given by the norm $||f|| = \sup_X |f(x)|$, and $C_c(X) = C(X)$ is a Banach space.

It is easily seen that any Borel measure μ defines a continuous linear functional $\mu(f) = \int_X f d\mu$ on $C_c(X)$. Moreover, $\mu(f) \geq 0$ for any nonnegative function $f \in C_c(X)$. Conversely, every such functional is given by a Borel measure.[10]

A measure which is positive on open subsets is called *strictly positive*. A measure μ on a Riemannian manifold is called *smooth* if it is equivalent to the measure given by the volume form (clearly, smooth measures are strictly positive).

THEOREM B.2 (Glimm, see [**Z**]). *The partition of a topological space X into orbits of a continuous action of a group F is measurable relative to a locally finite Borel measure $\iff$ almost all F-orbits are locally closed.*

THEOREM B.3 (Poincaré's recurrence theorem). *If a one-parameter flow preserves a finite Borel measure, then almost all its trajectories are recurrent in both directions.*

The *support* $\mathrm{supp}(\mu)$ of a Borel measure μ is the smallest closed set of full measure. A measure μ is strictly positive $\iff \mathrm{supp}(\mu) = X$.

LEMMA B.4 (Hedlund). *If a flow is ergodic relative to a strictly positive measure, then it is topologically transitive and, moreover, almost all its orbits are dense.*

Indeed, let U be an open subset in X. Then the set $M(U)$ of points $x \in X$ whose orbits do not intersect U is of zero measure (and is closed). Let $\{U_i\}$ be a countable base of open subsets in X. Then the family of nondense orbits is the countable union $\bigcup_i M(U_i)$ and hence is of zero measure (and is a first category set).

If a measure μ is finite and strictly positive, then measure-theoretic mixing implies topological mixing.

5. Uniform distribution. It is well known that the space $C_c(X)$ is dense in $L^1(X, \mu)$ (under all previous assumptions on X and μ). Let $(X, \mu, \phi_{\mathbb{R}})$ be an ergodic flow. One says that the orbit of a point $x \in \mathrm{supp}(\mu)$ is *uniformly distributed* relative to the measure μ if in notation of the Birkhoff–Khintchine theorem one has $s_f(T, x) \to \mu(f)$ for all $f \in C_c(X)$. Clearly, the closure of a uniformly distributed orbit is precisely $\mathrm{supp}(\mu)$. Since the space $C_c(X)$ is separable, μ-almost each trajectory of an ergodic flow is uniformly distributed (this strengthens Poincaré's theorem).

On the dual space $C_c^*(X)$ (that is, on the space of continuous linear functionals on $C_c(X)$) one defines the *$*$-weak topology*:

$$F_n \to F, \text{ if } F_n(f) \to F(f) \text{ for all } f \in C_c(X).$$

This topology makes $C_c^*(X)$ a locally convex linear topological space. In this space one distinguishes the convex subset $\mathcal{P}(X)$ of all Borel probability measures on X. A continuous action of a group F on X gives rise to a continuous affine action on $\mathcal{P}(X)$. In particular, the stabilizer $\mathrm{Stab}(\mu) \subset F$ of a measure $\mu \in \mathcal{P}(X)$ is a closed subgroup of F.

One knows that, given a compact space X, the set $\mathcal{P}(X)$ is also compact (relative to the weak* topology). By a theorem of Tychonoff and Schauder, a continuous action of $\mathbb{Z}$ (or $\mathbb{R}$) on a convex compact set (in a locally convex topological linear space) has a fixed point. Hence any one-generator action (or a flow) on a compact space always has an invariant Borel probability measure (theorem of Kryloff

[10]Therefore the set of Borel measures is a cone in the dual space $C_c^*(X)$.

and Bogoluboff). One can also find an invariant measure for an action on a compact space of any *amenable* group (see [**Z**]). We do not need a precise definition of amenability;[11] we only note that a connected Lie group is amenable if and only if it is a compact extension of a solvable Lie group.

If an F-invariant Borel probability measure is unique, the action (X, F) is said to be *uniquely ergodic.*[12] One example is a rectilinear ergodic flow on the torus (this follows because the subgroup $\mathrm{Stab}(\mu) \subset \mathbb{R}^n$ is closed and contains the dense subgroup $\mathbb{R}v\mathbb{Z}^n$).

Clearly, a uniquely ergodic action of a solvable Lie group F on a compact space is always minimal provided the unique invariant probability measure is strictly positive. The converse fails to hold: Furstenberg [**Fu61**] constructed an example of a minimal $\mathbb{Z}$-action by smooth diffeomorphisms on the torus $\mathbb{T}^2$ that is not uniquely ergodic.

To check a one-parameter flow for unique ergodicity one often applies the following observation: if all trajectories are uniformly distributed, then, as is easily seen, the flow is uniquely ergodic. If X is a compact space, the converse is also true. (The linear functional $s(T, x)$ on $C(X)$ gives rise to a Borel probability measure on X. Each limit measure in $\mathcal{P}(X)$ for the family $\{s(T, x),\ T \to \infty\}$ is invariant and hence is μ.)

One easily sees also that a uniformly continuous and topologically transitive flow on a compact space is uniquely ergodic. The following result is much less evident.

THEOREM B.5 (Furstenberg, [**Fu63**]). *A topologically transitive distal flow on a compact space is uniquely ergodic.*

6. Equivalence relations and suspension. Flows $(X, \phi_{\mathbb{R}})$ and $(X', \phi'_{\mathbb{R}})$ are said to be *topologically isomorphic* if there exists a homeomorphism $\alpha : X \to X'$ establishing their conjugacy:

$$\alpha(\phi_t x) = \phi'_t \alpha(x), \qquad \forall x \in X,\ \forall t \in \mathbb{R}.$$

If a homeomorphism α sends each $\phi_{\mathbb{R}}$-orbit to a $\phi'_{\mathbb{R}}$-orbit preserving the directions, the flows are said to be *topologically equivalent.*

Let A be a homeomorphism of a topological space X, and let $X' = (X \times \mathbb{R})/\sim$ be the quotient space of $X \times \mathbb{R}$ modulo the following equivalence relation:

$$(x, s) \sim (A(x), s - 1), \qquad x \in X,\ s \in \mathbb{R}.$$

Let a flow $A_{\mathbb{R}}$ on X' be defined as

$$A_t[(x, s)/\sim] = (x, s + t)/\sim .$$

It is easy to see that the flow $(X', A_{\mathbb{R}})$ is well-defined and continuous. The space X embeds naturally into X', and orbits of the flow $(X', A_{\mathbb{R}})$ intersect X precisely in orbits of the one-generator action given by $A \in \mathrm{Homeo}(X)$. The flow $(X', A_{\mathbb{R}})$ is

[11] Having a fixed point for every continuous affine action on a convex compact set (in a good enough topological linear space) is one of the definitions.

[12] In addition, if the action is minimal, it is called "strictly ergodic".

called the *standard suspension*[13] over a homeomorphism $A : X \to X$, whereas the original homeomorphism is called the *section* of the flow obtained.

C. Lie groups and Lie algebras

Here we use the information from [**VGO**].

1. General definitions. Let G be a real connected Lie group. In what follows, $1 \in G$ stands for the identity element, $\overline{M} \subset G$ is the closure of a subset $M \subset G$, and $H_0 \subset G$ is the connected component of the identity element of a closed subset $H \subset G$ containing $1 \in G$. Given subgroups $H, H' \subset G$, let $[H, H'] \subset G$ be their *commutator*, that is, the subgroup of G generated by elements

$$\{g, h\} = ghg^{-1}h^{-1}, \qquad g \in H, \quad h \in H'.$$

The notation $G = A \cdot B$ means that the group G is the *semidirect product* of subgroups A and B. More precisely, $A \cap B = 1$, A normalizes B (that is, $aBa^{-1} = B$, $a \in A$), and every element in G decomposes as $g = ab$, $a \in A$, $b \in B$.

In the sequel $Z(G) \subset G$ is the center of G, and $Z_H(a)$ is the centralizer of an element $a \in G$ in a subgroup $H \subset G$. $N_H(a)$ stands for the normalizer of $a \in G$ in $H \subset G$. Given an element $g \in G$, let $g^{\mathbb{Z}} \subset G$ be the cyclic subgroup $\{g^n,\ n \in \mathbb{Z}\}$.

The Lie algebra of a Lie group will be denoted by the corresponding gothic letter. Let $\mathfrak{g}$ be the Lie algebra of a group G. There exists a map $\exp : \mathfrak{g} \to G$ which establishes a homeomorphism of a neighborhood of $0 \in \mathfrak{g}$ with a neighborhood of $1 \in G$. The restriction of exp to any one-dimensional subspace $\mathbb{R}x$, $x \in \mathfrak{g}$, is a homomorphism which defines a one-parameter subgroup $\exp(\mathbb{R}x) \subset G$ with x as the unit tangent vector.

By the *adjoint representation* $\mathrm{Ad} : G \to \mathrm{Aut}(\mathfrak{g})$ we understand the differential of the conjugation $h \to ghg^{-1}$, that is,

$$\mathrm{Ad}_g(x) = \frac{dg\exp(tx)g^{-1}}{dt}\Big|_{t=0}.$$

Note that $\mathrm{Ad}_g \mathrm{Ad}_{g'} = \mathrm{Ad}_{gg'}$, and hence Ad is a representation of the group G in the automorphism group $\mathrm{Aut}(\mathfrak{g})$ of the algebra $\mathfrak{g}$.

Every connected Lie group G has the universal covering group $\widetilde{G}$. The simply connected group $\widetilde{G}$ has the same Lie algebra, and the covering map $\alpha : \widetilde{G} \to G$ factors $\widetilde{G}$ by a discrete central subgroup isomorphic to the fundamental group $\pi_1(G)$. A connected simply connected Lie group with the Lie algebra $\mathfrak{g}$ is unique up to isomorphism. Two Lie groups whose Lie algebras are isomorphic are called *locally isomorphic*. A connected normal subgroup of a simply connected Lie group is always closed.

An element $g \in G$ is called *unipotent* if the operator Ad_g on the Lie algebra $\mathfrak{g}$ is unipotent, that is, all its eigenvalues are equal to 1. This is equivalent to the property

$$(\mathrm{Ad}_g - E)^n \equiv 0 \text{ for some } n \in \mathbb{N}.$$

---

[13]There is a more general definition of the suspension over an automorphism of a Lebesgue space (X, μ) by means of a measurable function $f : X \to \mathbb{R}$. Our definition corresponds to the case $f \equiv 1$.

An element $g \in G$ is called *semisimple* or *reductive* if the operator Ad_g on the Lie algebra $\mathfrak{g}$ can be made diagonal over the field $\mathbb{C}$, and $\mathbb{R}$-*diagonal*[14] if it can be made diagonal over the field $\mathbb{R}$. Only an element of the center $Z(G)$ can be both semisimple and unipotent (this follows from the connectedness of G). A subgroup $H \subset G$ is called *unipotent* if all its elements are unipotent.

Each Lie group G carries a unique (up to rescaling) smooth G-right-invariant measure m called the *Haar measure*. It is given by a G-right-invariant Riemannian metric on G. A *modular function* is a character (continuous homomorphism) $\Delta : G \to \mathbb{R}^+$ for which

$$m(gA) = \Delta(g)m(A)$$

for any $g \in G$ and any measurable $A \subset G$. If the homomorphism Δ is trivial, the group G is called *unimodular*, and in this case G carries a bi-invariant Haar measure. A Lie group G is unimodular if and only if $\det(\mathrm{Ad}_g) = 1$ for all $g \in G$.

2. Nilpotent groups. Given a Lie group N, we consider the descending chain of normal subgroups

$$\{N_0 = N \supset \cdots \supset N_{i+1} = [N_i, N] \supset \cdots\}.$$

A Lie group N is called *nilpotent* if the chain degenerates, that is, if $N_i = 1$ for all $i \geq$ some k. Obviously, $N_{k-1} \subset Z(N)$, and hence the center of a nilpotent group is always nontrivial. Every operator Ad_g, $g \in N$, is unipotent, and hence all elements of a nilpotent group are unipotent.

If a group N is connected and simply connected, the map $\exp : \mathfrak{n} \to N$ is a homeomorphism. Let $V \subset \mathfrak{n}$ be a linear subspace complementary to the commutator $[\mathfrak{n}, \mathfrak{n}]$ (that is, as a vector space, $\mathfrak{n} = V + [\mathfrak{n}, \mathfrak{n}]$). Then V generates the Lie algebra $\mathfrak{n}$, whereas $\exp(V)$ generates the group N. In particular, any reductive automorphism $\alpha \in \mathrm{Aut}(N)$ inducing the trivial automorphism of the vector space $N/[N, N]$ is itself trivial. It is easily seen also that the normalizer of any proper connected subgroup $M \subset N$ contains M as a proper subgroup.

An example of a simply connected nilpotent Lie group is given by the group $\mathrm{U}(n, \mathbb{R})$ of strictly upper-triangular $n \times n$ matrices (with all diagonal elements 1). Every connected simply connected nilpotent Lie group embeds into $\mathrm{U}(n, \mathbb{R})$ for some n.

3. Solvable groups. For a Lie group R let us consider the descending chain of subgroups

$$\{R_0 = R \supset \cdots \supset R_{i+1} = [R_i, R_i] \supset \cdots\}.$$

Clearly, every subgroup R_{i+1} is normal in R_i. A Lie group R is called *solvable* if the chain degenerates, that is, if $R_i = 1$ for all $i \geq$ some k. Obviously, any nilpotent group is solvable. By $N \subset R$ we denote the *nilradical* in R, that is, the maximal connected normal nilpotent subgroup. This subgroup is unique, and the factor-group R/N is Abelian. The commutator $[R, R]$ is nilpotent and hence is a subgroup of the nilradical N. Any compact solvable group is commutative, hence is a torus. Any simply connected solvable group is homeomorphic to a vector space (although the map $\exp : \mathfrak{r} \to R$ need not be either injective or surjective). A

[14]It would be more correct to say that an element is Ad-unipotent, Ad-semisimple, etc. to avoid confusion with similar names for elements in algebraic groups. If there is a danger of ambiguity, we will do so.

connected subgroup of a simply connected solvable Lie group is simply connected and closed.

Let us give a classification of simply connected solvable Lie groups (see [**VGS**]). The type of a group R is defined by the set of eigenvalues of the adjoint operators Ad_g, $g \in R$, on the Lie algebra $\mathfrak{r}$.

A Lie group R is said to be of *class* (I) (for "imaginary") if every eigenvalue λ of every operator Ad_g, $g \in R$, lies on the unit circle: $|\lambda| = 1$.

A Lie group R is called *triangularizable* or of *class* (R) (for "real") if every eigenvalue λ of every operator Ad_g, $g \in R$, is purely real: $\lambda \in \mathbb{R}$.

A Lie group R is *class* (E) (for "exponential") if every eigenvalue λ of every operator Ad_g, $g \in R$, either equals 1, or does not lies on the unit circle: $|\lambda| \neq 1$.

Let us denote by (N) the class of nilpotent Lie groups, and by (I), (R), (E) the corresponding classes of solvable Lie groups. Then

$$(N) \subset (R) \subset (E), \qquad (N) \subset (I), \qquad (I) \cap (E) = (N).$$

Any closed subgroup and any factor-group of a solvable Lie group are the same class as the original group. In addition, the classes are closed under taking the direct product. The class (E) of exponential groups is precisely the class of all Lie groups R for which the exponential map $\exp : \mathfrak{r} \to R$ is a homeomorphism (Dixmier; see [**AGH**]). Another criterion is that there are no factor-groups containing a class (I) nonnilpotent subgroup.

Later we will give examples of solvable groups from all the classes. A standard example of a class (R) solvable group is the group $\mathrm{T}(n, \mathbb{R})$ of triangular matrices with determinant 1. It is the semidirect product

$$\mathrm{T}(n, \mathbb{R}) = \mathrm{D}(n, \mathbb{R}) \cdot \mathrm{U}(n, \mathbb{R}).$$

Here $\mathrm{U}(n, \mathbb{R})$ is the nilradical of $\mathrm{T}(n, \mathbb{R})$, and $\mathrm{D}(n, \mathbb{R})$ is the Abelian group of diagonal matrices with determinant 1, which is isomorphic to the product of $n-1$ copies of the multiplicative group $\mathbb{R}^* = \mathbb{R} - \{0\}$. Every connected simply connected triangularizable group embeds into $\mathrm{T}(n, \mathbb{R})$ for some $n \in \mathbb{N}$.

4. Semisimple groups. A Lie group L is called *semisimple* if it has no connected solvable normal subgroups, and *simple* if it has no connected normal subgroups. Any semisimple Lie group L decomposes into the product $L = \prod_{i=1}^n L_i$ of simple normal subgroups, where each intersection $L_i \cap L_j$, $i \neq j$, is discrete. In other words, L is an "almost direct" product of simple components (the product is direct if the center of L is trivial or L is simply connected).

The center of a connected semisimple group is discrete and is the kernel of the adjoint representation $\mathrm{Ad} : L \to \mathrm{Aut}(\mathfrak{l})$. The universal covering group of a compact semisimple group is itself compact. By $K \subset L$ we denote the product of all compact components, and by $S \subset L$ the product of all noncompact components. The decomposition $L = K \times S$ will be referred to as the *decomposition into compact and noncompact parts.*

Compact semisimple groups have no one-parameter unipotent subgroups; and, on the contrary, simple noncompact Lie groups are generated by such subgroups.

An example of a simple noncompact group is the group $\mathrm{SL}(n, \mathbb{R})$ of matrices with determinant 1, where $n \geq 2$. The same applies to the group $\mathrm{SO}(1, n) \subset \mathrm{SL}(n+1, \mathbb{R})$ of matrices preserving the quadratic form $x_1^2 - x_2^2 - \cdots - x_{n+1}^2$. On the contrary, the group $\mathrm{SO}(n,\mathbb{R})$ of orthogonal matrices is compact. For $n \le 2$ the group is Abelian; for $n \ge 3$ it is semisimple (and simple if $n \ne 4$).

A maximal connected Abelian subgroup $H \subset L$ consisting of semisimple elements is called a *Cartan subgroup*. Up to conjugation the group $\mathrm{SL}(2,\mathbb{R})$ contains two different Cartan subgroups: $\mathrm{SO}(2,\mathbb{R})$ and $\mathrm{D}(2,\mathbb{R})_0$.

Any Cartan subgroup decomposes into the direct product $H = T \times A$ of a torus T and an $\mathbb{R}$-diagonalizable subgroup A. Maximal $\mathbb{R}$-diagonalizable subgroups of L are conjugate, and their common dimension is called the $\mathbb{R}$-*rank* of the group L. For example, $\mathrm{rank}_{\mathbb{R}} \mathrm{SL}(n,\mathbb{R}) = n-1$, and $\mathrm{rank}_{\mathbb{R}} \mathrm{SO}(1,n) = 1$. The $\mathbb{R}$-rank of a compact group is 0.

Let C be a maximal connected compact subgroup of L. All such subgroups are conjugate in L. A maximal compact subgroup C of a simple Lie group L need not be simple (example: $L = \mathrm{SL}(2,\mathbb{R})$; note that $\widetilde{\mathrm{SL}(2,\mathbb{R})}$ contains no compact subgroups at all). Any simple group L with finite center is topologically the product of a maximal connected compact subgroup C and a maximal connected triangularizable subgroup B; that is, $C \cap B = 1$ and every element $g \in G$ decomposes as $g = cb$, $c \in C$, $b \in B$. Maximal triangularizable subgroups are conjugate in L. Each maximal triangularizable subgroup B splits into the semidirect product $B = A \cdot U$, where U is the nilradical of B (called a *maximal horospherical subgroup* of L) and A is a maximal connected $\mathbb{R}$-diagonalizable subgroup. Any connected unipotent subgroup is conjugate to a subgroup of U. The decomposition $L = CAU$ is called an *Iwasava decomposition*.

A standard Iwasava decomposition for $\mathrm{SL}(n,\mathbb{R})$ is arranged as follows:

$$\mathrm{SL}(n,\mathbb{R}) = \mathrm{SO}(n,\mathbb{R})\,\mathrm{D}(n,\mathbb{R})_0\,\mathrm{U}(n,\mathbb{R}),$$

where $A = \mathrm{D}(n,\mathbb{R})$ is the Abelian group of diagonal matrices and $U = \mathrm{U}(n,\mathbb{R})$ is the group of strictly upper-triangular matrices (note that $\mathrm{T}(n,\mathbb{R})_0 = \mathrm{D}(n,\mathbb{R})_0\,\mathrm{U}(n,\mathbb{R})$ is a maximal connected triangularizable subgroup).

Given a unipotent subgroup $u_{\mathbb{R}}$ of a semisimple group L, there exists a subgroup $H \subset L$ locally isomorphic to $\mathrm{SL}(2,\mathbb{R})$ and containing $u_{\mathbb{R}}$ (the Jacobson–Morozov lemma).

5. General groups. Let G be a connected simply connected Lie group. The maximal connected normal nilpotent (solvable) subgroup is called the *nilradical* (the *radical*) of G. Both these subgroups are unique. There is a *Levi decomposition* into the semidirect product $G = L \cdot R$ of the radical R and a maximal semisimple subgroup L (Levi subgroup), both subgroups being closed and simply connected. Any two Levi subgroups are conjugate by an element of the nilradical $N \subset R$ (Maltsev's theorem). It follows that the smallest normal subgroup G_∞ containing L is a subgroup of LN.

A subgroup $W \subset G$ is said to be *horospherical* if there exists an element $g \in G$ such that

$$W = \{h \in G : \ g^n h g^{-n} \to 1, \ n \to +\infty\}$$

(sometimes we also call W the contracting horospherical subgroup). Horospherical subgroups are connected and nilpotent.

D. Linear algebraic groups

Here we use [**Hu, Ma91c, VO, Z**].

1. Basic definitions. We introduce certain definitions from the theory of algebraic groups which are relevant to our purposes. A subset $H \subset \mathrm{SL}(n,\mathbb{R})$ is called (real-) *algebraic* or *Zariski closed* if it is the set of zeros of a polynomial in matrix entries with real coefficients. Such a definition supplies $\mathrm{SL}(n,\mathbb{R})$ and its algebraic subgroups with the *Zariski topology.* The usual topology on $\mathrm{SL}(n,\mathbb{R})$ will be referred to as the *Euclidean*[15] topology.

An algebraic group has a finite number of connected components relative to the Euclidean topology. An intersection of algebraic groups is algebraic. The centralizer $Z_G(H)$ and the normalizer $N_G(H)$ of a subgroup H of an algebraic group $G \subset \mathrm{SL}(n,\mathbb{R})$ are algebraic. The product $AB \subset \mathrm{SL}(n,\mathbb{R})$ of two algebraic subgroups $A, B \subset \mathrm{SL}(n,\mathbb{R})$ is locally closed in the Euclidean topology.

The smallest algebraic subgroup $\mathcal{A}(H) \subset \mathrm{SL}(n,\mathbb{R})$ containing $H \subset \mathrm{SL}(n,\mathbb{R})$ is called the *algebraic hull* or the *Zariski closure* of the group H (one also says that H is *Zariski dense* in $\mathcal{A}(H)$). The algebraic hull of an Abelian (nilpotent, solvable) group is Abelian (nilpotent, solvable). If H normalizes a connected subgroup $F \subset \mathrm{SL}(n,\mathbb{R})$ so does $\mathcal{A}(H)$.

Recall that an element $g \in \mathrm{SL}(n,\mathbb{R})$ is called *unipotent* if there exists a positive integer n such that $(g-E)^n = 0$, where E is the identity matrix (equivalently, all eigenvalues of g are equal to 1). If all eigenvalues of an element $g \in \mathrm{SL}(n,\mathbb{R})$ are of absolute value 1, g is said to be *quasi-unipotent.*

An element $g \in \mathrm{SL}(n,\mathbb{R})$ is said to be *semisimple* if it reduces to diagonal form over $\mathbb{C}$, and $\mathbb{R}$-*diagonalizable* if it does so over $\mathbb{R}$. An example of an $\mathbb{R}$-diagonalizable subgroup is given by the group $\mathrm{D}(n,\mathbb{R}) \subset \mathrm{SL}(n,\mathbb{R})$. Note that $\mathrm{D}(n,\mathbb{R})$ (like any Abelian group consisting of semisimple elements) contains at most countably many one-dimensional algebraic subgroups. The cylinder $\mathrm{D}(2,\mathbb{R}) \times \mathrm{SO}(2,\mathbb{R}) \subset \mathrm{SL}(4,\mathbb{R})$ has only two proper connected algebraic subgroups: $\mathrm{D}(2,\mathbb{R})$ and $\mathrm{SO}(2,\mathbb{R})$ (because the intersection of other one-parameter subgroups with $\mathrm{D}(2,\mathbb{R})$ is countable and hence nonalgebraic).

Any semisimple quasi-unipotent element in $\mathrm{SL}(n,\mathbb{R})$ is conjugate to an element of the maximal compact subgroup $\mathrm{SO}(n,\mathbb{R})$.

A subgroup $H \subset \mathrm{SL}(n,\mathbb{R})$ is said to be *unipotent* if all its elements are unipotent. The maximal normal unipotent subgroup of an algebraic group G is referred to as the *unipotent radical* of G.

Each element g in an algebraic group $H \subset \mathrm{SL}(n,\mathbb{R})$ is the product $g = g_s \times g_u$ of commuting semisimple and unipotent elements $g_s, g_u \in H$ (a *Jordan decomposition*). Any Abelian algebraic subgroup $A \subset H$ consisting of semisimple elements decomposes into the product $A = T \times V$ of a compact torus T and an $\mathbb{R}$-diagonalizable subgroup V. A solvable algebraic group R splits into the semidirect product $R = A \cdot U$ of the unipotent radical U and an Abelian subgroup A of semisimple elements (both subgroups being algebraic).

It is known that compact subgroups of $\mathrm{SL}(n,\mathbb{R})$, as well as connected unipotent subgroups, are algebraic. A semisimple subgroup of $\mathrm{SL}(n,\mathbb{R})$ is of finite index in its algebraic hull, which allows one to treat it as an algebraic group. Any semisimple Lie group L can be projected (with the discrete kernel $Z(L)$) onto the algebraic subgroup $\mathrm{Ad}\, L$ of the group $\mathrm{SL}(\mathfrak{l}) = \mathrm{SL}(n,\mathbb{R})$ of all unimodular linear transformations of the Lie algebra $\mathfrak{l}$, where $\dim \mathfrak{l} = n$.

---

[15]Sometimes it is called the Hausdorff topology, to emphasize the difference with the Zariski topology which does not have the Hausdorff property.

A maximal triangularizable subgroup B of a semisimple algebraic group L is itself algebraic. Clearly, $B = A \cdot U$, where A is $\mathbb{R}$-diagonalizable and U is the unipotent radical of B (and the nilradical of B). The group U is horospherical relative to an element $a \in A$ (called *regular*) and is maximal in the class of all horospherical subgroups of L. The normalizer $P_U = N_L(U)$ of the subgroup U contains B and is called a *minimal parabolic subgroup* of L. Since maximal triangularizable subgroups of L are conjugate, the same holds for maximal horospherical subgroups and minimal parabolic subgroups of L.

The normalizer $P_W = N_L(W)$ of an arbitrary horospherical subgroup $W \subset L$ is called a *parabolic subgroup* of L. Here W is the unipotent radical of P_W, and $P_W \supset P_U$, where $U \subset L$ is a maximal horospherical subgroup containing W.

By definition, a *reductive* subgroup $H \subset \mathrm{SL}(n, \mathbb{R})$ is an algebraic subgroup whose unipotent radical is trivial. We know that a reductive subgroup is the product of a semisimple group and an Abelian group of semisimple elements. The centralizer of a reductive subgroup is also reductive. A Levi decomposition of an algebraic group G is the decomposition $G = (L \times A) \cdot U$, where U is the unipotent radical, L is a Levi subgroup, and A is an Abelian subgroup of semisimple elements. Here $L \times A$ is a maximal reductive subgroup of G, and $A \cdot U$ is the radical of G (all the subgroups in the decomposition being algebraic).

2. Algebraicity over the field $\mathbb{Q}$. If a subgroup $H \subset \mathrm{SL}(n, \mathbb{R})$ is the set of zeros of a polynomial with rational coefficients, it is called a $\mathbb{Q}$-*defined* algebraic group (or simply a $\mathbb{Q}$-group). If $H \subset \mathrm{SL}(n, \mathbb{R})$ is defined over $\mathbb{Q}$, the group $H_{\mathbb{Q}} = H \cap \mathrm{SL}(n, \mathbb{Q})$ of its rational elements is dense in H. The algebraic hull of a subgroup $H \subset \mathrm{SL}(n, \mathbb{Q})$ is defined over $\mathbb{Q}$. The centralizer and the normalizer of a $\mathbb{Q}$-group $G \subset \mathrm{SL}(n, \mathbb{R})$ are also $\mathbb{Q}$-groups. The radical, the unipotent radical, the commutator, and the center of an algebraic $\mathbb{Q}$-group are $\mathbb{Q}$-groups as well. Given a $\mathbb{Q}$-group G, there exists a Levi decomposition $G = (L \times A) \cdot U$ such that L and A are also defined over $\mathbb{Q}$.

Given any element $g \in \mathrm{SL}(n, \mathbb{Q})$, the semisimple and unipotent parts of the Jordan decomposition of g are also elements of $\mathrm{SL}(n, \mathbb{Q})$.

By definition, a $\mathbb{Q}$-*character* of a $\mathbb{Q}$-group $H \subset \mathrm{SL}(n, \mathbb{R})$ is a homomorphism $\chi : H \to \mathbb{R}^*$ given by a polynomial over $\mathbb{Q}$ in matrix elements. Clearly, the kernel $\mathrm{Ker}(\chi)$ is a $\mathbb{Q}$-subgroup of H. Let $H_{\mathbb{Z}} = H \cap \mathrm{SL}(n, \mathbb{Z})$. Then $\chi(H_{\mathbb{Z}}) = 1$, because $\chi(H_{\mathbb{Z}}) \subset \frac{1}{N}\mathbb{Z}$ for some $N \in \mathbb{N}$ and $\chi(H_{\mathbb{Z}})$ is a multiplicative group. A unipotent $\mathbb{Q}$-group admits no nontrivial $\mathbb{Q}$-characters.

A $\mathbb{Q}$-*split* subgroup of a $\mathbb{Q}$-group H is an Abelian $\mathbb{Q}$-subgroup A such that $\bigcap \mathrm{Ker}(\chi) = \{1\}$, where the intersection is taken over all $\mathbb{Q}$-characters of A (then A is isomorphic over $\mathbb{Q}$ to the product of several copies of $\mathbb{R}^*$). A $\mathbb{Q}$-split subgroup is $\mathbb{R}$-diagonalizable. Example: the diagonal subgroup of $\mathrm{SL}(2, \mathbb{R})$ is $\mathbb{Q}$-split, whereas its conjugate subgroup containing a nontrivial element in $\mathrm{SL}(2, \mathbb{Z})$, is not.

Maximal $\mathbb{Q}$-split subgroups of a semisimple $\mathbb{Q}$-group $L \subset \mathrm{SL}(n, \mathbb{R})$ are conjugate by elements of $L_{\mathbb{Q}}$, and their common dimension is called the $\mathbb{Q}$-*rank* of L. Clearly, $0 \le \mathrm{rank}_{\mathbb{Q}} L \le \mathrm{rank}_{\mathbb{R}} L$. Here $\mathrm{rank}_{\mathbb{Q}} L = 0$ if and only if $L_{\mathbb{Q}}$ consists of semisimple elements. An algebraic semisimple group L admits different $\mathbb{Q}$-structures; in particular, $\mathrm{rank}_{\mathbb{Q}}$ can take either of the two extreme values (depending on the embedding of L into $\mathrm{SL}(N, \mathbb{R})$ as a $\mathbb{Q}$-group for some N). For instance, $\mathrm{SL}(4, \mathbb{R})$ contains a $\mathbb{Q}$-group L such that $L \simeq \mathrm{SL}(2, \mathbb{R})$ and $\mathrm{rank}_{\mathbb{Q}} L = 0$.

A semisimple $\mathbb{Q}$-group L with $\operatorname{rank}_{\mathbb{Q}} L \geq 1$ has a proper parabolic $\mathbb{Q}$-subgroup (which need not be a minimal parabolic subgroup of L). Note also that compact and noncompact parts of a semisimple $\mathbb{Q}$-group can fail to be $\mathbb{Q}$-subgroups.

3. Linear representations. Let $\rho : G \to \mathrm{SL}(N, \mathbb{R})$ be a finite-dimensional representation of a connected Lie group G. One says that a subgroup $H \subset G$ is *Zariski dense in the representation* ρ if the algebraic hulls of $\rho(H)$ and $\rho(G)$ are the same. The representation is said to be *unipotent* if the subgroup $\rho(G) \subset \mathrm{SL}(N, \mathbb{R})$ is unipotent.

Let $G \subset \mathrm{SL}(n, \mathbb{R})$ be an algebraic group. A representation $\rho : G \to \mathrm{SL}(N, \mathbb{R})$ is said to be *rational* if it is given by a collection of polynomials in matrix entries. If in addition the polynomials have all their coefficients in $\mathbb{Q}$, then ρ is called a *representation over* $\mathbb{Q}$. One easily verifies that in this case the orbit $\rho(G_{\mathbb{Z}})v$ is discrete for any $v \in \mathbb{Q}^N$. In what follows, we make use of the following result.

THEOREM D.1 (Chevalley, [**Z**]). *Given an algebraic group $G \subset \mathrm{SL}(n, \mathbb{R})$, there exist a rational representation $\rho : \mathrm{SL}(n, \mathbb{R}) \to \mathrm{SL}(N, \mathbb{R})$ and a vector $v \in \mathbb{R}^N$ such that*

$$G = \{g \in \mathrm{SL}(n, \mathbb{R}) : \ \rho(g)(\mathbb{R}v) = \mathbb{R}v\}.$$

Moreover, if the group G is unipotent or reductive, there exist ρ and $v \in \mathbb{R}^N$ such that

$$G = \{g \in \mathrm{SL}(n, \mathbb{R}) : \ \rho(g)v = v\}.$$

If G is a $\mathbb{Q}$-group, then in both these results one can find a representation ρ over $\mathbb{Q}$ and a vector $v \in \mathbb{Q}^N$.

In other words, an algebraic group is the stabilizer in a projective representation. Moreover, unipotent and reductive groups are stabilizers in a linear representation. In particular, the image of G in $\mathrm{SL}(n, \mathbb{R})/\mathrm{SL}(n, \mathbb{Z})$ is closed provided G is a reductive or unipotent $\mathbb{Q}$-group. In general this is not the case: the product $\mathrm{T}(2, \mathbb{R})\,\mathrm{SL}(2, \mathbb{Z})$ is dense in $\mathrm{SL}(2, \mathbb{R})$.

Note that any orbit $U(v) \subset \mathbb{R}^N$ of a unipotent algebraic subgroup $U \subset \mathrm{SL}(N, \mathbb{R})$ is closed (Rosenlicht).

E. Finite volume spaces

Unless otherwise stated, all the results given below can be found in [**Rag**] and [**VGS**].

1. Basic definitions. In what follows, G stands for a *connected* Lie group. Given any closed subgroup $D \subset G$, a right-invariant Haar measure on G induces a smooth volume measure ν on the right homogeneous space G/D. If both G and D are unimodular, the measure ν is G-invariant under the left action[16] (in general this is not the case). If the measure ν is finite and G-invariant under the left action of G, one says that G/D is a *space of finite volume* (we write $\mathrm{vol}(G/D) < \infty$). If $D \subset \Lambda \subset G$, then

$$\mathrm{vol}(G/D) < \infty \Longleftrightarrow \mathrm{vol}(G/\Lambda) < \infty, \ \mathrm{vol}(\Lambda/D) < \infty.$$

In particular, if a subgroup $H \subset G$ is normal, $\mathrm{vol}(G/D) < \infty$, and the product HD is closed, then $H/D \cap H$ is of finite volume.

[16]Sometimes the G-invariant measure ν on G/D will be also called Haar measure.

If a subgroup Γ is discrete and $\nu(G/\Gamma) < \infty$, then the measure ν is always G-invariant, and Γ is called a *lattice* in G.

A subgroup $D \subset G$ is said to be *uniform* if the space G/D is compact. A compact homogeneous space G/D can fail to be of finite volume (example: D is a maximal triangularizable subgroup in a semisimple Lie group G), and vice versa (example: $\mathrm{SL}(n, \mathbb{Z})$ is a nonuniform lattice in $\mathrm{SL}(n, \mathbb{R})$). However, a uniform discrete subgroup is always a lattice.

A subgroup $D \subset G$ will be referred to as a *quasilattice* in G if $\mathrm{vol}(G/D) < \infty$ and D has no nontrivial connected subgroups normal in G. A lattice is said to be *faithful* if it has no nontrivial subgroups normal in G (equivalently, the intersection with the center $Z(G)$ is trivial).

Subgroups $\Gamma, \Gamma' \subset G$ are said to be *commensurable* if the intersection $\Gamma \cap \Gamma'$ is of finite index in Γ and Γ'. If Γ is a lattice in G, then every subgroup commensurable with it, is a lattice. A simple (but very useful) result says that the lattices $\Gamma = \mathrm{SL}(n, \mathbb{Z})$ and $g\Gamma g^{-1}$ in $\mathrm{SL}(n, \mathbb{R})$ are commensurable for every $g \in \mathrm{SL}(n, \mathbb{Q})$.

2. Nilpotent homogeneous spaces. Let N be a simply connected nilpotent Lie group. Recall that there exists an embedding $N \subset \mathrm{U}(n, \mathbb{R})$ which allows one to treat N as an algebraic group all of whose connected subgroups are algebraic subgroups. Let D be a subgroup of N, and $H = \mathcal{A}(D)$. The groups N and H are unimodular, and hence the space N/H carries an N-invariant measure. On the other hand, it is diffeomorphic to a vector space. It follows that N/D and N/H can be of finite volume only if D is Zariski dense in N. It turns out that this is a sufficient condition as well.

THEOREM E.1 (Maltsev). *If N is a simply connected nilpotent Lie group and $D \subset N$ is a closed subgroup, then the following conditions are equivalent:*

a) $\mathrm{vol}(N/D) < \infty$.

b) *The space N/D is compact.*

c) *D is Zariski dense in all finite-dimensional unipotent representations of N.*

d) *N has no proper connected subgroups containing D.*

As a corollary, we deduce that any quasilattice $D \subset N$ is a lattice. Indeed, the adjoint representation $\mathrm{Ad} : N \to \mathrm{Aut}(\mathfrak{n})$ is unipotent, and $\mathrm{Ad}(D)$ normalizes the Lie algebra $\mathfrak{d} \subset \mathfrak{n}$ of its identity component D_0. It follows that $\mathrm{Ad}(N)$ normalizes $\mathfrak{d}$, and hence D_0 is a normal subgroup of N, whence $D_0 = 1$. Therefore while studying nilpotent spaces of finite volume, it suffices to work with lattices $\Gamma \subset N$.

Let Γ be a lattice in N. Equip N with the structure of an algebraic group (recall that all connected subgroups of N are algebraic). Then

$$\mathcal{A}([\Gamma, \Gamma]) = [\mathcal{A}(\Gamma), \mathcal{A}(\Gamma)],$$

and hence $[\Gamma, \Gamma]$ is a lattice in $[N, N]$. It follows that N/Γ fibers over the torus $N/\Gamma[N, N]$.

Also, the product $Z(N)\Gamma$ is closed. This follows because $Z(N)\Gamma$ normalizes Γ and hence $\overline{Z(N)\Gamma}$ normalizes Γ as well. One concludes that the connected group $(\overline{Z(N)\Gamma})_0$ normalizes the discrete group Γ and hence commutes with the latter. But since $\Gamma \subset N$ is Zariski dense, the group $(\overline{Z(N)\Gamma})_0$ commutes with all of N. It follows that $(\overline{Z(N)\Gamma})_0 \subset Z(N)$, which means that the product $Z(N)\Gamma$ is closed.

Not every nilpotent Lie group contains a lattice (such examples exist for $\dim N > 6$). If a subgroup $N \subset \mathrm{U}(n, \mathbb{R})$ is defined over $\mathbb{Q}$, the group of rational points

$N_{\mathbb{Q}} = N \cap \mathrm{U}(n,\mathbb{Q})$ is Zariski dense in N. On the other hand, some power of any element in $\mathrm{U}(n,\mathbb{Q})$ belongs to $\mathrm{U}(n,\mathbb{Z})$, and hence the group of integer points $N_{\mathbb{Z}} = N \cap \mathrm{U}(n,\mathbb{Z})$ is also Zariski dense in N. Consequently, the $\mathbb{Q}$-algebraic group N contains $N_{\mathbb{Z}}$ as a lattice. It turns out that the converse holds true as well.

THEOREM E.2 (Maltsev). *A simply connected nilpotent Lie group contains a lattice $\Longleftrightarrow$ it admits the structure of an algebraic group over $\mathbb{Q}$.*

Note that the $\mathbb{Q}$-algebraicity condition is equivalent to the existence of a basis in $\mathfrak{n}$ with all structure constants rational.

3. Solvable homogeneous spaces.

THEOREM E.3 (Mostow). *If G is a solvable Lie group, then a space G/D is of finite volume $\Longleftrightarrow$ it is compact. The identity component D_0 of a quasilattice $D \subset G$ belongs to the nilradical $N \subset R$, and the space $N/D \cap N$ is compact.*

As is seen, a compact solvable manifold R/D always fibers over a torus of smaller dimension (either over R/ND if R is not nilpotent, or over $R/[R,R]D$ if $R = N$).

If a group R is class (R), then any uniform subgroup $D \subset R$ is Zariski dense in the Ad-representation (in particular, any quasilattice is a lattice). This is not the case for classes (I) and (E) (see §4 for more detail).

4. Semisimple homogeneous spaces. One knows that every semisimple Lie group contains uniform as well as nonuniform lattices (Borel). One of the basic results is the following density theorem.

THEOREM E.4 (Borel). *Let $L = K \times S$ be the decomposition of a semisimple Lie group into compact and noncompact parts, and let* $\mathrm{vol}(L/D) < \infty$ *and* $L = \overline{SD}$. *Then D is Zariski dense in all finite-dimensional representations of L.*

In particular, if $L = \overline{S\Gamma}$ then any quasilattice $\Gamma \subset L$ is a lattice. Moreover, the product $Z(L)\Gamma$ is closed. In the general case both statements fail to hold. For instance, the group $\mathrm{SL}(2,\mathbb{R})$ has a (nonuniform) lattice isomorphic to the free group F_2 with two generators, and hence the universal covering group $\widetilde{\mathrm{SL}(2,\mathbb{R})}$ contains a lattice $F_2 \times Z$, where $Z \simeq \mathbb{Z}$ is the center of $\widetilde{\mathrm{SL}(2,\mathbb{R})}$. Take a cyclic subgroup $Z(a) \subset Z \times \mathrm{SO}(2,\mathbb{R})$ such that $\overline{ZZ(a)} = Z \times \mathrm{SO}(2,\mathbb{R})$. Then $\Gamma = F_2 \times Z(a)$ is a lattice in $L = \widetilde{\mathrm{SL}(2,\mathbb{R})} \times \mathrm{SO}(3,\mathbb{R})$, the product of which with the center Z is not closed.[17]

We will also need the following result.

THEOREM E.5 (Prasad–Raghunathan [**PR**]). *Let Γ be a lattice in a semisimple Lie group L, and let $H \subset L$ be a Cartan subgroup. Then the set*

$$\{g \in L : \; g\Gamma g^{-1} \cap H \text{ is a lattice in } H\}$$

is dense in L.

One possible way to construct lattices in a semisimple Lie group is as follows.

[17]This example was communicated to the author by È. B. Vinberg.

THEOREM E.6 (Borel and Harish-Chandra). 1) *If* $H \subset \mathrm{SL}(n, \mathbb{R})$ *is a* $\mathbb{Q}$*-defined group, then the intersection* $H_{\mathbb{Z}} = \mathrm{SL}(n, \mathbb{Z}) \cap H$ *is a lattice in* $H \Longleftrightarrow H$ *admits no nontrivial* $\mathbb{Q}$*-characters.*

2) *Given a semisimple* $\mathbb{Q}$*-group* L*, the lattice* $L_{\mathbb{Z}}$ *is uniform* $\Longleftrightarrow L_{\mathbb{Q}}$ *consists of semisimple elements.*

In particular, $H_{\mathbb{Z}}$ is a lattice in H whenever the $\mathbb{Q}$-group H is either semisimple or unipotent.

Lattices $L_{\mathbb{Z}} \subset L = SK$ and $KL_{\mathbb{Z}} \cap S \subset S$ given by the $\mathbb{Q}$-structure on L provide examples of the so-called arithmetic lattices. More specifically, a lattice Γ in a semisimple Lie group L is called *arithmetic* if there exist a $\mathbb{Q}$-defined group $H \subset \mathrm{SL}(n, \mathbb{R})$ and an epimorphism $\rho : H \to \mathrm{Ad}\, L$ with compact kernel such that $\rho(H_{\mathbb{Z}})$ and $\mathrm{Ad}\,\Gamma$ are commensurable.

A lattice Γ in a semisimple Lie group S with no compact components is said to be *irreducible* if $\overline{S'\Gamma} = S$ for each nontrivial connected normal subgroup $S' \subset S$. We know that, given a lattice Γ in a semisimple Lie group S with no compact components, there exist a subgroup of finite index $\Gamma' \subset \Gamma$ and a decomposition $S = \prod_{i=1}^{n} S_i$ into the product of normal subgroups such that $\Gamma' = \prod_{i=1}^{n}(S_i \cap \Gamma')$ and $S_i \cap \Gamma'$ is an irreducible lattice in S_i for every $i = 1, \ldots, n$. In fact, if $S = S' \times S''$ and $\Gamma \cap S'$ is a lattice in S', then the connected group $(\overline{S''\Gamma})_0$ normalizes the lattice $\Gamma \cap S'$ and hence commutes with it. Since $\Gamma \cap S' \subset S'$ is a Zariski dense lattice, the group $(\overline{S''\Gamma})_0$ commutes with S'. Hence the product $S''\Gamma$ is closed or, equivalently, $\Gamma \cap S''$ is a lattice in S''.

The central result in the theory of arithmetic lattices is as follows.

THEOREM E.7 (Margulis; see [**Ma91c**]). *If* Γ *is an irreducible lattice in a semisimple Lie group* S *with no compact components and with* $\mathrm{rank}_{\mathbb{R}} S \geq 2$*, then* Γ *is arithmetic.*

Note that the first examples of nonarithmetic lattices in $\mathrm{SO}(1, n)$, $n = 3, 4, 5$, were constructed by Makarov and Vinberg (see [**VGS**]).

5. General homogeneous spaces. In what follows, G is a connected simply connected Lie group with a Levi decomposition $G = L \cdot R$, and $L = S \times K$ is the decomposition of the Levi subgroup L into noncompact and compact parts.

THEOREM E.8 (Auslander). *If* $\mathrm{vol}(G/D) < \infty$ *and the identity component* D_0 *is solvable, then there exists a torus* $T \subset L$ *such that* $TR = (\overline{RD})_0$ *and the space* $TR/D \cap TR$ *is compact.*

A particular case of the theorem is given by the following result.

THEOREM E.9 (Bieberbach). *Given a discrete uniform group* Γ *of Euclidean motions of* $\mathbb{R}^n$*, there exists a finite index subgroup* $\Gamma_0 \subset \Gamma$ *consisting of pure translations.*

In fact, Γ as above is a lattice in the group $\mathrm{O}(n, \mathbb{R}) \cdot \mathbb{R}^n$. By Auslander's theorem, Γ contains a finite index subgroup $\Gamma' \subset T \cdot \mathbb{R}^n$, and by Mostow's theorem, the intersection $\Gamma_0 = \Gamma' \cap \mathbb{R}^n$ is of finite index in Γ'.

Another corollary of Auslander's theorem is the following useful statement.

THEOREM E.10. *If* $G = (S \times K) \cdot R$ *is a Levi decomposition, then the intersection of a lattice* $\Gamma \subset G$ *with the normal subgroup* KR *is a uniform lattice in* KR*.*

Indeed, let $q : G = L \cdot R \to L$ be the projection of the group G along the radical R onto the Levi subgroup $L = S \times K$, and let $T = (\overline{q(\Gamma)})_0$. Since $\mathrm{vol}(G/\overline{R\Gamma}) < \infty$, we have $\mathrm{vol}(L/\overline{q(\Gamma)}) < \infty$. Note that $\overline{KR\Gamma} = K\overline{R\Gamma}$, whence $\mathrm{vol}(S/K\overline{q(\Gamma)} \cap S) < \infty$. By Borel's theorem, the subgroup $KT \cap S = (K\overline{q(\Gamma)} \cap S)_0$ is normal in S, and by Auslander's theorem it is solvable. It follows that T is a torus in K and the product $KR\Gamma$ is closed. □

In particular, if $G = S \cdot R$ (that is, if Levi subgroups of G have no compact components), then the intersection of a lattice $\Gamma \subset G$ with the radical R is a lattice in R. It follows that the projection $q(\Gamma) \subset S$ is a lattice in S (Wang's theorem [**Rag**]). Therefore in this case the structure of a lattice Γ is consistent with a Levi decomposition of G.

The presence of the compact part K of a Levi subgroup L significantly complicates the study of homogeneous spaces G/D. For instance, the intersection of a (quasi)lattice D with the radical R fails to be a (quasi)lattice in R (that is, sometimes $\mathrm{vol}(R/D \cap R) = \infty$), and hence the image $q(D) \subset L$ need not be closed. Example: $G = \mathrm{SO}(3, \mathbb{R}) \times \mathbb{R}$ and Γ is a cyclic subgroup of the cylinder $\mathrm{SO}(2, \mathbb{R}) \times \mathbb{R}$ such that $\overline{\mathbb{R}\Gamma} = \mathrm{SO}(2, \mathbb{R}) \times \mathbb{R}$. Moreover, even if $G = K \cdot R$ and the compact group K acts effectively on R (that is, with trivial kernel), the intersection $D \cap R$ need not be a (quasi)lattice in R; an example is given in [**St84b**].

This explains why the study of G/D may not reduce to solvable and semisimple cases. However, under certain conditions on the quasilattice $D \subset G$, life is much easier.

THEOREM E.11 (Witte [**Wi87**], Starkov [**St87a**]). *If D is a quasilattice in G, and $G = \overline{SRD}$, then the intersection $D \cap R$ is a quasilattice in the radical R, and the projection $q(D)$ is a lattice in L.*

If $\Gamma \subset G$ is a lattice, the proof is similar to that of Theorem E.10.

If one does not impose any restrictions on a lattice $\Gamma \subset G$, the following generalization of Mostow's theorem holds:

THEOREM E.12 (Starkov [**St93**]). *If Γ is a lattice in a Lie group G with a Levi decomposition $G = L \cdot R$ and $N \subset R$ is the nilradical of G, then the intersection $\Gamma \cap LN$ is a lattice in LN.*

The proof relies on a statement from Witte's paper [**Wi87**] which directly implies the theorem in the case $G = \overline{SR\Gamma}$.

The density theorems of Maltsev and Borel can be generalized for arbitrary Lie groups as follows.

THEOREM E.13 (Dani [**Da80a**]). *Let $\rho : G \to \mathrm{SL}(n, \mathbb{R})$ be a finite-dimensional representation of a connected Lie group G, and let $\mathrm{vol}(G/D) < \infty$. Then the algebraic hull $\rho(D)$ contains all unipotent one-parameter subgroups of $\rho(G)$.*

F. Mahler's criterion

Consider the action of $\mathrm{SL}(n, \mathbb{R})$ on $\mathbb{R}^n$. The lattice $\mathrm{SL}(n, \mathbb{Z}) \subset \mathrm{SL}(n, \mathbb{R})$ keeps the lattice $\mathbb{Z}^n \subset \mathbb{R}^n$ invariant, and for every $g \in \mathrm{SL}(n, \mathbb{R})$ the lattice $g(\mathbb{Z}^n) \subset \mathbb{R}^n$ is *unimodular*, that is, it has a fundamental domain of unit volume. The noncompact space of finite volume $\mathrm{SL}(n, \mathbb{R}) / \mathrm{SL}(n, \mathbb{Z})$ is homeomorphic to the space of unimodular lattices in $\mathbb{R}^n$. The main link between number theory and the theory of homogeneous flows is the following criterion of Mahler's.

THEOREM F.1 (see [**Rag**]). *A sequence* $g_i \operatorname{SL}(n, \mathbb{Z})$ *exits to infinity in the space* $\operatorname{SL}(n, \mathbb{R}) / \operatorname{SL}(n, \mathbb{Z})$ *if and only if there exists a sequence of nonzero vectors* $e_i \in \mathbb{Z}^n$ *such that* $g_i(e_i) \to 0$ *as* $i \to \infty$.

The theorem has a stronger formulation in which the speed of exiting to infinity of the sequence $g_i \operatorname{SL}(n, \mathbb{Z})$ is controlled by the rate of approaching to $0 \in \mathbb{R}^n$ by elements of the lattices $g_i(\mathbb{Z}^n)$ (see §30).

CHAPTER 1

Ergodicity and Mixing of Homogeneous Flows

The study of ergodic properties of homogeneous flows on finite volume spaces involves the theory of unitary representations of Lie groups. Recall that unitary representations were first used by Gelfand and Fomin [**GF**] in the study of the spectrum of a geodesic flow on a manifold of constant curvature -1 (ergodicity and mixing of the flow had already been established by Hopf and Hedlund). Quite soon, Parasjuk [**Pa**] found the spectrum of the horocycle flow and proved its mixing property. Both papers relied heavily on the (previously known) classification of irreducible unitary representations of $\mathrm{SO}(1,n)$.

Mautner [**Mau**] applied the theory of unitary representation to the study of ergodic properties of geodesic flows on locally symmetric spaces. Actually, he noticed that, given a unitary representation of the corresponding Lie group, any fixed vector for the subgroup inducing the geodesic flow is also fixed by the horospherical subgroups inducing the contracting and expanding foliations. This enables one to derive ergodicity of many partially hyperbolic flows by elementary methods (that is, without knowing all irreducible unitary representations of the Lie group involved). Subsequently this argument was used by Moore [**Mo66**] and also by Auslander and Green [**AG**] in the study of homogeneous flows in the semisimple and solvable cases (in full scope, the spectrum of a flow on a semisimple homogeneous space was found by Stepin [**Ste73**]).

The ergodicity criterion and the spectrum of nilflows were first found by Green [**G**], [**AGH**].[1] He also used the structure of irreducible unitary representations (this time, of nilpotent Lie groups). Refining Green's results, Stepin [**Ste69**] and Safonov [**Saf**] completed the computation of spectra (including multiplicities) of nilflows and flows on solvable manifolds.

The search for ergodicity and mixing criteria in the general case was also based on the computation of the spectrum of homogeneous flows. Here the main step was provided by Dani's paper [**Da77**], where the general case was reduced to the semisimple and solvable cases, which were already well understood. The strongest technical result in the theory of unitary representations needed for the computation of spectra of homogeneous flows was proved by Moore [**Mo80**], who found the most general formulation[2] of the so-called *Mautner phenomenon*. The phenomenon says that fixed vectors for a subgroup $F \subset G$ are also fixed under a certain normal subgroup $M(F) \subset G$ (for any unitary representation of the Lie group G).

Along with the subgroup $M(F)$ (we call it the *Moore normal subgroup*) we use the *Auslander normal subgroup* $A(F)$ and the *Dani normal subgroup* $J(F)$ (note that $A(F) \subset J(F) \subset M(F)$). The Auslander normal subgroup plays a key role in

[1] In §11 we correct a gap in his original proof.

[2] Actually, Moore was concerned with one-parameter subgroups F. The observation that all the results hold true for an arbitrary subgroup F was made in [**Ma91b**].

the reduction of many problems to the case of a quasi-unipotent subgroup $g_{\mathbb{R}}$ (in other words, in the reduction to a homogeneous flow with zero entropy), whereas the Dani and Moore subgroups prove useful in finding the ergodic decomposition and the mixing criterion.

Later, Margulis [**Ma91b**] outlined a proof of the Mautner phenomenon which used only elementary arguments from the theory of representations. This approach was developed in [**MT96**] and plays a key role in our study of the Mautner phenomenon in §9. Thus, we do not use spectral information for the proof of ergodicity and weak mixing. Spectral theory is presented in §11.

The study of solvable flows relies heavily on Maltsev's construction of a *semisimple splitting* of solvable Lie groups and its modification by Auslander [**AG**], [**Au**]. Roughly speaking, the construction enables one to embed a solvable Lie group into a real-algebraic group and to use well-developed techniques of algebraic groups. For general Lie groups this approach was refined by Auslander and Brezin [**AB68**], Brezin and Moore [**BM**], and the author [**St84a,89**]. In essence, it consists in simulating an algebraic structure on an arbitrary Lie group (the idea is that the factor group by a certain discrete central subgroup can be embedded into a "true" algebraic group)[3]. This will be shown in §6.

Another technical tool for studying homogeneous flows is purely algebraic information on the structure of finite volume spaces G/D. Apart from classical results due to Maltsev, Mostow, Borel, Auslander, and others, one needs extra effort to reduce the general case to the semisimple and solvable cases. For instance, the ergodicity criterion was originally found by Dani [**Da77**] and by Brezin and Moore [**BM**] for the so-called admissible spaces of finite volume. It was proved independently by the author [**St86a**] and Witte [**Wi87**] that all finite volume spaces are admissible. All related results on the structure of finite volume spaces are presented without proofs in §0. The ergodicity and weak mixing criteria for arbitrary homogeneous flows will be obtained in §9 along the lines of [**Da77, BM, St87a**].

In studying homogeneous flows we will see that many homogeneous spaces G/D do not carry an ergodic flow $(G/D, g_{\mathbb{R}})$. Thus, there naturally arises the problem of constructing the ergodic decomposition. The author has shown in a series of papers [**St83,86b,89**] that components of the ergodic decomposition E for a flow $(G/D, g_{\mathbb{R}})$ are certain closed invariant submanifolds $E_h(g_{\mathbb{R}})$, each carrying a finite smooth ergodic measure μ_h (see §1). Hence the study of a nonergodic homogeneous flow on a finite volume space reduces to the study of a family of smooth ergodic flows.

If the subgroup $g_{\mathbb{R}}$ is unipotent, all the ergodic submanifolds are homogeneous and have finite volume (this will be used in Chapter 2). In the general case an ergodic submanifold $E_h(g_{\mathbb{R}})$ fails to be a homogeneous subspace of G/D (see §4) and *a priori* in the course of the reduction one loses the homogeneous nature of the flow. Fortunately, one can prove that each flow $(E_h(g_{\mathbb{R}}), g_{\mathbb{R}})$ is finitely covered by an ergodic flow on the quotient space of a Lie group by a lattice (§10). Hence, from both measure-theoretic and topological points of view, the study of flows on homogeneous spaces of finite volume is reduced to the situation, where $(G/\Gamma, g_{\mathbb{R}})$ is an ergodic flow and Γ is a lattice in a Lie group G. Exactly such flows will be studied in the next chapters.

[3]Note that this idea has been independently exploited for other purposes in many papers; we mention [**HM**] and [**Go**].

§1. General constructions in the theory of homogeneous flows

1. Given a closed subgroup D of a connected Lie group G, let us consider the homogeneous space G/D. With respect to the transitive action of G by translations, the subgroup D is the *isotropy subgroup* (that is, the stationary subgroup of the point $D \in G/D$). G-right-invariant Haar measure on G induces a smooth volume measure ν on G/D. Note that such a measure arises also as the volume measure for a Riemannian metric on G/D defined by a G-right-invariant metric on G. A *homogeneous action* $(G/D, F)$ induced by a subgroup $F \subset G$, is the left action of F on the right homogeneous space G/D equipped with the measure ν.

First we make a few immediate observations. Let $(G/D, F)$ be the action by left translations of a subgroup $F \subset G$ on a homogeneous space of a Lie group G. Then without loss of generality the group G may be assumed to be simply connected. In fact, let $\alpha : \widetilde{G} \to G$ be the universal covering map, and let $\widetilde{D} = \alpha^{-1}(D)$ and $\widetilde{F} = \alpha^{-1}(F)$. Then the actions $(G/D, F)$ and $(\widetilde{G}/\widetilde{D}, \widetilde{F})$ are measure-theoretically and topologically isomorphic (though, formally speaking, the groups F and $\widetilde{F}$ may differ). If we want the subgroup F to be connected (for instance, one-parameter), in place of $\widetilde{F}$ we can take the identity component $(\alpha^{-1}(F))_0$.

Without loss of generality we can agree that the isotropy subgroup $D \subset G$ has no connected subgroups normal in G. In fact, if $H \subset D$ is the maximal connected subgroup normal in G, and

$$\alpha : G \to G' = G/H, \;\; D' = D/H, \;\; F' = \alpha(F),$$

then the actions $(G/D, F)$ and $(G'/D', F')$ are isomorphic. In particular, if $\mathrm{vol}(G/D) < \infty$, then D may be assumed to be a quasilattice.

The left action of F on the right homogeneous space G/D is isomorphic to the right action of F on the left homogeneous space $D\backslash G$ via the map $gD \to Dg^{-1}$. Hence it is not important which kind of actions (right or left) we are studying. Almost everywhere we will work with left actions; we turn to right actions only in §§14–16 while studying flows on surfaces of constant negative curvature.

Finally, conjugate subgroups induce isomorphic actions. Indeed, the actions $(G/D, F)$ and $(G/D, gFg^{-1})$ are isomorphic via the left translation $\beta_g(hD) = ghD$ because

$$\beta_g(fhD) = gfhD = gfg^{-1}\beta_g(hD) \quad \text{for all } h \in G, \; f \in F.$$

2. We fix an element $g \in G$ and consider the linear operator Ad_g on the Lie algebra $\mathfrak{g}$. With respect to the operator, the complexified Lie algebra $\mathfrak{g}_{\mathbb{C}} = \mathfrak{g} \bigotimes_{\mathbb{R}} \mathbb{C}$ decomposes into finitely many generalized root subspaces $\mathfrak{g}_{\mathbb{C}}^{\lambda}$, where

$$\mathfrak{g}_{\mathbb{C}}^{\lambda} = \{x \in \mathfrak{g}_{\mathbb{C}} : \; \exists n : \; (\mathrm{Ad}_g - \lambda E)^n x = 0\}$$

and E stands for the identity operator. A number $\lambda \in \mathbb{C}$ is called an *eigenvalue* of the operator Ad_g if $\mathfrak{g}_{\mathbb{C}}^{\lambda} \neq 0$ (then there exists $x \in \mathfrak{g}_{\mathbb{C}}^{\lambda}$, $x \neq 0$, such that $\mathrm{Ad}_g x = \lambda x$). The dimension (over $\mathbb{C}$) of the root subspace $\mathfrak{g}_{\mathbb{C}}^{\lambda}$ is called the *multiplicity* of the eigenvalue λ.

The set of all eigenvalues will be denoted by Σ. Since $\mathrm{Ad}_g[x, y] = [\mathrm{Ad}_g x, \mathrm{Ad}_g y]$, we have $[\mathfrak{g}_{\mathbb{C}}^{\lambda}, \mathfrak{g}_{\mathbb{C}}^{\theta}] \subset \mathfrak{g}_{\mathbb{C}}^{\lambda\theta}$. For real λ we define $\mathfrak{g}^{\lambda} = \mathfrak{g}_{\mathbb{C}}^{\lambda} \cap \mathfrak{g}$, and for a pair of complex conjugate eigenvalues λ and $\overline{\lambda}$ we set

$$\mathfrak{g}^{\lambda} = \mathfrak{g}^{\overline{\lambda}} = (\mathfrak{g}_{\mathbb{C}}^{\lambda} + \mathfrak{g}_{\mathbb{C}}^{\overline{\lambda}}) \cap \mathfrak{g}.$$

We have obtained the decomposition $\mathfrak{g} = \sum_{\lambda \in \Sigma} \mathfrak{g}^{\lambda}$ into the sum of Ad_g-invariant subspaces. Note that

$$[\mathfrak{g}^{\lambda}, \mathfrak{g}^{\theta}] \subset \mathfrak{g}^{\lambda\theta} + \mathfrak{g}^{\lambda\overline{\theta}}.$$

We define

$$\mathfrak{g}^{+} = \sum_{|\lambda|>1} \mathfrak{g}^{\lambda}, \qquad \mathfrak{g}^{0} = \sum_{|\lambda|=1} \mathfrak{g}^{\lambda}, \qquad \mathfrak{g}^{-} = \sum_{|\lambda|<1} \mathfrak{g}^{\lambda}.$$

It is easy to see that $\mathfrak{g}^{+}$ and $\mathfrak{g}^{-}$ are nilpotent subalgebras which are normalized by the subalgebra $\mathfrak{g}^{0}$.

Let G^{+}, Q, G^{-} be the analytic (that is, connected) subgroups of G corresponding to the subalgebras $\mathfrak{g}^{+}, \mathfrak{g}^{0}, \mathfrak{g}^{-}$. The subgroups G^{+} and G^{-} will be called the *expanding* and *contracting horospherical* subgroups (relative to the element g). The subgroup Q is said to be the *neutral* subgroup. It is easy to check that

$$G^{+} = \{h \in G : \; g^{-n} h g^{n} \to 1, \; n \to +\infty\},$$
$$G^{-} = \{h \in G : \; g^{n} h g^{-n} \to 1, \; n \to +\infty\},$$

and both the subgroups are nilpotent and are normalized by Q. The subgroup $A \subset G$ generated by G^{+} and G^{-} will be referred to as the *Auslander normal subgroup*[4] for the element $g \in G$.

An element g is said to be *partially hyperbolic* if the subgroup $A = A(g)$ is nontrivial. Otherwise (if $Q = G$, that is, if all eigenvalues of Ad_g are of absolute value 1), the element g is called *quasi-unipotent.* Recall that an element g is called *unipotent* if all eigenvalues of the operator Ad_g are equal to 1. If the operator Ad_g is diagonalizable over $\mathbb{C}$, the element g is *semisimple*, and if it is diagonalizable over $\mathbb{R}$, the element g is $\mathbb{R}$-*diagonalizable.* Note that an element g may be semisimple and quasi-unipotent simultaneously if and only if the operator Ad_g belongs to a compact subgroup of $\mathrm{Aut}(\mathfrak{g})$.

3. It is easily seen that the Auslander normal subgroup A is the smallest normal subgroup of G such that the element $gA \in G/A$ is quasi-unipotent in the factor group G/A.

The *Dani normal subgroup*[5] J for an element $g \in G = LR$ (where as usual R is the radical of the group G and L is a Levi subgroup) is the smallest normal subgroup of G such that $gJ \in G/J$ is quasi-unipotent in G/J, and $gJR \in G/JR$ is semisimple in the semisimple Lie group G/JR.

Finally, the *Moore normal subgroup*[6] M for an element $g \in G$ is the smallest normal subgroup $M \subset G$ such that $gM \in G/M$ is both quasi-unipotent and semisimple in G/M.

Obviously, $A \subset J \subset M$. If the group G is nilpotent then $A = J = 1$. If G is solvable then $A = J$. If G is semisimple then $J = M$.

In the sequel we use the constructive definition of the Dani subgroup J. Given the projection $q : G = L \cdot R \to L$, we take a Jordan decomposition $q(g_t) = a_t \times u_t$

[4] In [**AG**] this subgroup was called the unstable normal subgroup for the element g.

[5] We call it so because it was Dani [**Da77**] who introduced it for the first time.

[6] In [**Mo80**] it was called the Ad-compact subgroup, and in [**Ma91b**] the Mautner subgroup for $g \in G$.

into reductive and unipotent subgroups $a_{\mathbb{R}}$, $u_{\mathbb{R}} \subset L$. By L' we denote the smallest normal subgroup of L containing $u_{\mathbb{R}}$. Let us consider the (completely reducible) action of L' by conjugations on the nilradical $N \subset R$. Let $N' \subset N$ be the subgroup generated by all nontrivial irreducible L'-submodules, and define $F = L' \cdot N'$. Clearly, the subgroups N' and F are normal in G, and $J = AF$ is the Dani normal subgroup for $g_{\mathbb{R}} \subset G$.

We emphasize an important difference between the group M and its subgroups A and J. Assume that a closed subgroup $G' \subset G$ contains $g_{\mathbb{R}}$ and A. Then $G^+, G^- \subset G'$, and hence A remains the Auslander normal subgroup for $g_{\mathbb{R}}$ if one considers G' instead of G. As follows from the constructive definition, the Dani normal subgroup J has the same property of "self-preservation". But if $g_{\mathbb{R}} M \subset G'$ then the Moore normal subgroup M' for $g_{\mathbb{R}} \subset G'$ does not necessarily coincide with M (although it belongs to the latter). This will be clearly seen in the case of a nilpotent Lie group G.

Now let F be a subgroup of G. The subgroup $A = A(F)$ generated by the Auslander normal subgroups for elements $g \in F$ is called the Auslander normal subgroup for F. In a similar way one defines the Dani normal subgroup and the Moore normal subgroup for F. Note that for a one-parameter group $g_{\mathbb{R}}$ all three subgroups coincide with the corresponding subgroups for a single element g_s, $s \neq 0$. For instance, $A(g_{\mathbb{R}}) = A(g_s)$, $s \neq 0$.

Let us make another remark. Assume that $g_t = c_t \times h_t$, where the subgroup $c_{\mathbb{R}}$ is reductive and quasi-unipotent (that is, $M(c_{\mathbb{R}}) = 1$). Then

$$A(g_{\mathbb{R}}) = A(h_{\mathbb{R}}), \qquad J(g_{\mathbb{R}}) = J(h_{\mathbb{R}}), \qquad M(g_{\mathbb{R}}) = M(h_{\mathbb{R}}).$$

4. When G/D has a finite G-invariant measure ν, the ergodic properties of homogeneous dynamical systems on G/D are fairly well studied by means of the theory of representations. To demonstrate the connection, let us consider the Hilbert space $L^2(G/D)$ of complex-valued functions f on G/D which are square-integrable relative to ν. The space is equipped with the scalar product

$$(f, f') = \int_{G/D} f(x)\overline{f'(x)} d\nu_x.$$

The left action of G on G/D induces a strongly continuous unitary representation $\rho : G \to \mathbb{U}(L^2(G/D))$ by the formula

$$\rho(g) f(hD) = f(g^{-1}hD).$$

In the sequel we often use the following simple argument (see [**Mo66, Da77**]).

LEMMA 1.1. *Let H be a normal subgroup of G, and let G/D be a space of finite volume. Then the measurable hull ξ of the foliation of G/D into H-orbits is* (mod 0) *the smooth fibration into closed leaves $g\overline{HD}$, $g \in G$. The space*

$$\mathrm{Fix}(H) = \{f \in L^2(G/D) : f(hx) = f(x),\ h \in H,\ x \in G/D\}$$

is isomorphic to the Hilbert space $L^2(G/\overline{HD})$.

Proof. First we prove the second part. Let $f \in L^2(G/D)$ be constant along H-orbits. Then the function $\widetilde{f}$ on G defined as $\widetilde{f}(g) = f(gD)$ is locally integrable (relative to the right Haar measure m on G) and invariant under the right action of the subgroup $HD \subset G$. Any open subset $O \subset G$ with compact closure gives rise to a seminorm on the space $L^2_{\text{loc}}(G)$ by $\|\phi\|_O = \int_O |\phi(g)| dm_g$. The action of G is continuous relative to this seminorm, and hence $\widetilde{f}$ is $\overline{HD}$-invariant. Consequently, f is constant along leaves of the fibration $G/D \to G/\overline{HD}$, and therefore defines an element in $L^2(G/\overline{HD})$.

Now let ξ be the measurable hull of the foliation of G/D into H-orbits. Then there exists a countable system of functions $\{f_i \in L^2(G/D),\ i \in \mathbb{N}\}$ which are constant along the partition ξ and such that $y \in \xi_x \iff f_i(x) = f_i(y),\ i \in \mathbb{N}$. Since each function $f_i \in L^2(G/D)$ is constant along ξ, it is constant along H-orbits and hence along leaves of the fibration $G/D \to G/\overline{HD}$. It is obvious that the fibration is a measurable partition and hence is none other than ξ. □

5. *In what follows, we always assume that D is a quasilattice in G.*

For a moment we restrict ourselves to the consideration of a one-parameter flow $(G/D, g_{\mathbb{R}})$. Let d_G be a right-invariant metric on G. Then, given any $h \in G^-$, one has

$$d_G(g_t, g_t h) = d(1, g_t h g_{-t}) \to 0, \qquad t \to +\infty,$$

and hence orbits of G^- on G/D form the contracting foliation for the flow $(G/D, g_{\mathbb{R}})$ in the metric d on G/D induced by the metric d_G (and the rate of contracting is uniform and exponential)[7].

Likewise, G^+-orbits on G/D form the smooth expanding foliation for our flow. If one of the foliations on G/D is nontrivial, so are the others (because the volume measure on G/D is invariant), and our flow is *uniformly* partially hyperbolic. More specifically, there exist constants $C \geq 1$ and $\alpha < 1$ such that for any $h \in G^-$, $h' \in G^+$ and $x \in G/D$,

$$d(g_t x, g_t h x) < C\alpha^t d(x, hx),\ t > 0, \quad \text{and} \quad d(g_t x, g_t h' x) < C\alpha^{|t|} d(x, h'x),\ t < 0.$$

To show this, it suffices to consider the finite set Σ of eigenvalues of the operator Ad_{g_1} on the Lie algebra $\mathfrak{g}$ and to select $\alpha < 1$ such that

$$\alpha > \max\{|\lambda| < 1,\ \lambda \in \Sigma\} \quad \text{and} \quad \alpha > \max\{\tfrac{1}{|\lambda|} < 1,\ \lambda \in \Sigma\}$$

(We cannot *a priori* take $C = 1$, because matrix entries of the one-parameter group Ad_{g_t} on the root space $\mathfrak{g}_\lambda$ may involve polynomials in t.) For a quasi-unipotent flow, the rate of divergence of close trajectories is at most polynomial.

6. Now we turn to calculating the entropy of homogeneous flows. It is known (Sinai [**Si66**], Parry [**P69**]) that the Pinsker partition π of a smooth uniformly partially hyperbolic flow $(M, g_{\mathbb{R}})$ on a manifold with finite invariant measure is such that almost every element π_x contains the expanding and contracting leaves of the point $x \in M$. Applying this to a homogeneous flow $(G/D, g_{\mathbb{R}})$, one proves that the partition π is invariant under both G^+ and G^-, and hence under the Auslander normal subgroup $A \subset G$. Lemma 1.1 now ensures that the Pinsker partition for the flow $(G/D, g_{\mathbb{R}})$ consists of leaves of the fibration $G/D \to G/\overline{AD}$.

[7] If the isotropy subgroup D is discrete, then the foliation into G^--orbits is obviously smooth. But given an arbitrary subgroup D, the dimension of G^--orbits on G/D may vary. However, if $\operatorname{vol}(G/D) < \infty$, then all G^--orbits are of the same dimension [**St91**] and we can talk about smooth foliation.

Since D is a quasilattice in G, the Pinsker partition is nontrivial (and hence the entropy of the flow is positive) provided that $A \neq 1$.

The quotient flow $(G/\overline{AD}, g_{\mathbb{R}})$ is quasi-unipotent, and hence the rate of divergence of its close trajectories is at most polynomial. There is a general result due to Kushnirenko which says that a smooth flow on a compact manifold with polynomial rate of divergence of its orbits has zero entropy. As was shown in [**Da77**], the proof can be adapted for a homogeneous flow on a (not necessarily compact) space of finite volume. Hence the following holds.

THEOREM 1.2 (Dani [**Da76b, Da77**]). *The flow $(G/\overline{AD}, g_{\mathbb{R}})$ is the maximal quotient flow with zero entropy for $(G/D, g_{\mathbb{R}})$. In particular, $(G/D, g_{\mathbb{R}})$ has positive entropy $\iff$ it is partially hyperbolic (that is, $A \neq 1$), and is a K-flow $\iff G = \overline{AD}$.*

The exact formula for the entropy of a homogeneous flow is as follows.

THEOREM 1.3 (Dani [**Da77**]). *Let D be a lattice in G. Then the entropy of a homogeneous flow $(G/D, g_{\mathbb{R}})$ is given by the formula*

$$e(g_1) = \sum_{|\lambda_i|>1} m_i \log |\lambda_i|,$$

where $\lambda_1, \ldots, \lambda_n$ are eigenvalues of the operator Ad_{g_1} on $\mathfrak{g}$, and $m_1, \ldots, m_n$ the corresponding multiplicities.

Note that it is not difficult to prove that the entropy of a partially hyperbolic homogeneous flow is positive. The entropy formula in the compact case was found by Bowen [**Bo71**] and follows from a general Pesin–Millionshchikov formula [**Pe, Mi76**] which expresses the entropy of a smooth dynamical system in terms of Lyapunov exponents. Dani extended these results to the case when the space G/D is of finite volume but not necessarily compact.

The normal subgroups $J \subset M \subset G$ also play a key role while studying ergodic properties of a flow $(G/D, g_{\mathbb{R}})$ relative to a finite G-invariant measure ν. More specifically, as we will see below, the quotient flow $(G/\overline{JD}, g_{\mathbb{R}})$ is distal and the flow $(G/\overline{MD}, g_{\mathbb{R}})$ is uniformly continuous (and each of the two is maximal among all quotient flows of homogeneous origin with the corresponding property).

§2. The Mautner phenomenon and the ergodic decomposition

1. Let $\rho : G \to \mathbb{U}(L^2(G/D))$ stand for the natural unitary representation given as

$$\rho(g)f(hD) = f(g^{-1}hD).$$

Recall that $f \in L^2(G/D)$ is called an *eigenfunction* for a flow $(G/D, g_{\mathbb{R}})$ if there exists $\lambda \in \mathbb{C}$, $|\lambda| = 1$, such that

$$\rho(g_t)f = \lambda^t f \quad \text{for all } t \in \mathbb{R}.$$

The equality $\rho(g_t)f = f$, $t \in \mathbb{R}$, stands for an *invariant* function.

In this setting, ergodicity of a flow is clearly equivalent to all invariant functions in $L^2(G/D)$ being constant. Weak mixing means that any eigenfunction is constant, that is, invariant under the entire group G. Clearly, while checking these properties of a flow it is very important to know the maximal normal subgroup of G keeping all eigenfunctions of the flow fixed. The following result is crucial in the ergodic theory of homogeneous flows.

THEOREM 2.1 [**Mo80**]. *Let $\rho : G \to \mathbb{U}(\mathbb{H})$ be a continuous unitary representation of a group G in a Hilbert space $\mathbb{H}$, and let $v \in \mathbb{H}$ be a fixed vector for a subgroup $g_{\mathbb{R}} \subset G$. Then $\rho(M)v \equiv v$, where $M \subset G$ is the Moore normal subgroup for $g_{\mathbb{R}}$.*

Given a subgroup $F \subset G$, let $\mathrm{Fix}(F) \subset \mathbb{H}$ denote the closed subspace of F-fixed vectors in $\mathbb{H}$. In this notation, Theorem 2.1 ensures that $\mathrm{Fix}(g_{\mathbb{R}}) \subset \mathrm{Fix}(M)$ for each continuous unitary representation of G. This property of the subgroup M is called the *Mautner phenomenon* for unitary representations, since it was first discovered (in a particular case) by Mautner [**Mau**].

Actually, a somewhat stronger result holds: if $v \in \mathbb{H}$ is an eigenvector for a subgroup $g_{\mathbb{R}}$, then $v \in \mathrm{Fix}(M)$. Indeed, assume that $\rho(g_t)v = \lambda^t v$ and $G' = G \times \mathbb{R}$. One extends the representation ρ onto G' as follows:

$$\rho(t)w = \lambda^{-t}w \qquad \text{for all } t \in \mathbb{R},\ w \in \mathbb{H}.$$

Then $v \in \mathrm{Fix}(h_{\mathbb{R}})$, where $h_t = g_t \times t$. Clearly, M is the Moore normal subgroup for $h_{\mathbb{R}} \subset G'$, and hence $v \in \mathrm{Fix}(M)$.

It is also straightforward that the one-parameter subgroup $g_{\mathbb{R}}$ in the theorem may be replaced by any connected[8] subgroup $F \subset G$ if one takes the corresponding Moore subgroup $M = M(F)$.

As was noticed by Margulis [**Ma91b**], the proof of Theorem 2.1 is, roughly speaking, based on the following simple argument.

LEMMA 2.2. *Assume that $g \in \overline{FO(1)F}$ for any neighborhood $O(1) \subset G$. Then $\mathrm{Fix}(F) \subset \mathrm{Fix}(g)$ for every continuous unitary representation of the group G. Similarly, if there exist $g_i \in G$, $g_i \to 1$, and $f_i, f_i' \in F$ such that $g_i f_i g_i^{-1} f_i' \to g$, then $\mathrm{Fix}(F) \subset \mathrm{Fix}(g)$.*

Proof. Given $v \in \mathrm{Fix}(F)$, we introduce a function $f(h) = (\rho(h)v, v)$ on G. Note that since the representation ρ is unitary, the function f is constant along the sets FhF, $h \in G$. But $f(1) = (v, v)$ and ρ is continuous, so $f(g) = (v, v)$. Finally, because of the unitary property, $f(g) = (v, v) \iff \rho(g)v = v$.

Similarly,

$$f(g_i f_i g_i^{-1} f_i') = (v_i + \rho(f_i)\varepsilon_i, v_i + \varepsilon_i) \to (v, v), \qquad \text{where } \rho(g_i^{-1}v) = v_i + \varepsilon_i,$$

and

$$v_i \in \mathrm{Fix}(F),\ v_i \to v, \qquad \text{and} \qquad \varepsilon_i \in \mathbb{H} \ominus \mathrm{Fix}(F),\ \varepsilon_i \to 0.$$

We again conclude that $f(g) = (v, v)$. □

Theorem 2.1 will be completely proved in §9 (and in §11 it will be derived from a general result in spectral theory). However, we can already claim that $g_{\mathbb{R}}$-fixed vectors are also A-fixed. In fact, since $h_n = g_n h g_{-n} \to 1$ as $n \to +\infty$ for any $h \in G^-$ and $h = g_{-n} h_n g_n$, it follows that any element $h \in G^-$ satisfies the hypothesis of Lemma 2.2. Hence $\rho(G^-)v = v$ whenever $\rho(g_{\mathbb{R}})v = v$. Likewise, $\rho(G^+)v = v$, and hence $\rho(A)v = v$.

Thus, replacing G by G/A and $\mathbb{H}$ by the closed G-invariant subspace $\mathrm{Fix}(A)$, we reduce the proof of Theorem 2.1 to the quasi-unipotent case.

[8] Actually, Theorem 2.1 holds for any (not necessarily connected) subgroup $F \subset G$, but this is not relevant to our purposes.

2. Now we apply Theorem 2.1 to the study of ergodic properties of a flow $(G/D, g_{\mathbb{R}})$.

COROLLARY 2.3. *The flows* $(G/D, g_{\mathbb{R}})$ *and* $(G/\overline{MD}, g_{\mathbb{R}})$ *are either both ergodic or both nonergodic.*

Proof. Assume that $\rho(g_{\mathbb{R}})f \equiv f$ for a function $f \in L^2(G/D)$. Then $\rho(M)f \equiv f$, and since the subgroup M is normal, the function f is constant along almost all leaves $gMD \subset G/D$. By Lemma 1.1, f is constant along almost all closed leaves $g\overline{MD} \subset G/D$ and therefore defines a function $f' \in L^2(G/\overline{MD})$. If the quotient flow on $G/\overline{MD}$ is ergodic, then $f' \equiv$ const. It follows that $f \equiv$ const, and hence the original flow is also ergodic. □

In a similar fashion one proves the following.

COROLLARY 2.4. *If* $G = \overline{MD}$, *then the flow* $(G/D, g_{\mathbb{R}})$ *is weak mixing.*

Hence in order to check a flow for ergodicity or mixing, one passes from the space G/D to the quotient space $G/\overline{MD}$. From the dynamical viewpoint this passage consists in "killing" the divergence of close trajectories, thus obtaining a uniformly continuous (and even isometric) flow. Indeed, assume that $g_{\mathbb{R}}$ has trivial Moore subgroup; that is, the closure $K = \overline{\mathrm{Ad}(g_{\mathbb{R}})} \subset \mathrm{Aut}(\mathfrak{g})$ is compact. One can select a scalar product on the Lie algebra $\mathfrak{g}$ that is invariant under the compact group K. This gives rise to a G-right-invariant metric d_G on the Lie group G. The metric d_G is invariant under left translations of the subgroup $g_{\mathbb{R}}$, and hence our flow keeps fixed the metric d on G/D induced by the metric d_G.

3. Recall that a measurable partition ξ on a measure space (M, ν) has a system of conditional measures $\{\nu_C, C \in M/\xi\}$, each of which is supported on the corresponding element $C \in \xi$, and which being glued together give the measure ν. A partition ξ invariant under a flow on (M, ν) is called the *ergodic decomposition* if almost all conditional measures are ergodic. It is known [**Ro**] that the ergodic decomposition is unique (mod 0).

Actually, to construct the ergodic decomposition of a homogeneous flow $(G/D, g_{\mathbb{R}})$ it suffices to handle the quotient flow $(G/\overline{MD}, g_{\mathbb{R}})$. In fact, consider the fibration of finite volume spaces $\pi : G/D \to G/\overline{MD}$ with $\overline{MD}/D$ as a fiber. Let ν be the G-invariant probability measure on G/D, let ν' be the G-invariant probability measure on $G/\overline{MD}$, and let μ be the $\overline{MD}$-invariant probability measure on $\overline{MD}/D$. Then the measure ν is the *semidirect product* [**We**] of measures μ and ν'. More specifically, given any function $f \in C_c(G/D)$, one has

$$\int_{G/D} f(gD)d\nu_{gD} = \int_{G/\overline{MD}} d\nu'_{g\overline{MD}} \int_{\overline{MD}/D} f(ghD)d\mu_{hD}.$$

Let $\{\nu'_{C'},\ C' \in (G/\overline{MD})/\xi'\}$ be the ergodic decomposition of the measure ν'. Let $\xi = \pi^{-1}(\xi')$ be the corresponding partition of the space G/D into subsets $C = \pi^{-1}(C')$, $C' \in \xi'$. Denote by ν_C the system of $\overline{g_{\mathbb{R}}M}$-invariant measures on G/D obtained by semidirect products of measures $\nu'_{C'}$ with the measure μ. Now we can prove that this system of measures forms the ergodic decomposition of the measure ν.

In fact, it suffices to show that almost all measures ν_C are ergodic. This means that, given any $g_{\mathbb{R}}$-invariant bounded measurable function f on G/D, one has

$f \equiv$ const relative to ν_C for almost all $C \in \xi$. But by the Mautner phenomenon, every invariant function is constant along almost all leaves $g\overline{MD}$ and so determines a $g_{\mathbb{R}}$-invariant function f' on $G/\overline{MD}$. Then $f' \equiv$ const relative to almost every measure $\nu'_{C'}$, and hence $f \equiv$ const relative to almost every measure ν_C.

Recall that the quotient flow $(G/\overline{MD}, g_{\mathbb{R}})$ is uniformly continuous. It follows that the closure of any orbit of it is a minimal set and the flow is uniquely ergodic on each orbit closure (and, as we will see later, is isomorphic to a rectilinear flow on the torus). Consequently, the ergodic decomposition of the flow is the partition into orbit closures. This implies that the ergodic decomposition $\widetilde{E}$ of the original flow is the partition into closed invariant subsets $\widetilde{E}_h(g_{\mathbb{R}}) = \overline{g_{\mathbb{R}}MhD} \subset G/D$, $h \in G$, and every such set carries a $\overline{g_{\mathbb{R}}M}$-invariant measure $\widetilde{\mu}_h$. For almost all $h \in G$ the measure is ergodic. Let us summarize the above.

THEOREM 2.5. *The ergodic decomposition $\widetilde{E}$ of a homogeneous flow $(G/D, g_{\mathbb{R}})$ relative to a finite G-invariant measure ν is the partition of G/D into closed sets*

$$\widetilde{E}_h(g_{\mathbb{R}}) = \overline{g_{\mathbb{R}}MhD} \subset G/D, \qquad h \in G.$$

4. However, one can construct another version of the ergodic decomposition of the original flow. Consider the quotient flow $(G/\overline{JD}, g_{\mathbb{R}})$. We will see below that orbit closures of the flow are also minimal sets, and the flow on each of them is uniquely ergodic (and isomorphic to an ergodic nilflow). So they form a partition of the space $G/\overline{JD}$ which is the ergodic decomposition. The construction of the semidirect product of measures provides a partition E of the space G/D into closed subsets

$$E_h(g_{\mathbb{R}}) = \overline{g_{\mathbb{R}}JhD} \subset G/D,$$

equipped with $\overline{g_{\mathbb{R}}J}$-invariant measures μ_h. But since the Dani normal subgroup in the group $G' = \overline{g_{\mathbb{R}}J}$ is again J, in view of the Mautner phenomenon *each* measure μ_h is ergodic. This reveals a difference between two constructions of the ergodic decomposition of a homogeneous flow.

In order to distinguish E from $\widetilde{E}$, we call E the *ergodic partition*. Clearly, $E_h(g_{\mathbb{R}}) \subset \widetilde{E}_h(g_{\mathbb{R}})$ for all $h \in G$. Since the ergodic decomposition is unique (in the measure-theoretic sense), it follows that $E_h(g_{\mathbb{R}}) = \widetilde{E}_h(g_{\mathbb{R}})$ for almost all $h \in G$. In the sequel we will see that all subsets $E_h(g_{\mathbb{R}})$ and $\widetilde{E}_h(g_{\mathbb{R}})$ are submanifolds, and the corresponding measures are smooth. Note that although ergodic components in the abstract ergodic decomposition are defined on a conull set, they are defined *everywhere* for a homogeneous flow.

Thus, we have reduced the study of a nonergodic homogeneous flow to that of a family of smooth ergodic flows. But such a reduction is not satisfactory: we have lost the homogeneous structure. Fortunately (as we will see later), every flow on $\widetilde{E}_h(g_{\mathbb{R}})$ is finitely covered by a homogeneous flow, and so is every flow on $E_h(g_{\mathbb{R}})$.

§3. Nilpotent homogeneous flows

1. We start with the Abelian case and consider a connected simply connected Abelian Lie group A with a lattice $\Gamma \subset A$. Clearly, the group A is isomorphic to a real vector space $\mathbb{R}^n$. In addition, the lattice Γ will be isomorphic to $\mathbb{Z}^n$ in the basis formed by n generators of Γ. Since the Lie algebra of A is again $\mathbb{R}^n$, an Abelian homogeneous flow is a rectilinear flow on a torus $(\mathbb{R}^n/\mathbb{Z}^n, \mathbb{R}v)$, where $v = (v_1, \dots, v_n) \in \mathbb{R}^n$ is the directing vector. The natural measure for such a flow is the $\mathbb{R}^n$-invariant smooth Lebesgue measure ν which is unique up to rescaling. We already know Kronecker's ergodicity criterion for such a flow:

THEOREM 3.1. *A flow* $(\mathbb{R}^n/\mathbb{Z}^n, \mathbb{R}v)$ *is ergodic if and only if the numbers* $v_1, \ldots, v_n$ *are linearly independent over the field* $\mathbb{Q}$ *of rational numbers.*

Let us look more closely at orbits of a rectilinear flow. Since $\mathbb{R}vg\mathbb{Z}^n = g\mathbb{R}v\mathbb{Z}^n$, the closure of any orbit is obtained from the closure of the "marked" orbit $\mathbb{R}v\mathbb{Z}^n$ via translation by an element $g \in \mathbb{R}^n$. It follows that our flow is minimal if and only if it is topologically transitive; more specifically, $\mathbb{R}^n = \overline{\mathbb{R}v\mathbb{Z}^n}$. Since the identity component $(\overline{\mathbb{R}v\mathbb{Z}^n})_0$ is a $\mathbb{Q}$-defined vector subspace $V \subset \mathbb{R}^n$, the criteria of minimality, topological transitivity and ergodicity for a rectilinear flow on the torus are the same.

On the other hand, let μ be an ergodic invariant probability measure for $(\mathbb{R}^n/\mathbb{Z}^n, \mathbb{R}v)$. Then the stabilizer $\Lambda(\mu) \subset \mathbb{R}^n$ is closed and contains $\mathbb{R}v$. But it contains $\mathbb{Z}^n$ as well, because the left action of the lattice $\mathbb{Z}^n$ on $\mathbb{R}^n/\mathbb{Z}^n$ is trivial. It follows that $\Lambda(\mu) = \overline{\mathbb{R}v\mathbb{Z}^n}$, and hence the measure μ is different from the smooth measure ν only if the flow is nonergodic relative to ν. We have derived the following result.

COROLLARY 3.2. *For a rectilinear flow on the torus, the properties of topological transitivity, minimality, ergodicity and unique ergodicity are equivalent.*

Suppose that the subspace $V = (\overline{\mathbb{R}v\mathbb{Z}^n})_0$ is proper in $\mathbb{R}^n$, that is, the flow is nonergodic. Then the torus $\mathbb{T}^n = \mathbb{R}^n/\mathbb{Z}^n$ fibers into subtori

$$E_g(v) = gV\mathbb{Z}^n \subset \mathbb{T}^n$$

having the same dimension $k = \dim V$, and each of them carries a smooth V-invariant ergodic measure μ_g. Every subflow $(gV\mathbb{Z}^n, \mathbb{R}v)$ is uniquely ergodic and minimal, and the smooth $\mathbb{R}^n$-invariant measure ν is glued from the measures μ_g. Hence the fibration E of the torus $\mathbb{T}^n$ into subtori $E_g(v)$, $g \in \mathbb{R}^n$, is the ergodic decomposition of the original nonergodic flow.

2. Now we consider a connected simply connected nilpotent Lie group N with a connected subgroup $U \subset N$. In this case the Moore subgroup $M = M(U) \subset N$ is the smallest normal subgroup of N such that the group $F = UM$ is normal in N as well. Let us prove the Mautner phenomenon in the nilpotent case.

THEOREM 3.3. *Let* $M \subset N$ *be the Moore normal subgroup for a connected subgroup* $U \subset N$. *Then* $\mathrm{Fix}(U) \subset \mathrm{Fix}(M)$ *for any continuous unitary representation of the group* N.

For a demonstration we consider the simplest example of a nilpotent group.

Example 3.4. Let $N = \mathrm{U}(3, \mathbb{R})$ be the connected simply connected three-dimensional Heisenberg group, that is, the nilpotent Lie group of strictly upper-triangular matrices

$$h = \begin{pmatrix} 1 & a & b \\ 0 & 1 & c \\ 0 & 0 & 1 \end{pmatrix}, \qquad \text{where } a, b, c \in \mathbb{R}.$$

Any one-parameter subgroup of N is of the form

$$g_t = \begin{pmatrix} 1 & \alpha_1 t & \alpha_2 t + \frac{\alpha_1\alpha_3}{2}t^2 \\ 0 & 1 & \alpha_3 t \\ 0 & 0 & 1 \end{pmatrix}.$$

The one-dimensional subgroup

$$\left\{ \begin{pmatrix} 1 & 0 & b \\ 0 & 1 & 0 \\ 0 & 0 & 1 \end{pmatrix}, \ b \in \mathbb{R} \right\}$$

is simultaneously the commutator and the center of the group N. It is easily seen that, given another subgroup $h_{\mathbb{R}} \subset N$, one has

$$g_t h_{1/t} g_{-t} = h_{1/t} f \to f, \qquad t \to \infty, \qquad \text{where } f \in Z(N).$$

Here $f \neq 0$ if $g_{\mathbb{R}}$ and $h_{\mathbb{R}}$ do not commute, and for $f \in Z(N)$ one can take any element of $Z(N)$ (it suffices to replace h_t by h_{kt} with suitable $k \in \mathbb{R}$). By virtue of Lemma 2.2, we conclude that $\mathrm{Fix}(g_{\mathbb{R}}) \subset \mathrm{Fix}(Z(N))$ if $g_{\mathbb{R}}$ does not coincide with the center $Z(N)$. Note that $Z(N)$ is precisely the Moore normal subgroup for all subgroups $g_{\mathbb{R}}$ different from $Z(N)$.

Let us say that a subgroup $F \subset N$ has the *Mautner property* with respect to a subgroup $U \subset N$ if $\mathrm{Fix}(U) \subset \mathrm{Fix}(F)$ for every (continuous) unitary representation of the group N. By $F(U) \subset N$ we denote the maximal connected subgroup having the Mautner property with respect to U. It is not difficult to verify that Theorem 3.3 is equivalent to the following.

THEOREM 3.5. *Let U be a connected subgroup of N. Then the subgroup $F(U)$ is normal in N.*

Proof (cf. [**MT96**]). It is well known that, given two connected subgroups $H, V \subset N$, the space N/H is diffeomorphic to $\mathbb{R}^n$, and all V-orbits in N/H are closed and diffeomorphic to $\mathbb{R}^k$, $k \leq n$. We set $H = F(U)$. Then $U \subset H$ and $F(U) = F(H)$. If the subgroup H is not normal in N, then the normalizer $N_N(H)$ is a proper subgroup of N, and hence one can find a sequence of elements $g_i \in N - N_N(H)$ convergent to $1 \in N$. Let $S \subset N/H$ be a sphere centered at the point $H \in N/H$. The orbit $g_i H g_i^{-1} H \subset N/H$, being closed and noncompact, must interesect the sphere S at a point $s_i \in S$. Passing, if necessary, to a subsequence, we can assume that $s_i \to s \in S$, $i \to \infty$. Let $a \in N$ be an element such that $s = aH$. Then there exist $h_i, h_i' \in H$ such that $g_i h_i g_i^{-1} h_i' \to a$. By Lemma 2.2, the element $a \in N - H$ has the Mautner property with respect to H and hence with respect to U. Clearly, the element $a \in N - H$ can be made arbitrarily close to 1 (taking, if necessary, a sphere $S \subset N/H$ of smaller radius). Since the group $F(H)$ is closed, we arrive at a contradiction with its maximality. Hence the group $H = F(U)$ is normal in N. □

Now let Γ be a lattice in N. It follows from Theorem 3.3 that the flows $(N/\Gamma, U)$ and $(N/\overline{M\Gamma}, U)$ are both ergodic or both nonergodic. If $N = \overline{U[N, N]\Gamma}$, then N has no proper subgroup $H \subset N$ such that $U \subset H$ and the product $H\Gamma$ is closed (otherwise $N = H[N, N]\Gamma$ and so $N = H$). In particular, $N = \overline{UM\Gamma}$, and we have proved Green's ergodicity criterion for nilflows:

THEOREM 3.6 [**AGH**]. *A nilflow $(N/\Gamma, U)$ is ergodic if and only if the rectilinear quotient flow on the torus $(N/[N,N]\Gamma, U)$ is ergodic.*

3. Let us comment on the behavior of a nilflow relative to the metric d on N/Γ induced by a right-invariant metric d_N on N. The following holds.

THEOREM 3.7 (Auslander [**AGH**]). *Any nilflow $(N/\Gamma, U)$ is distal.*

Proof [**AB71**]. We apply induction on $\dim N$. If $\dim N = 1$, the group N is Abelian and there is nothing to prove. Take $Z = Z(N)$ and $N' = N/Z$. Then the product $Z\Gamma$ is closed and $\dim N' < \dim N$. Pick $x, y \in N/\Gamma$ and let $u_n x, u_n y \to q \in N/\Gamma$ for a sequence $u_n \in U$. By the inductive hypothesis, $x = zy$ for some $z \in Z$. But $u_n zy = zu_n y \to zq$, and so $zq = q$. Hence $z \in Z \cap \Gamma$, whence $x = y$. □

Since locally
$$d(g_t h\Gamma, g_t\Gamma) = d_N(1, g_t h g_{-t})$$
and the right-hand function is polynomial in t, the flow $(N/\Gamma, U)$ may be uniformly continuous only if the subgroup U is central in N.

Look again at the Heisenberg group $N = \mathrm{U}(3, \mathbb{R})$ with the lattice $\Gamma = \mathrm{U}(3, \mathbb{Z})$. Let $F \subset N$ be the two-dimensional Abelian group of matrices
$$\begin{pmatrix} 1 & a & b \\ 0 & 1 & 0 \\ 0 & 0 & 1 \end{pmatrix}.$$
Assume that a subgroup $g_{\mathbb{R}}$ does not belong to F; that is, the parameters $\alpha_1, \alpha_2, \alpha_3$ determining $g_{\mathbb{R}}$ (see §2) are such that $\alpha_3 \neq 0$. Then the flow $(N/\Gamma, g_{\mathbb{R}})$ is the standard suspension over the *skew shift*[9] A of the torus $\mathbb{T}^2 = F/\Gamma \cap F$, where
$$A(a, b) = (a + \beta_1, b - a + \beta_2 - \beta_1/2), \qquad \beta_1 = \alpha_1/\alpha_3,\ \beta_2 = \alpha_2/\alpha_3.$$

As follows from Weyl's theorem (see §7) on uniform distribution of the fractional parts of values of a polynomial, the skew shift A is ergodic, and even uniquely ergodic if the number β_1 is irrational. This implies unique ergodicity of the flow $(N/\Gamma, g_{\mathbb{R}})$. Originally, unique ergodicity of nilflows (in particular cases) was established in precisely this way. However, as we will see in a moment, unique ergodicity can be proved by purely ergodic methods.

The example of the Heisenberg group is a particular case of Ellis' Theorem B.1 and Furstenberg's Theorem B.5, which claim that a topologically transitive distal flow is minimal and uniquely ergodic. For nilflows the proof of the theorems is too elegant to be omitted. Subsequently we derive the Weyl theorem.

THEOREM 3.8 [**Fu63, P69**]. *For nilflows, the properties of topological transitivity, minimality, ergodicity and unique ergodicity are equivalent.*

Proof. Clearly, it suffices to derive unique ergodicity of a flow $(N/\Gamma, g_{\mathbb{R}})$ from its ergodicity. We apply induction on the dimension of N/Γ. If $\dim(N/\Gamma) = 2$, the space N/Γ is a torus and we are done. Let us make the inductive step. Suppose that $\dim(N/\Gamma) = n$. We have to prove that a $g_{\mathbb{R}}$-invariant ergodic probability measure μ on N/Γ is none other than the N-invariant measure ν. Let $\pi : N/\Gamma \to$

[9] So called because A is a composite of a (unipotent) automorphism of the torus $\mathbb{T}^2$ and a translation.

$N/\Gamma Z(N)$ be the projection. By $E_\nu \subset N/\Gamma$ we denote the family of all uniformly distributed orbits of the flow $(N/\Gamma, g_\mathbb{R})$ relative to the measure ν. Likewise, we define the family E_μ. By the Birkhoff–Khintchine ergodic theorem, $\nu(E_\nu) = 1$ and $\mu(E_\mu) = 1$. The quotient flow $(N/\Gamma Z(N), g_\mathbb{R})$ is ergodic, and by the inductive hypothesis it is uniquely ergodic. It follows that $\pi_*\mu = \pi_*\nu$, and hence the sets $\pi(E_\nu)$ and $\pi(E_\mu)$ coincide almost everywhere. On the other hand, every element of $Z(N)$ sends a uniformly distributed trajectory relative to ν into another such, and hence $E_\nu = \pi^{-1}(\pi(E_\nu))$. This implies that at least one point $x \in N/\Gamma$ belongs simultaneously to E_ν and E_μ. But then

$$\int_{N/\Gamma} f\,d\nu = \lim_{T\to\infty} \frac{1}{T}\int_0^T f(g_t x)dt = \int_{N/\Gamma} f\,d\mu, \qquad f \in C(N/\Gamma),$$

and hence the measures μ and ν are the same. □

4. Suppose that a nilflow $(N/\Gamma, U)$ is nonergodic. What can we say about the structure of its orbits? To answer, we let $F_1 = (\overline{U\Gamma})_0$. Let us prove that F_1 is a subgroup of N. Indeed, let F' be the smallest subgroup of N such that $U \subset F'$ and the intersection $\Gamma \cap F'$ is a lattice in F'. Then by Green's criterion, the flow $(F'/\Gamma \cap F', U)$ is ergodic and hence minimal. It follows that $F_1 = F'$, and the closure

$$E_1(U) = \overline{U\Gamma} = F_1\Gamma \subset N/\Gamma$$

is a homogeneous subspace isomorphic to $F_1/\Gamma \cap F_1$.

Likewise, the closure

$$E_h(U) = \overline{Uh\Gamma} \subset N/\Gamma = h\overline{h^{-1}Uh\Gamma}$$

is isomorphic to a compact nilspace $F_h/\Gamma \cap F_h$, where $F_h = (\overline{h^{-1}Uh\Gamma})_0$ is a proper connected subgroup of N.

Since every flow $(E_h(U), U)$ is minimal, the ergodic submanifolds $E_h(U)$ and $E_{h'}(U)$ either do not intersect or coincide. Therefore we have obtained a partition E of the space N/Γ.

Example 3.9. Look again at the Heisenberg group $N = \mathrm{U}(3, \mathbb{R})$ with the lattice $\Gamma = \mathrm{U}(3, \mathbb{Z})$. The subgroup

$$g_t = \begin{pmatrix} 1 & t & 0 \\ 0 & 1 & 0 \\ 0 & 0 & 1 \end{pmatrix}$$

induces a nonergodic flow on N/Γ. The orbit closure of a point $h\Gamma \in N/\Gamma$ is $E_h(g_\mathbb{R}) = hF_h\Gamma$, where F_h is either the two-dimensional normal Abelian subgroup

$$F = \left\{ \begin{pmatrix} 1 & a & b \\ 0 & 1 & 0 \\ 0 & 0 & 1 \end{pmatrix} \; a, b \in \mathbb{R} \right\},$$

or a certain one-dimensional subgroup $F_h \subset F$. Indeed,

$$\text{if } h = \begin{pmatrix} 1 & a & b \\ 0 & 1 & c \\ 0 & 0 & 1 \end{pmatrix}, \text{ then } h^{-1}g_t h = \begin{pmatrix} 1 & t & tc \\ 0 & 1 & 0 \\ 0 & 0 & 1 \end{pmatrix}$$

and hence $F_h = F$ for matrices $h \in N$ with irrational $c \in \mathbb{R}$, whereas $F_h \subset F$ for $h \in N$ with rational $c \in \mathbb{R}$.

Define

$$\widetilde{E}_h(U) = hF\Gamma = Fh\Gamma \subset N/\Gamma.$$

Then the fibration $\widetilde{E}$ of the nilspace N/Γ into two-dimensional invariant subtori has the property that almost every flow $(\widetilde{E}_h(U), U)$ is ergodic, and on the remaining subtori it is nonergodic and has periodic orbits $E_h(U)$. Clearly, every subtorus $\widetilde{E}_h(U) = Fh\Gamma$ carries a smooth F-invariant measure $\widetilde{\mu}_h = h\widetilde{\mu}_1$, and the N-invariant measure ν on N/Γ is glued from the measures $\widetilde{\mu}_h$. Consequently, $\widetilde{E}$ is the ergodic decomposition of the original flow. The partition E of the space N/Γ into homogeneous subspaces $E_h(U) = hF_h\Gamma$ coincide almost everywhere with the fibration $\widetilde{E}$. Hence from the measure-theoretic viewpoint it is an ergodic decomposition as well. Every subspace

$$E_h(U) = hF_h\Gamma = hF_hh^{-1}h\Gamma$$

carries a smooth hF_hh^{-1}-invariant ergodic measure μ_h. The partition E is not a smooth fibration even locally, because periodic orbits of the flow form a dense family in N/Γ, due to the density of rational numbers in $\mathbb{R}$.

5. Let us prove (without using the Mautner phenomenon) that the same situation occurs in the general case.

THEOREM 3.10 [**St84a, Le**]. *The ergodic decomposition $\widetilde{E}$ of a nilflow $(N/\Gamma, U)$ is a smooth fibration into closed orbits of some normal subgroup $F \subset N$. The orbit closure of any point*

$$E_h(U) = \overline{Uh\Gamma} \subset N/\Gamma$$

is a homogeneous subspace, and $E_h(U) = Fh\Gamma$ for almost all $h \in G$.

Proof. We have already constructed the partition E of the space N/Γ into homogeneous subspaces $E_h(U) = \overline{Uh\Gamma} = hF_h\Gamma$, where $F_h = (\overline{h^{-1}Uh\Gamma})_0$ is a connected subgroup of N. Let F' be the normal subgroup of N generated by the set $\{h^{-1}Uh,\ h \in N\}$, and $F = (\overline{F'\Gamma})_0$. The subgroup F is normalized by the lattice $\Gamma \subset N$ and hence is normal in N. It suffices to prove that almost every subflow $(Fh\Gamma, U)$ is ergodic, that is, $F = (\overline{h^{-1}Uh\Gamma})_0$ for almost all $h \in N$.

The connected simply connected group F has countably many connected subgroups H for which the intersection $\Gamma \cap H$ is a lattice in H. Hence the proof of the theorem is immediate modulo the following assertion.

LEMMA 3.11 [**Da80b**]. *Let G be a connected Lie group with a connected closed subgroup $H \subset F$. Given an element $f \in G$, define*

$$B(f) = \{g \in G:\ g^{-1}fg \in H\}.$$

Then either $B(f) = G$, or $B(f)$ is a closed set of zero measure in G.

Proof. Introduce a diffeomorphism $\alpha : G \times G \to G \times G$ as follows:

$$\alpha(g, h) = (g, ghg^{-1}) \quad \text{for all } g, h \in G.$$

It is known that a connected Lie group admits the structure of an analytic manifold in which the multiplication and taking the inverse element are analytic maps and connected closed subgroups are analytic submanifolds. It follows that the diffeomorphism α is an analytic diffeomorphism, and $\alpha(G\times H)$ is an analytic submanifold in $G \times G$. In view of the definition of $B(f)$ and α, one has

$$B(f) \times f = \alpha(G \times H) \cap (G \times f).$$

But the connected analytic submanifold $G \times f$ either is contained in the analytic submanifold $\alpha(G \times H)$, or intersects the latter in a closed subset of zero measure in $G \times f$. This precisely means that either $B(f) = G$, or $B(f)$ is a closed set of zero measure. □

It is easily seen as well that the submanifold $E_h(U) = \overline{Uh\Gamma}$ is the maximal invariant submanifold which contains the point $h\Gamma \in N/\Gamma$ and on which the flow is ergodic relative to a smooth measure. Indeed, if $S_h \subset N/\Gamma$ is an invariant submanifold with a smooth ergodic measure μ_h, then due to the smoothness there exists at least one dense orbit: $S_h = \overline{Uh'\Gamma} = E_{h'}(U)$ for some point $h'\Gamma \in S_h$. But then $E_h(U) \subset E_{h'}(U)$, and hence $E_h(U) = E_{h'}(U) = S_h$.

So, in the nilpotent case the ergodic decomposition $\widetilde{E}$ is still a smooth fibration into closed orbits of a normal subgroup $F \subset N$. But, generally speaking, not every subflow

$$\widetilde{E}_h(U) = hF\Gamma = Fh\Gamma \subset N/\Gamma$$

is ergodic relative to a smooth F-invariant measure. Hence the decomposition $\widetilde{E}$ need not coincide with the partition E into maximal ergodic submanifolds $E_h(U)$. Every submanifold $E_h(U)$ is a homogeneous subspace in N/Γ which coincides with $\widetilde{E}_h(U)$ for almost all $h \in N$. However, on a null subset of N/Γ the dimension of the subspace $E_h(U)$ may drop, and hence E may fail to be a smooth fibration. Every ergodic flow $(E_h(U), U)$ is minimal.

Note that the subgroup F which determines the ergodic decomposition $\widetilde{E}$ is of the form $F = (\overline{UM\Gamma})_0 \subset N$, where M is the Moore normal subgroup for U. In fact, F is the smallest subgroup of N which contains all the conjugate subgroups hUh^{-1}, $h \in N$, and intersects Γ in a lattice. But by the definition of the Moore subgroup, in the nilpotent case UM is the smallest (normal) subgroup containing all the conjugate subgroups hUh^{-1}, $h \in N$. It follows that

$$\widetilde{E}_h(U) = \overline{UMh\Gamma} \subset N/\Gamma \quad \text{for all } h \in N$$

(which can be immediately derived from the Mautner phenomenon).

Note also that, similarly to the Abelian case, almost all one-parameter subgroups $g_{\mathbb{R}} \subset N$ induce ergodic flows. Moreover, ergodicity or nonergodicity of a flow $(N/\Gamma, g_{\mathbb{R}})$ only depends on the mutual positions of the subgroup $g_{\mathbb{R}}$ and the lattice Γ.

6. To conclude the study of nilflows, let us describe U-invariant ergodic probability measures for a nilflow $(N/\Gamma, U)$. Let μ be such a measure, and let $\widetilde{E}$ be the ergodic decomposition relative to the smooth measure ν. We know that $\widetilde{E}$ is the smooth fibration into orbits of the normal subgroup $F = (\overline{UM\Gamma})_0 \subset N$. The image μ^* of the measure μ in $N/F\Gamma$ is a U-ergodic measure. But the action of U on $N/F\Gamma$ is trivial, and hence the measure μ^* is concentrated at a point in $N/F\Gamma$. It follows that the measure μ is supported on a compact homogeneous space $Fg\Gamma \subset N/\Gamma$. If the flow $(Fg\Gamma, U)$ is nonergodic relative to a smooth Haar measure on $Fg\Gamma$, we again construct the ergodic fibration and prove that the measure μ is supported on a smaller subspace. Eventually, we will find a compact homogeneous subspace $Hg\Gamma$ which supports the measure μ and on which the action of U is ergodic relative to an H-invariant measure. But such a flow is uniquely ergodic, and hence μ is H-invariant. We have proved that all U-ergodic measures are of algebraic origin.

THEOREM 3.12. *Every U-invariant ergodic measure μ on N/Γ is supported on a closed orbit $Hg\Gamma \subset N/\Gamma$ and is H-invariant.*

7. As an application of nilflows to number theory, we prove Weyl's theorem on uniform distribution of the fractional parts of values of a polynomial. A sequence $\{x_n,\ n \in \mathbb{Z}\} \subset [0,1]$ is called *uniformly distributed* on the segment $[0,1]$ if for every $f \in C[0,1]$,

$$\frac{1}{2N+1}\sum_{i=-N}^{N} f(x_i) \to \int_0^1 f(x)dx, \qquad N \to +\infty.$$

THEOREM 3.13 (WEYL). *Let $p(t) = a_0t^n + \cdots + a_{n-1}t + a_n$ be a polynomial having at least one of the first n coefficients irrational. Then the sequence $p(\mathbb{Z})$ mod 1 is uniformly distributed on the segment $[0,1]$.*

Note that for polynomials of the first degree the statement implies that an irrational rotation of the circle is uniquely ergodic (that is, all its orbits are uniformly distributed). Another particular case is a theorem of Hardy and Littlewood: if a number a_0 is irrational and $n \in \mathbb{N}$, then the fractional parts of powers $a_0m^n,\ m \in \mathbb{Z}$, are dense in the segment $[0,1]$.

Originally this result was proved by Weyl back in 1919 by using trigonometric series. A proof along the lines of [**Fu61**] using skew shifts on the torus was given in the book [**KSF**]. In our opinion, the usage of nilflows simplifies the proof significantly. In essence, we only use the following two results: Green's ergodicity criterion and Furstenberg's theorem on uniform distribution, both being applied to a rather special example of a nilflow (cf. [**AGH**]).

First we note that it suffices to consider polynomials $p(t) = a_0t^n + \cdots + a_n$ with leading coefficient a_0 irrational (see [**KSF**] for the proof).

Now let a nilpotent Lie group N consist of matrices

$$g_{a,b} = \begin{pmatrix} 1 & a & a^2/2 & \dots & a^{n-1}/(n-1)! & b_1 \\ 0 & 1 & a & \dots & a^{n-2}/(n-2)! & b_2 \\ \cdot & \cdot & \cdot & \cdot & \cdot & \cdot \\ 0 & 0 & \dots & 1 & a & b_{n-1} \\ 0 & 0 & 0 & \dots & 1 & b_n \\ 0 & 0 & 0 & \dots & 0 & 1 \end{pmatrix},$$

where $a \in \mathbb{R}, b = (b_1, \dots, b_n) \in \mathbb{R}^n$. Note that $\dim N = n+1$ and N contains a normal Abelian subgroup M isomorphic to $\mathbb{R}^n$ (M consists of matrices $g_{0,b}$). Let $K = (n-1)!$. The subgroup $\Gamma \subset N$ of matrices $g_{a,b}$, where $a \in K\mathbb{Z}, b \in \mathbb{Z}^n$, is a lattice in N. It is easy to check that the family

$$\{u_t = g_{t,\lambda t^n/n!,\dots,\lambda t^2/2,\lambda t},\ t \in \mathbb{R}\}$$

is a one-parameter subgroup of N. Green's ergodicity criterion combined with Furstenberg's theorem ensures that the flow $(N/\Gamma, u_{\mathbb{R}})$ is uniquely ergodic for every irrational λ. This flow is the suspension over a skew shift A of the torus $\mathbb{T}^n =$

$M/\Gamma\cap M$. It follows that all orbits of the A-action on $\mathbb{T}^n$ are uniformly distributed as well. The explicit expression for A is found from the equation

$$u_K g_{0,b}\Gamma = g_{0,A(b)}\Gamma,$$

whence

$$A\begin{pmatrix} b_1 \\ b_2 \\ \dots \\ b_n \end{pmatrix} = \begin{pmatrix} b_1 + b_2K + \dots + b_nK^{n-1}/(n-1)! + \lambda K^n/n! \\ b_2 + b_3K + \dots + b_nK^{n-2}/(n-2)! + \lambda K^{n-1}/(n-1)! \\ \dots \\ b_n + \lambda K \end{pmatrix}.$$

Now we fix $b_1, \dots, b_n, \lambda$ and introduce the polynomial

$$q(t) = b_1 + b_2t + \dots + b_nt^{n-1}/(n-1)! + \lambda t^n/n!.$$

By applying Taylor's formula

$$q(t+s) = q(t) + q'(t)s + q''(t)s^2/2 + \dots + q^{(n)}(t)s^n/n!,$$

we deduce that

$$A^l\begin{pmatrix} q(0) \\ q'(0) \\ \dots \\ q^{(n-1)}(0) \end{pmatrix} = \begin{pmatrix} q(lK) \\ q'(lK) \\ \dots \\ q^{(n-1)}(lK) \end{pmatrix}, \qquad l \in \mathbb{Z}.$$

This, in particular, implies the uniform distribution of the sequence $q(K\mathbb{Z}) \mod 1$. All we need now is to find coefficients of the polynomial q from the relation $p(t) = q(Kt)$ and to note that λ is irrational whenever a_0 is. □

§4. Toral automorphisms and examples of solvable flows

Here we provide examples of solvable flows and study their ergodic properties and orbit closures. We will see that, in contrast to the nilpotent case: 1) a solvable manifold does not necessarily carry an ergodic homogeneous flow; 2) the ergodic decomposition $\widetilde{E}$ need not be a smooth fibration; 3) the ergodic submanifolds $E_g(x)$ and $\widetilde{E}_g(x)$ can fail to be homogeneous subspaces; and 4) an orbit closure need not be a manifold.

Arbitrary solvable flows will be studied in §7, after we introduce the construction of semisimple splitting.

Let R be a connected simply connected solvable Lie group. Recall that the type of R is determined as follows. On the complex plane $\mathbb{C}$, mark all eigenvalues of operators Ad_g, $g \in R$, on the Lie algebra $\mathfrak{r}$. If all of them are equal to 1, the group R is nilpotent; if all of them belong to the unit circle $\mathbb{C}_1 = \{|z| = 1\}$, then R is of class (I); if all of them are real, the group is class (R); if apart from 1 there are no other eigenvalues on $\mathbb{C}_1$, then R is of class (E).

1. We start with simple examples of class (I) groups taken from [**AGH**] and [**St84a**].

Example 4.1. Let $O_{\mathbb{R}}$ be the additive group $\mathbb{R}$ equipped with the orthogonal action on the plane $\mathbb{R}^2$. More specifically, the action of an element O_t in the basis (e_1, e_2) of $\mathbb{R}^2$ is given by the orthogonal matrix

$$O(t) = \begin{pmatrix} \cos t & \sin t \\ -\sin t & \cos t \end{pmatrix}.$$

The action enables one to take the semidirect product:

$$R = O_{\mathbb{R}} \cdot \mathbb{R}^2, \text{ that is, } O_t v O_{-t} = O(t)(v).$$

Then R is a connected simply connected three-dimensional solvable Lie group with center $O_{2\pi\mathbb{Z}}$ and nilradical $N = [R, R] = \mathbb{R}^2$. Let us prove that it is of class (I).

Clearly, all operators Ad_g, $g \in \mathbb{R}^2$, are unipotent, because $(\mathrm{Ad}_g - E)^2 = 0$. Pick an element $g = O_t u$, where $u \in \mathbb{R}^2$. Let us show that whenever t is not an integer multiple of 2π, the element $g = O_t u$ is conjugate to the element O_t. In fact, it suffices to find an element $v \in \mathbb{R}^2$ such that

$$v O_t u v^{-1} = O_t, \text{ that is, } O_{-t} v O_t v^{-1} = u^{-1}.$$

But the operator $v \to O_{-t} v O_t v^{-1}$ is given by the matrix $O(-t) - E$, which is nondegenerate provided that $t \notin 2\pi\mathbb{Z}$. It follows that the equation is solvable for $t \notin 2\pi\mathbb{Z}$, and hence every element $g \notin O_{2\pi\mathbb{Z}} \times \mathbb{R}^2$ is conjugate to an element from $O_{\mathbb{R}}$. Hence for such g, the adjoint operator Ad_g is semisimple and has 3 eigenvalues:

$$\lambda_1 = 1, \quad \lambda_2 = \cos t + i \sin t, \quad \lambda_3 = \cos t - i \sin t.$$

As for $g \in O_{2\pi\mathbb{Z}} \times \mathbb{R}^2$, the operator Ad_g is obviously unipotent. Thus, all eigenvalues belong to the unit circle and R is of class (I). The exponential map $\exp : \mathfrak{r} \to R$ is neither injective (because any one-parameter subgroup which is not a subgroup of the nilradical contains the center $O_{2\pi\mathbb{Z}} \subset R$), nor surjective (because no element of the form $g = O_{2\pi k} u$, $k \in \mathbb{Z} - \{0\}$, $u \neq 0$, belongs to a one-parameter subgroup of R).

The simplest example of a lattice in the group R is the direct product of the center $O_{2\pi\mathbb{Z}} \subset R$ with the integer lattice $\mathbb{Z}^2 \subset \mathbb{R}^2$. We denote it by Γ_1. Let us prove that the space R/Γ_1 is diffeomorphic to the 3-torus. To do this, we look at the factorization $\alpha : O_{\mathbb{R}} \to S = O_{\mathbb{R}}/O_{2\pi\mathbb{Z}}$. The actions of $O_{\mathbb{R}}$ and S on $\mathbb{R}^2$ commute, and we can form the semidirect product[10]

$$R_s = (O_{\mathbb{R}} \times S) \cdot \mathbb{R}^2.$$

The cylinder $O_{\mathbb{R}} \times S$ contains the "anti-diagonal" subgroup

$$\Delta = \{(O_t, \alpha(O_t)^{-1}),\ t \in \mathbb{R}\},$$

which is both the kernel of the action on $\mathbb{R}^2$ and the center of R_s. Hence the nilradical N_R of the group R_s is $N_R = \Delta \times \mathbb{R}^2$. Let

$$\Pi : R_s = \Delta \times S \cdot \mathbb{R}^2 \to \Delta \times \mathbb{R}^2$$

be the projection along S. Then the restriction $\Pi : R \to N_R$ is a diffeomorphism of the group R onto the three-dimensional Abelian Lie group $\Delta \times \mathbb{R}^2$ (for this reason, N_R is called the *nilshadow* of the group R). Note that Π *is not a group homomorphism.* The map Π keeps the Abelian lattice Γ_1 fixed, because

$$\Gamma_1 = O_{2\pi\mathbb{Z}} \times \mathbb{Z}^2 \subset \Delta \times \mathbb{R}^2.$$

It follows that the diffeomorphism Π induces a diffeomorphism Π^* of the manifold R/Γ_1 onto the 3-torus N_R/Γ_1.

Let us turn to homogeneous flows on R/Γ_1. Clearly, if $g_{\mathbb{R}}$ is a subgroup of the nilradical $\mathbb{R}^2 \subset R$, we obtain a nonergodic flow which determines the fibration of

[10]Later we will see that R_s is none other than the *semisimple splitting* of the group R.

R/Γ_1 into invariant 2-tori $\mathbb{R}^2 h\Gamma_1$. Note that $\mathbb{R}^2$ is the Moore subgroup $M(g_{\mathbb{R}})$ for $g_{\mathbb{R}} \subset \mathbb{R}^2$, $g_{\mathbb{R}} \neq 1$. A flow $(\mathbb{R}^2 h\Gamma_1, g_{\mathbb{R}})$ is isomorphic to the flow $(\mathbb{R}^2/\mathbb{Z}^2, h^{-1}g_{\mathbb{R}}h)$. Since the set of conjugate subgroups $\{h^{-1}g_{\mathbb{R}}h,\ h \in R\}$ covers all directions in $\mathbb{R}^2$, the flow is ergodic for almost all $h \in R$. Hence the fibration is the ergodic decomposition $\widetilde{E}$ of our flow. The ergodic partition E differs from $\widetilde{E}$ on a dense null set. More precisely, E divides nonergodic tori into periodic orbits.

Assume now that the subgroup $g_{\mathbb{R}}$ does not belong to the nilradical $\mathbb{R}^2$. We have already proved that $g_{\mathbb{R}}$ is conjugate to the subgroup $O_{\mathbb{R}} \subset R$, and hence we can assume from the beginning that $g_{\mathbb{R}} = O_{\mathbb{R}}$. But this flow is periodic! Indeed, the "marked" orbit $g_{\mathbb{R}}\Gamma_1$ is periodic because $O_{2\pi\mathbb{Z}} \subset g_{\mathbb{R}} \cap \Gamma_1$. On the other hand, any conjugate group $h^{-1}g_{\mathbb{R}}h$ also contains the center $O_{2\pi\mathbb{Z}}$, and hence any orbit $g_{\mathbb{R}}h\Gamma_1$ is periodic. It follows that in this case the ergodic partitions E and $\widetilde{E}$ coincide and are the partition of R/Γ_1 into periodic orbits (here $M(g_{\mathbb{R}}) = 1$). Since the intersection $h^{-1}g_{\mathbb{R}}h \cap \Gamma_1$ is equal to $O_{2\pi\mathbb{Z}}$ for any $h \in R$, the partition is a smooth fibration.

We emphasize that the space R/Γ_1 carries no homogeneous ergodic flows. This reveals a difference with the nilpotent case where almost all homogeneous flows are ergodic.

Example 4.2. Let us consider the same Lie group R with another lattice

$$\Gamma_2 = O_{\pi\mathbb{Z}} \cdot \mathbb{Z}^2 \subset R.$$

Note that the element $O_\pi \in O_{\mathbb{R}}$ inverts the plane $\mathbb{R}^2$ and hence keeps the lattice $\mathbb{Z}^2 \subset \mathbb{R}^2$ invariant. It follows that the lattice Γ_2 is not Abelian and contains Γ_1 as a subgroup of index 2. The diffeomorphism $\Pi : R \to N_R$ induces a two-fold covering of the manifold R/Γ_2 by the 3-torus N_R/Γ_1. Flows on R/Γ_2 induced by unipotent subgroups $g_{\mathbb{R}} \subset \mathbb{R}^2$ are arranged in the same way as the corresponding flows on R/Γ_1. But if $g_{\mathbb{R}} = O_{\mathbb{R}}$, there is a difference. This flow is also periodic, but here the length of orbits is not constant because the intersection $h^{-1}g_{\mathbb{R}}h \cap \Gamma_2$ is $O_{\pi\mathbb{Z}}$ for $h \in O_{\mathbb{R}}$ and $O_{2\pi\mathbb{Z}}$ for $h \notin O_{\mathbb{R}}$. (Thus, the "marked" orbit is half as long as the others.) For this reason, the partition into periodic orbits (which also serves as the ergodic decomposition) is not a smooth fibration. Indeed, the orbit space $M = O_{\mathbb{R}}\backslash R/\Gamma_2$ is the quotient of the plane $O_{\mathbb{R}}\backslash R \simeq \mathbb{R}^2$ by orbits of the action of Γ_2 on $\mathbb{R}^2$. The action of Γ_2 on $\mathbb{R}^2$ is generated by translations along $\mathbb{Z}^2$ and the inversion O_π. It is easily seen that the space M is homeomorphic to the two-dimensional sphere S^2. If the map $R/\Gamma_2 \to S^2$ were a smooth fibration (with S^1 as a fiber), we would derive from the homotopy sequence of the fibration that $\pi_3(M) = 0$, which contradicts the well-known formula $\pi_3(S^2) = \mathbb{Z}$.

Example 4.3. We take the same Lie group R with a new lattice

$$\Gamma_3 = Z(a) \times \mathbb{Z}^2, \qquad \text{where } a = O_{2\pi} \times (\sqrt{2}, 0)$$

and $Z(a)$ is the cyclic group generated by the element a. Since $\Gamma_3 \subset \Delta \times \mathbb{R}^2$, the diffeomorphism $\Pi : R \to \Delta \times \mathbb{R}^2$ induces a diffeomorphism

$$\Pi^* : R/\Gamma_3 \to (\Delta \times \mathbb{R}^2)/\Gamma_3.$$

Let

$$g_t = O_t \quad \text{and} \quad u_t = \Pi(g_t) = O_t \times \alpha(O_t)^{-1} \subset \Delta.$$

Let us prove that the flow $(R/\Gamma_3, g_{\mathbb{R}})$ is measure-theoretically and smoothly isomorphic to the flow $((\Delta \times \mathbb{R}^2)/\Gamma_3, u_{\mathbb{R}})$. Indeed,

$$\Pi(g_t h\Gamma_3) = \Pi(\alpha(g_t)\Pi(g_t)\alpha(h)\Pi(h)\Gamma_3) = \Pi(\alpha(g_t)\alpha(h)u_t\Pi(h)\Gamma_3) = u_t\Pi(h\Gamma_3),$$

because $u_{\mathbb{R}} = \Pi(g_{\mathbb{R}}) = \Delta$ commutes with $S = \alpha(g_{\mathbb{R}})$. Let $e_0 = O_1 \times \alpha(O_1)^{-1}$ be the generator element in Δ. Relative to the basis (e_0, e_1, e_2) of the three-dimensional vector space $\Delta \times \mathbb{R}^2$, the lattice Γ_3 is generated by the elements

$$(2\pi, \sqrt{2}, 0), \ (0,1,0), \ (0,0,1),$$

and the directing vector of the flow is $(1,0,0)$. Thus the flow $((\Delta \times \mathbb{R}^2)/\Gamma_3, u_{\mathbb{R}})$ is nonergodic and the subgroup $(\overline{u_{\mathbb{R}}\Gamma_3})_0$ is $\Delta \times \mathbb{R}'$, where $\mathbb{R}' = \mathbb{R}e_1 \subset \mathbb{R}^2$. Hence the set $(\overline{g_{\mathbb{R}}\Gamma_3})_0$ coincides with the product $O_{\mathbb{R}}\mathbb{R}'$. But the latter is not a subgroup of R, because $O_{\mathbb{R}}$ is not normalized by $\mathbb{R}'$. Therefore the orbit closure $\overline{g_{\mathbb{R}}\Gamma_3} \subset R/\Gamma_3$ is a submanifold in R/Γ_3 but is not a homogeneous subspace (that is, no subgroup of R acts transitively on it).

Likewise, given an element $h \in G$, let

$$\Pi_h : R_s = \Delta \times h^{-1}Sh \cdot \mathbb{R}^2 \to \Delta \times \mathbb{R}^2$$

be the projection along $h^{-1}Sh$. Then Π_h establishes a diffeomorphism

$$\Pi_h^* : R/\Gamma_3 \to (\Delta \times \mathbb{R}^2)/\Gamma_3$$

and sends the left action of $h^{-1}g_{\mathbb{R}}h = h^{-1}O_{\mathbb{R}}h$ to the left action of Δ. It follows that

$$\overline{h^{-1}g_{\mathbb{R}}h\Gamma_3} = h^{-1}O_{\mathbb{R}}h\mathbb{R}'\Gamma_3,$$

and so

$$\overline{g_{\mathbb{R}}h\Gamma_3} = h\overline{h^{-1}g_{\mathbb{R}}h\Gamma_3} = hh^{-1}O_{\mathbb{R}}h\mathbb{R}'\Gamma_3 = O_{\mathbb{R}}h\mathbb{R}'\Gamma_3.$$

Thus, every orbit closure of the flow $(R/\Gamma_3, g_{\mathbb{R}})$ is a smooth nonhomogeneous submanifold in R/Γ_3.

Define

$$E_h(g_{\mathbb{R}}) = \overline{g_{\mathbb{R}}h\Gamma_3} = O_{\mathbb{R}}h\mathbb{R}'\Gamma_3.$$

Then every flow

$$(E_h(g_{\mathbb{R}}), g_{\mathbb{R}}) \simeq (E_1(hg_{\mathbb{R}}h^{-1}), hg_{\mathbb{R}}h^{-1})$$

is smoothly isomorphic to the flow $((\Delta \times \mathbb{R}')\Gamma_3, u_{\mathbb{R}})$ and hence is uniquely ergodic relative to the smooth measure μ_g obtained from a $\Delta \times \mathbb{R}'$-invariant measure on the 2-torus $(\Delta \times \mathbb{R}')\Gamma_3$. Since every flow on $E_h(g_{\mathbb{R}})$ is minimal, any two submanifolds $E_h(g_{\mathbb{R}})$ and $E_f(g_{\mathbb{R}})$ either coincide or do not intersect each other. Clearly, the partition E into these submanifolds coincides with the ergodic decomposition $\widetilde{E}$ of the flow $(R/\Gamma_3, g_{\mathbb{R}})$.

Hence any ergodic submanifold $E_h(g_{\mathbb{R}})$ is the orbit closure $\overline{g_{\mathbb{R}}h\Gamma_3} = O_{\mathbb{R}}h\mathbb{R}'\Gamma_3$ and therefore is not homogeneous. At the same time, every flow $(E_h(g_{\mathbb{R}}), g_{\mathbb{R}})$ is smoothly isomorphic to a rectilinear ergodic flow $((\Delta \times \mathbb{R}')\Gamma_3, u_{\mathbb{R}})$ on the 2-torus. This presents a new effect, compared to the nilpotent case where all ergodic submanifolds are homogeneous.

Example 4.4. One might get the impression that whatever the lattice Γ in our Lie group $R = O_{\mathbb{R}} \cdot \mathbb{R}^2$ is, the space R/Γ carries no ergodic flows. But this is not the case. Let

$$\Gamma_4 = Z(a) \times \mathbb{Z}^2, \qquad \text{where } a = O_{2\pi} \times (\sqrt{2}, \sqrt{3})$$

and $Z(a)$ is the cyclic group generated by the element a. Then $\Gamma_4 \subset \Delta \times \mathbb{R}^2$, and hence the diffeomorphism $\Pi : R \to \Delta \times \mathbb{R}^2$ induces a diffeomorphism of R/Γ_4 with the 3-torus $(\Delta \times \mathbb{R}^2)/\Gamma_4$. Define $g_{\mathbb{R}} = O_{\mathbb{R}}$. Then Π induces a smooth isomorphism of flows

$$\Pi^* : (R/\Gamma_4, g_{\mathbb{R}}) \to ((\Delta \times \mathbb{R}^2)/\Gamma_4, u_{\mathbb{R}}),$$

where $u_{\mathbb{R}} = \Delta$. If $e_0 = O_1 \times \alpha(O_1)^{-1}$ is the generator in Δ, then relative to the basis (e_0, e_1, e_2) of the three-dimensional vector space $\Delta \times \mathbb{R}^2$, the lattice Γ_3 is generated by the elements

$$(2\pi, \sqrt{2}, \sqrt{3}), \ (0, 1, 0), \ (0, 0, 1),$$

and the directing vector of the subgroup $u_{\mathbb{R}}$ is $(1, 0, 0)$. Consequently the flow $((\Delta \times \mathbb{R}^2)/\Gamma_4, u_{\mathbb{R}})$ is ergodic and minimal. The same goes for the original flow $(R/\Gamma_4, g_{\mathbb{R}})$.

We conclude that a compact solvable manifold R/Γ may admit or not admit an ergodic homogeneous flow, depending on the choice of the lattice $\Gamma \subset R$.

2. So far, all the flows studied have been quasi-unipotent. Now we consider the opposite type of solvable flows. More precisely, as we will see, the standard suspension over a toral automorphism is a one-parameter homogeneous flow on a compact solvable manifold. If the automorphism is ergodic, the dynamics of the obtained flow differs drastically from the previous examples. In particular, such a flow is partially hyperbolic and may have rather "exotic" orbits with their closures nowhere locally connected.

Let $X = G/\Gamma$, where Γ is a lattice in a Lie group G, and let $A \in \operatorname{Aut}(G, \Gamma)$ be an automorphism of G keeping the lattice Γ invariant. Then A determines an automorphism of the space X by the formula $A(g\Gamma) = A(g)\Gamma$. It turns out that the standard suspension over the action $A \in \operatorname{Aut}(X)$ quite often can be viewed as a homogeneous flow.

More specifically, suppose that the group $\operatorname{Aut}(G)$ contains a one-parameter subgroup $A_{\mathbb{R}} \subset \operatorname{Aut}(G)$ such that $A_1 = A$ (in general, the element $A \in \operatorname{Aut}(G)$ does not embed into a one-parameter subgroup, but some power of it does). Define $G' = A_{\mathbb{R}} \cdot G$ and $\Gamma' = A_{\mathbb{Z}} \cdot \Gamma$. There is a natural homeomorphism $G'/\Gamma' \to X' = (X \times \mathbb{R})/\sim$, defined as

$$A_s g\Gamma' \to (g\Gamma, s)/\sim, \qquad g \in G, \ s \in \mathbb{R}.$$

It is easy to see that the homogeneous flow $(G'/\Gamma', A_{\mathbb{R}})$ is precisely the suspension over the one-generator action $(G/\Gamma, A_{\mathbb{Z}})$. Hence the theory of solvable flows involves the study of toral automorphisms.

The main series of examples arises as follows. Let

$$G/\Gamma = \mathbb{R}^n/\mathbb{Z}^n = \mathbb{T}^n \qquad \text{and} \qquad A \in \mathrm{SL}(n, \mathbb{Z}) \subset \operatorname{Aut}(\mathbb{R}^n, \mathbb{Z}^n).$$

If $A = \exp(a)$ for some element $a \in \mathrm{sl}(n, \mathbb{R})$, then the suspension over the automorphism A of the torus $\mathbb{T}^n$ is a flow $(G'/\Gamma', \exp(\mathbb{R}a))$, where $G' = \exp(\mathbb{R}a) \cdot \mathbb{R}^n$ is a solvable Lie group with the lattice

$$\Gamma' = \exp(\mathbb{Z}a) \cdot \mathbb{Z}^n = A^{\mathbb{Z}} \cdot \mathbb{Z}^n.$$

Remark. The map $\exp : \mathrm{sl}(n, \mathbb{R}) \to \mathrm{SL}(n, \mathbb{R})$ is neither surjective nor injective. Therefore the suspension over a toral automorphism cannot always be realized as a homogeneous flow; and if it can, there may exist infinitely many such realizations. If two distinct one-parameter subgroups $A_{\mathbb{R}}, A'_{\mathbb{R}} \subset \mathrm{SL}(n, \mathbb{R})$ contain an element $A \in \mathrm{SL}(n, \mathbb{Z})$, the Lie groups obtained $G = A_{\mathbb{R}} \cdot \mathbb{R}^n$ and $G' = A'_{\mathbb{R}} \cdot \mathbb{R}^n$ may not be isomorphic (although both of them contain the lattice $\Gamma = A^{\mathbb{Z}} \cdot \mathbb{Z}^n$). However, from the dynamical point of view, the flows $(G/\Gamma, A_{\mathbb{R}})$ and $(G'/\Gamma, A'_{\mathbb{R}})$ are smoothly and measure-theoretically isomorphic (which is not surprising, because both of them are suspensions over the same toral automorphism).

3. Let us look more closely at automorphisms of the torus $\mathbb{T}^n$. By making use of Fourier series we will prove a particular case of Halmos' ergodicity criterion for automorphisms of a compact Abelian group.

So, let $A \in \mathrm{SL}(n, \mathbb{Z})$ be an automorphism of the torus $\mathbb{T}^n = \mathbb{R}^n/\mathbb{Z}^n$. We will treat elements of $\mathbb{Z}^n$ as multi-indices $I = (i_1, \ldots, i_n) \in \mathbb{Z}^n$. Also we agree that $\mathbb{Z}^n$ is generated by the standard basis $e_1, \ldots, e_n$ in $\mathbb{R}^n$, and $\mathbb{R}^n$ is equipped with the scalar product $(x, y) = \sum_1^n x_i y_i$ relative to the basis.

Every complex-valued function $f \in L^2(\mathbb{T}^n)$ is given by its Fourier series

$$f(\overline{x}) = \sum_I f_I e^{2\pi i(x,I)}, \qquad \text{where } x \in \mathbb{R}^n,\ \overline{x} = x + \mathbb{Z}^n \in \mathbb{T}^n, \text{ and } \sum_{I \in \mathbb{Z}^n} |f_I|^2 < \infty.$$

Note that

$$\begin{aligned} f(A\overline{x}) &= \sum_I f_I e^{2\pi i(Ax,I)} = \sum_I f_I e^{2\pi i(x,A^{\intercal} I)} \\ &= \sum_I f_{(A^{\intercal})^{-1} I} e^{2\pi i(x,I)} = \sum_I f_{\mathcal{A} I} e^{2\pi i(x,I)}, \end{aligned}$$

where $\mathcal{A} = (A^{\intercal})^{-1} : \mathbb{R}^n \to \mathbb{R}^n$ stands for the contragredient matrix for A. We observe that orbits of A and $\mathcal{A}$ on $\mathbb{Z}^n$ are in one-to-one correspondence.

LEMMA 4.5. *Given $A \in \mathrm{SL}(n, \mathbb{Z})$, the following conditions are equivalent*:
1) *The automorphism A is ergodic on $\mathbb{T}^n$.*
2) *A and $\mathcal{A}$ have no periodic orbits in $\mathbb{Z}^n - \{0\}$.*
3) *No eigenvalue of A is a root of* 1.
4) *The automorphism A is mixing on $\mathbb{T}^n$.*

Proof. 2) $\Rightarrow$ 1). Let $f = \sum_I f_I e^{2\pi i(x,I)}$ be an A-invariant function. Then $f_{\mathcal{A}I} = f_I$ for all $I \in \mathbb{Z}^n$. If the orbit $\mathcal{A}^{\mathbb{Z}} I$ is infinite, then $f_I = 0$ because $\sum_I |f_I|^2 < \infty$.

1) $\Rightarrow$ 2). If $\mathcal{A}^k I = I$, let $f_{\mathcal{A}^l I} = 1$, $l \in \mathbb{Z}$, and 0 for other indices. We obtain an A-invariant function $f \in L^2(\mathbb{T}^n)$.

2) $\Rightarrow$ 3). If $A^k x = x$ for some $k \in \mathbb{N}$ and $x \in \mathbb{C}^n$, then the space $\mathrm{Fix}(A^k) = \{x \in \mathbb{C}^n : A^k x = x\}$ is defined over $\mathbb{Q}$ and hence intersects $\mathbb{Z}^n$ in a lattice. It follows that A and $\mathcal{A}$ have a periodic orbit in $\mathbb{Z}^n - \{0\}$.

2) $\Rightarrow$ 4). The space of functions $f \in L^2(\mathbb{T}^n)$ with $c(f) = \sum_I |f_I| < \infty$ is dense in $L^2(\mathbb{T}^n)$. Hence it suffices to check the mixing property on functions f, g such that $f_0 = 0$ and $c(f), c(g) < \infty$. We have

$$\Big|\sum_k (A^k f, g)\Big| = \Big|\sum_k \sum_I f_{\mathcal{A}^k I}\bar{g}_I\Big| \le \sum_I \sum_k |f_{\mathcal{A}^k I}||g_I| \le c(f)c(g) < \infty,$$

because $\mathcal{A}$ has no periodic orbits in $\mathbb{Z}^n - \{0\}$. It follows that $(A_k f, g) \to 0,\ k \to \infty$.

The implications 3) $\Rightarrow$ 2) and 4) $\Rightarrow$ 1) are obvious. □

Let $A = A_s \times A_u$ be a Jordan decomposition of a matrix $A \in \mathrm{SL}(n, \mathbb{Z})$ into semisimple and unipotent parts. Then $A_s, A_u \in \mathrm{SL}(n, \mathbb{Q})$, and hence there exists $m \in \mathbb{N}$ such that $A_u^m \in \mathrm{SL}(n, \mathbb{Z})$. It follows that $A^m = A_s^m \times A_u^m$ is a Jordan decomposition of the matrix A^m inside $\mathrm{SL}(n, \mathbb{Z})$. The lemma implies that A, A^m and A_s^m as toral automorphisms are either all ergodic or all nonergodic. If the matrix A is quasi-unipotent, then all eigenvalues of A and A_s belong to the unit circle in $\mathbb{C}$ and hence all orbits of A_s in $\mathbb{R}^n$ are bounded. It follows that all orbits of A_s^m on $\mathbb{Z}^n$ are periodic. As a corollary, we deduce that only a partially hyperbolic automorphism $A \in \mathrm{SL}(n, \mathbb{Z})$ can be ergodic.

This remark allows one to prove another (less effective) ergodicity criterion. Let $U \subset \mathbb{R}^n$ be the Auslander unstable subspace relative to the automorphism A (that is, no eigenvalue of the restriction of A onto U belongs to the unit circle in $\mathbb{C}$, whereas all eigenvalues for the quotient action of A on $\mathbb{R}^n/U$ belong to the unit circle). As was already proved in §2, the automorphism A is ergodic on $\mathbb{T}^n$ if $\mathbb{T}^n = \overline{U\mathbb{Z}^n}$. Conversely, the automorphism A on the factor-space $\mathbb{T}^n/\overline{U\mathbb{Z}^n}$ is quasi-unipotent and hence in view of ergodicity the factor-space is trivial.

By the way, as was mentioned in the introduction, the condition $\mathbb{T}^n = \overline{U\mathbb{Z}^n}$ ensures that A is a K-automorphism (and, in particular, is mixing of all orders).

Let $\mathrm{Per}(A) \subset \mathbb{T}^n$ stand for the set of periodic points of an automorphism A on the torus $\mathbb{T}^n$. Note that $\mathbb{Q}^n/\mathbb{Z}^n \subset \mathrm{Per}(A)$. In fact, $\mathbb{T}^n$ can be viewed as an additive group with the zero element $\mathbb{Z}^n \in \mathbb{R}^n/\mathbb{Z}^n$. If $x \in \mathbb{Q}^n/\mathbb{Z}^n$ then $mx = 0$ for some $m \in \mathbb{N}$, and then $mA(x) = A(mx) = 0$. It follows that A preserves the finite subgroup $\Sigma_m = \{x \in \mathbb{T}^n : mx = 0\}$, and hence the point x is periodic. One can provide one more ergodicity criterion.

LEMMA 4.6. *An automorphism $A \in \mathrm{SL}(n, \mathbb{Z})$ is ergodic* $\Longleftrightarrow \mathrm{Per}(A) = \mathbb{Q}^n/\mathbb{Z}^n$.

Proof. Suppose that $A^k x = x$ for some $x \in \mathbb{T}^n$, $k \in \mathbb{N}$, where $x = v + \mathbb{Z}^n \in \mathbb{R}^n/\mathbb{Z}^n$ for some $v \in \mathbb{R}^n$. Then $(A^k - E)v \in \mathbb{Z}^n$. In view of ergodicity, the matrix $A^k - E$ is invertible and its inverse belongs to $\mathrm{GL}(n, \mathbb{Q})$. It follows that $v \in \mathbb{Q}^n$.

Conversely, if A is nonergodic, then by Lemma 4.5, for some $k \in \mathbb{N}$, the kernel L of the operator $A^k - E$ on $\mathbb{R}^n$ is nontrivial. But then the subtorus $L/\mathbb{Z}^n \cap L \subset \mathbb{T}^n$ consists entirely of periodic points. □

Now we prove the ergodicity criterion for automorphisms of a nilmanifold.

LEMMA 4.7 (Parry [**P69**]). *Let $A \in \mathrm{Aut}(N, \Gamma)$ be an automorphism of a compact nilmanifold N/Γ and let $U \subset N$ be the Auslander normal subgroup relative to A. Then the following conditions are equivalent.*

1) *A is ergodic on the nilmanifold N/Γ.*

2) *A is ergodic on the torus $N/\Gamma[N, N]$.*

3) $N = \overline{U\Gamma}$.

Proof. The implications 1) $\Rightarrow$ 2) and 3) $\Rightarrow$ 1) are obvious. To prove 2) $\Rightarrow$ 3), define $L = (\overline{U\Gamma})_0$. Then L is a normal subgroup of N and the product $L\Gamma$ is closed. It follows that the product $L[N,N]\Gamma$ is closed as well. On the other hand, in view of ergodicity, $N = \overline{U[N,N]\Gamma}$ and hence $L[N,N] = N$. Thus, $L = N$, as desired. □

4. Now we proceed to consider orbits of an ergodic toral automorphism. We turn to the simplest example: $A = \begin{pmatrix} 2 & 1 \\ 1 & 1 \end{pmatrix} \in \mathrm{SL}(2,\mathbb{Z})$. Let us agree that the lattice $\mathbb{Z}^2 \in \mathbb{R}^2$ has the standard basis $b_1, b_2 \in \mathbb{R}^2$, whereas eigenvectors of the matrix A generate another (orthogonal) basis $e_1, e_2 \in \mathbb{R}^2$. More specifically, $\mathbb{R}e_1$ is the expanding direction for A, and $\mathbb{R}e_2$ is contracting. Clearly, these directions generate transversal dense windings on the torus $\mathbb{T}^2$ (a closed trajectory cannot contract or expand uniformly at all points).

Let $x \in \mathbb{T}^2$, $x \neq 0$, be an intersection point of the contracting and expanding leaves of the fixed point $0 \in \mathbb{T}^2$. Such a point is called *homoclinic.* Then $A^k x \to 0$ as $k \to \pm\infty$. This means that the orbit of the point x is locally closed and nonclosed. In particular, its closure is not a manifold.

Pick a small square centered at $0 \in \mathbb{T}^2$ with its sides formed by segments of the contracting and expanding foliations and of length $\delta > 0$. The interior of the square is a neighborhood P_0 of the point $0 \in \mathbb{T}^2$ which contains all but a finite number of the points of the orbit $A^{\mathbb{Z}}x$. We surround the orbit points $x_1, \ldots, x_n$ exterior to P_0 by square neighborhoods of the same size δ and denote them by $P_1, \ldots, P_n$. Define

$$P = P_0 \cup P_1 \cup \cdots \cup P_n \quad \text{and} \quad M = \bigcap_{i \in \mathbb{Z}} A^i P.$$

Let us show that from the topological viewpoint the invariant set M is arranged as a two-dimensional Cantor set.

In fact, under the action of A, each square P_k is sent to a rectangle, and for a large enough $i > 0$, the rectangle $A^i P_k$ is a band which gets thinner and longer as i increases (keeping its area constant). Since every expanding leaf is everywhere dense in $\mathbb{T}^2$, the band gets more and more uniformly distributed in $\mathbb{T}^2$. Hence the intersection $P_0 \cap A^l P_k$ is a family of thin parallel bands of length δ. We call them horizontal bands. Likewise, $A^{-l}P_k \cap P_0$ is a family of thin vertical bands, whereas $A^{-l}P_k \cap P_0 \cap A^l P_k$ is a family of many small squares inside P_0, which form a fairly uniform net. As l increases, every such small square gives rise to a net of smaller squares, and this picture describes the local structure of the set $\bigcap_{i=-l}^{l} A^i P$.

A rigorous proof of the fact that M is a topologically transitive locally disconnected perfect set can be given by the methods of symbolic dynamics (see [**Al**] or [**Bo79**]). Actually, M can be a minimal set of positive Hausdorff dimension (see [**Be**]). In the sequel we only need the following: the automorphism A (as well as any Anosov diffeomorphism of a compact manifold) has various orbits: periodic, locally closed, and those with locally disconnected perfect closures.

On the other hand, according to Sinai [**Si72**] and Bowen [**Bo75**], any Anosov automorphism has uncountably many ergodic measures: those with their supports not locally connected, and those that are strictly positive (but singular to a smooth volume measure). The simplest example of a singular strictly positive ergodic measure was constructed by Anosov [**An67**]. In this example one takes our hyperbolic automorphism $A : \mathbb{T}^2 \to \mathbb{T}^2$ and constructs a continuous family of smooth perturbations $A_s : \mathbb{T}^2 \to \mathbb{T}^2$, $A_0 = A$, preserving the Lebesgue measure l on $\mathbb{T}^2$. Here the entropy of A_s varies continuously and reaches a local maximum at the point $s = 0$. By Anosov's structural stability theorem [**An67**], given small enough s, there exists a homeomorphism $\phi_s : \mathbb{T}^2 \to \mathbb{T}^2$ such that $\phi_s A_s = A\phi_s$. It is easily seen that each measure $(\phi_s)_*(l)$ is strictly positive, ergodic, and different from l by entropy considerations.

An ergodic toral automorphism need not be Anosov (that is, totally hyperbolic). Given such an automorphism, it is easy to prove the existence of a locally closed nonperiodic orbit (a counterpart of a homoclinic orbit), but in the general case the structure of orbit closures and minimal sets is not well understood. Let us give an example of such an automorphism (due to Auslander) that was used in [**St84b,94**].

Example 4.8. Let

$$A = \begin{pmatrix} 0 & 1 & 0 & 0 \\ 0 & 0 & 1 & 0 \\ 0 & 0 & 0 & 1 \\ -1 & 8 & -10 & 8 \end{pmatrix}.$$

It is easy to verify that A has two pairs of eigenvalues:

$$\lambda_{1,2} = e^{\pm i\theta},\ \theta \in \mathbb{R}, \qquad \text{and} \qquad \lambda_{3,4} = e^{\pm\alpha}, \alpha > 1.$$

Hence the space $\mathbb{R}^4$ decomposes into the sum of two invariant planes $\mathbb{R}^4 = V + V'$, where the action of A on V is orthogonal, and on V' hyperbolic. So we may view the matrix A as an element of the cylinder $\mathrm{SO}(V) \times \mathrm{SH}(V') \subset \mathrm{SL}(4, \mathbb{R})$, where $\mathrm{SO}(V)$ stands for the group of orthogonal rotations of V, and $\mathrm{SH}(V')$ for the group of unimodular hyperbolic transformations of V'. Here it is of interest that the ratio θ/π is irrational and hence the numbers $\lambda_{1,2}$ are not roots of 1. It follows that A determines an ergodic automorphism of the torus $\mathbb{T}^4$.

On the other hand, the matrix A belongs to countably many distinct one-parameter subgroups $A_{\mathbb{R}}(k)$, $k \in \mathbb{Z}$, of the cylinder $\mathrm{SO}(V) \times \mathrm{SH}(V')$. Hence each of the flows

$$(G_k/\Gamma, A_{\mathbb{R}}(k)), \qquad \text{where } G_k = A_{\mathbb{R}}(k) \cdot \mathbb{R}^4 \text{ and } \Gamma = A^{\mathbb{Z}} \cdot \mathbb{Z}^4,$$

is the suspension over the automorphism A of the torus $\mathbb{T}^4$. The groups G_k are mutually nonisomorphic, and hence from the algebraic viewpoint the flows are different (though topologically and measure-theoretically they are isomorphic). The lattice $\Gamma \subset G_k$ has interesting rigidity properties (see [**St94**]).

5. We return to the class of solvable flows. The suspension over the Anosov automorphism $A = \left(\begin{smallmatrix} 2 & 1 \\ 1 & 1 \end{smallmatrix}\right)$ of the torus $\mathbb{T}^2$ is a flow $(R/\Gamma, A_{\mathbb{R}})$, where

$$\Gamma = A^{\mathbb{Z}} \cdot \mathbb{Z}^2, \qquad R = A_{\mathbb{R}} \cdot \mathbb{R}^2,$$

where $A_{\mathbb{R}} \subset \mathrm{SL}(2, \mathbb{R})$ is the one-parameter subgroup such that $A = A_1$ (in this case such a subgroup is unique). Then $A_{\mathbb{R}}$ is conjugate to the diagonal subgroup

$$\left\{ H_t = \begin{pmatrix} e^{st} & 0 \\ 0 & e^{-st} \end{pmatrix} \right\}, \qquad \text{where } e^s + e^{-s} = 3.$$

Clearly, R is a connected simply connected three-dimensional unimodular class (R) solvable group, and the flow $(R/\Gamma, A_{\mathbb{R}})$ is Anosov and hence ergodic. Note that unlike the automorphism A, this flow (like any solvable flow) is not mixing: the rotation group of the circle is its quotient flow. Apart from countably many periodic orbits, it has locally closed nonperiodic orbits, and also orbits whose closures are not locally connected. If a subgroup $g_{\mathbb{R}} \subset R$ does not belong to $\mathbb{R}^2$, it is (up to rescaling of time) conjugate to the subgroup $A_{\mathbb{R}}$ and hence induces a flow isomorphic to the suspension.

But if $g_{\mathbb{R}} \subset \mathbb{R}^2$, the flow $(R/\Gamma, g_{\mathbb{R}})$ is nonergodic and determines the fibration of R/Γ into invariant 2-tori $\mathbb{R}^2 h\Gamma$. Every flow $(\mathbb{R}^2 h\Gamma, g_{\mathbb{R}})$ is isomorphic to the flow $(\mathbb{R}^2\Gamma, h^{-1}g_{\mathbb{R}}h)$, and one has to distinguish two cases. The group R contains precisely two normal one-parameter subgroups $g_{\mathbb{R}}$ spanned by eigenvectors e_1 and e_2 of A. Each of them determines an irrational winding on the torus $\mathbb{R}^2/\mathbb{Z}^2$, and all subflows $(\mathbb{R}^2 h\Gamma, g_{\mathbb{R}})$, $h \in R$, are ergodic and are obtained from each other by linear time change. In this case the ergodic decompositions E and $\widetilde{E}$ coincide with the fibration of the space R/Γ into invariant 2-tori $\mathbb{R}^2 h\Gamma$. If the subgroup $g_{\mathbb{R}} \subset \mathbb{R}^2$ is not normal in R, then the conjugate subgroups $\{h^{-1}g_{\mathbb{R}}h,\ h \in R\}$ cover a cone in $\mathbb{R}^2$, and almost all of them induce ergodic flows. It follows that the fibration of the space R/Γ into invariant 2-tori $\mathbb{R}^2 h\Gamma$ is the ergodic decomposition $\widetilde{E}$ for our flow. The ergodic partition E differs from $\widetilde{E}$ on a dense null set. More specifically, E breaks down nonergodic subtori into periodic orbits.

Finally, we note that, in contrast to the nilpotent case, the ergodicity of a flow on R/Γ does not depend on the mutual positions of the one-parameter subgroup and the lattice.

§5. Construction of the semisimple splitting

1. Construction of the semisimple splitting (Maltsev's splitting) was first applied to the study of solvable homogeneous flows by Auslander and Green [**AG**]. Subsequently, the study of the construction was continued by Auslander and Brezin [**AB68**]. The general idea consists in embedding solvable and general simply connected Lie groups into split groups (for nilpotent and semisimple groups the construction gives nothing new) and is due to Maltsev [**M**]. Note that even more powerful and complex techniques (covering all connected Lie groups) was developed by Hochschild and Mostow [**HM**]; however we do not need it for studying homogeneous flows. It is more suitable to introduce construction of the semisimple splitting for Lie algebras, and then to transfer the results to the class of connected simply connected Lie groups.

Let $\mathfrak{r}$ be a solvable Lie algebra and let $\mathrm{Der}(\mathfrak{r})$ be the algebraic Lie algebra of all derivations of $\mathfrak{r}$. Due to the algebraicity of $\mathrm{Der}(\mathfrak{r})$, there is a decomposition

$$\mathrm{Der}(\mathfrak{r}) = (\mathfrak{p} \oplus \mathfrak{t}) \dotplus \mathfrak{u},$$

where $\mathfrak{p}$ is a maximal semisimple subalgebra in $\mathrm{Der}(\mathfrak{r})$, $\mathfrak{t}$ is an Abelian subalgebra of semisimple elements which commutes with $\mathfrak{p}$, and $\mathfrak{u}$ is the nilradical (the ideal containing all nilpotent elements). The subalgebra $\mathrm{ad}(\mathfrak{r})$ of all inner derivations of $\mathfrak{r}$ is a solvable ideal in $\mathrm{Der}(\mathfrak{r})$, and hence its algebraic hull belongs to the radical $\mathfrak{t} \dotplus \mathfrak{u}$. Let $\theta : \mathfrak{t} \dotplus \mathfrak{u} \to \mathfrak{t}$ be the projection and let $\mathfrak{t}_r = \theta(\mathrm{ad}(\mathfrak{r}))$. Since $\mathfrak{t}_r$ is a Lie algebra of derivations of $\mathfrak{r}$, one can take the semidirect sum $\mathfrak{r}_s = \mathfrak{t}_r \dotplus \mathfrak{r}$.

The Lie algebra

$$\mathfrak{r}_s = \mathfrak{t}_r \dotplus \mathfrak{r}$$

constructed above is called the *semisimple splitting* of the solvable Lie algebra $\mathfrak{r}$, and its nilradical $\mathfrak{n}_r$ is called the *nilshadow* of $\mathfrak{r}$.

It is easy to check [**AB68**] that

$$\mathfrak{r}_s = \mathfrak{t}_r \dotplus \mathfrak{n}_r, \text{ that is, } \mathfrak{r}_s = \mathfrak{t}_r + \mathfrak{n}_r \text{ and } \mathfrak{t}_r \cap \mathfrak{n}_r = 0.$$

In addition, $\mathfrak{r}_s = \mathfrak{r} + \mathfrak{n}_r$. One can prove [**AB68**] that the two properties provide an equivalent nonconstructive definition.

More precisely, let $\mathfrak{t}_r$ be an Abelian Lie algebra of nontrivial semisimple derivations of a solvable Lie algebra $\mathfrak{r}$. Define $\mathfrak{r}_s = \mathfrak{t}_r \dotplus \mathfrak{r}$, and let $\mathfrak{n}_r$ be the nilradical of $\mathfrak{r}_s$. If $\mathfrak{r}_s = \mathfrak{t}_r \dotplus \mathfrak{n}_r$ and $\mathfrak{r}_s = \mathfrak{r} + \mathfrak{n}_r$, the Lie algebra $\mathfrak{r}_s$ is called the *semisimple splitting* of the solvable Lie algebra $\mathfrak{r}$.

Note that $\dim \mathfrak{r} = \dim \mathfrak{n}_r$, and the ideal $\mathfrak{n} = \mathfrak{r} \cap \mathfrak{n}_r$ is the nilradical of $\mathfrak{r}$.

Let us give some examples of the semisimple splitting. If a Lie algebra $\mathfrak{r}$ is nilpotent, by definition $\mathfrak{r}_s = \mathfrak{r}$. Suppose now that $\mathfrak{r}$ is a split solvable Lie algebra, that is, $\mathfrak{r} = \mathfrak{a} \dotplus \mathfrak{n}$, where $\mathfrak{n}$ is the nilradical of $\mathfrak{r}$, and $\mathfrak{a}$ is a Lie subalgebra consisting of nontrivial semisimple derivations of $\mathfrak{n}$. Since the center of $\mathfrak{r}$ belongs to $\mathfrak{n}$, it follows that $\mathfrak{a}$ consists of nontrivial semisimple derivations of the nilradical $\mathfrak{n}$, and hence the representation $\mathrm{ad} : \mathfrak{a} \to \mathrm{Der}(\mathfrak{r})$ is faithful. Define $\mathfrak{t}_r = \mathrm{ad}(\mathfrak{a})$. It is easily seen that the Lie algebra $\mathfrak{r}_s = \mathfrak{t}_r \dotplus \mathfrak{r}$ is the semisimple splitting of $\mathfrak{r}$, and the nilshadow $\mathfrak{n}_r$ is as follows: $\mathfrak{n}_r = \Delta \oplus \mathfrak{n}$. Here $\Delta = \{(\mathrm{ad}(a), -a),\ a \in \mathfrak{a}\}$ is the "anti-diagonal" in $\mathrm{ad}(\mathfrak{a}) \oplus \mathfrak{a}$, and Δ belongs to the center of $\mathfrak{r}_s$.

Recall that the nilradical (the maximal nilpotent ideal) is a characteristic ideal and hence unique. As for the maximal reductive subalgebra $\mathfrak{t}_r \subset \mathfrak{r}_s$, one can select it in many ways (all of them are conjugate via inner automorphisms of $\mathfrak{r}_s$ of the form $\exp(\mathrm{ad}_m) = \mathrm{Ad}_{\exp(m)}$, where m is an element of the nilradical $\mathfrak{n}_r \subset \mathfrak{r}_s$). The Lie algebra $\mathfrak{r}_s$ is split (and in terms of [**AB68**] almost algebraic) and admits a Jordan decomposition for every element $x \in \mathfrak{r}_s$ into commuting semisimple and nilpotent parts. Hence the following holds.

LEMMA 5.1. *Given any element $x \in \mathfrak{r}_s$, there exists a maximal Abelian subalgebra of semisimple elements $\mathfrak{t}_r \subset \mathfrak{r}_s$ such that $x = a + n$, where $a \in \mathfrak{t}_r$, $n \in \mathfrak{n}_r$ and $[a, n] = 0$.*

Denote by

$$p : \mathfrak{r}_s = \mathfrak{t}_r \dotplus \mathfrak{n}_r \to \mathfrak{t}_r, \qquad \pi : \mathfrak{r}_s = \mathfrak{t}_r \dotplus \mathfrak{n}_r \to \mathfrak{n}_r$$

the projections along $\mathfrak{n}_r$ and $\mathfrak{t}_r$. Then the restriction $p : \mathfrak{r} \to \mathfrak{t}_r$ is an epimorphism of Lie algebras, and the restriction $\pi : \mathfrak{r} \to \mathfrak{n}_r$ is an isomorphism of $\mathfrak{r}$ and $\mathfrak{n}_r$ as linear spaces but not as Lie algebras (for this reason, $\mathfrak{n}_r$ is called the nilshadow of $\mathfrak{r}$). In addition, the epimorphism $p : \mathfrak{r} \to \mathfrak{t}_r$ is the composite $\theta \circ \mathrm{ad} : \mathfrak{r} \to \mathrm{Der}(\mathfrak{r}) \to \mathfrak{t}_r$.

2. Now we investigate how the semisimple splitting $\mathfrak{r}_s$ behaves relative to such natural operations over the original algebra $\mathfrak{r}$ as localization onto an ideal, factorization by an ideal, and extension by means of an Abelian algebra of semisimple derivations.

LEMMA 5.2. *If $\mathfrak{i}$ is an ideal in $\mathfrak{r}$, then $\mathfrak{i}_s = \mathfrak{t}_i \dotplus \mathfrak{n}_I$, where $\mathfrak{t}_i = p(\mathfrak{i})$ and $\mathfrak{n}_i = \pi(\mathfrak{i})$.*

Proof. Define $\mathfrak{t}_i = p(\mathfrak{i}) = \theta(\mathrm{ad}(\mathfrak{i}))$. Since $\mathfrak{i}$ is an ideal in $\mathfrak{r}$, then $\mathrm{ad}(\mathfrak{i})$ is an ideal in $\mathrm{ad}(\mathfrak{r})$. Hence the algebraic hull of $\mathrm{ad}(\mathfrak{i})$ is an ideal in the algebraic hull of $\mathrm{ad}(\mathfrak{r})$. It follows that $\mathfrak{t}_i = \theta(\mathrm{ad}(\mathfrak{i}))$ belongs to the algebraic hull of $\mathrm{ad}(\mathfrak{i})$, and hence $[\mathfrak{t}_i, \mathfrak{r}] \subset \mathfrak{i}$. On the other hand, $\mathfrak{t}_r$ is a subalgebra of the algebraic hull of $\mathrm{ad}(\mathfrak{r})$, and hence $[\mathfrak{t}_r, \mathfrak{i}] \subset \mathfrak{i}$. Define $\mathfrak{i}_s = \mathfrak{t}_i + \mathfrak{i}$. Then

$$[\mathfrak{i}_s, \mathfrak{r}_s] = [\mathfrak{t}_i + \mathfrak{i}, \mathfrak{t}_r + \mathfrak{r}] \subset [\mathfrak{t}_i, \mathfrak{r}] + [\mathfrak{i}, \mathfrak{t}_r] + \mathfrak{i} \subset \mathfrak{i}.$$

It follows that $\mathfrak{i}$ and $\mathfrak{i}_s$ are ideals in $\mathfrak{r}_s$, and $\mathfrak{n}_i = \pi(\mathfrak{i}) = \mathfrak{i}_s \cap \mathfrak{n}_r$ is an ideal in $\mathfrak{r}_s$. Since $\mathfrak{t}_i$ is a reductive algebra of nontrivial derivations of $\mathfrak{r}$, and $[\mathfrak{t}_i, \mathfrak{r}] \subset \mathfrak{i}$, it follows that $\mathfrak{t}_i$ consists of nontrivial derivations of $\mathfrak{i}$, and $\mathfrak{n}_i$ is the nilradical of $\mathfrak{i}_s$. Now it is clear that $\mathfrak{i}_s = \mathfrak{t}_i + \mathfrak{i} = \mathfrak{t}_i \dotplus \mathfrak{n}_i$ satisfies all the requirements of the nonconstructive definition, and hence is the semisimple splitting of the ideal $\mathfrak{i} \subset \mathfrak{r}$. □

LEMMA 5.3. *Let $\mathfrak{i}$ be an ideal in $\mathfrak{r}$ and let $\mathfrak{r}' = \mathfrak{r}/\mathfrak{i}$. Then*

$$\mathfrak{r}'_s = \mathfrak{t}'_r \dotplus \mathfrak{n}'_r,$$

where $\mathfrak{t}'_r = \mathfrak{t}_r/\mathfrak{t}(\mathfrak{i})$, $\mathfrak{n}'_r = \mathfrak{n}_r/\mathfrak{n}_i$, *and* $\mathfrak{t}(\mathfrak{i}) = \{x \in \mathfrak{t}_r : \ [x, \mathfrak{r}] \subset \mathfrak{i}\}$.

Proof. Since $[\mathfrak{t}(\mathfrak{i}) \dotplus \mathfrak{n}_i, \mathfrak{r}_s] \subset \mathfrak{i} + \mathfrak{n}_i = \mathfrak{i}_s = \mathfrak{t}_i \dotplus \mathfrak{n}_i$ and $\mathfrak{t}_i \subset \mathfrak{t}(\mathfrak{i})$, the subspace $\mathfrak{t}(\mathfrak{i}) \dotplus \mathfrak{n}_i$ is an ideal in $\mathfrak{r}_s$. Also, since $(\mathfrak{t}_r \dotplus \mathfrak{n}_i) \cap \mathfrak{r} = \mathfrak{i}$, then $(\mathfrak{t}(\mathfrak{i}) \dotplus \mathfrak{n}_i) \cap \mathfrak{r} = \mathfrak{i}$. Define $\mathfrak{r}'_s = \mathfrak{r}_s/(\mathfrak{t}(\mathfrak{i}) \dotplus \mathfrak{n}_i)$. Then $\mathfrak{r}' \subset \mathfrak{r}'_s$, and $\mathfrak{t}'_r = \mathfrak{t}_r/\mathfrak{t}(\mathfrak{i})$ is an Abelian algebra of nontrivial semisimple derivations of $\mathfrak{r}'$. Also $\mathfrak{n}'_r = \mathfrak{n}_r/\mathfrak{n}_i$ is the nilradical of $\mathfrak{r}'_s$, and $\mathfrak{r}'_s$ satisfies all the requirements of the nonconstructive definition of the semisimple splitting of the Lie algebra $\mathfrak{r}'$. □

Note that $\mathfrak{r}'_s \neq \mathfrak{r}_s/\mathfrak{i}_s$, because $\mathfrak{t}_i$ is a subgroup of $\mathfrak{t}(\mathfrak{i})$.

Now let $\rho : \mathfrak{t}_0 \to \mathrm{Der}(\mathfrak{r})$ be a homomorphism of an Abelian Lie algebra $\mathfrak{t}_0$ to the algebraic Lie algebra $\mathrm{Der}(\mathfrak{r})$, where $\rho(\mathfrak{t}_0)$ consists of semisimple derivations of $\mathfrak{r}$. Since all maximal Abelian subalgebras of $\mathrm{Der}(\mathfrak{r}) = (\mathfrak{p} \oplus \mathfrak{t}) \dotplus \mathfrak{u}$ of semisimple elements are conjugate, we can select $\mathfrak{t}$ in such a way that $\rho(\mathfrak{t}_0) \subset \mathfrak{t}$, and hence the actions of $\rho(\mathfrak{t}_0)$ and $\mathfrak{t}_r$ on $\mathfrak{r}$ commute.

LEMMA 5.4. *Let $\mathfrak{r}' = \mathfrak{t}_0 \dotplus \mathfrak{r}$, where the semidirect sum is given by the representation ρ. Then*

$$\mathfrak{r}'_s = \mathfrak{t}'_r \dotplus \mathfrak{r}' = \mathfrak{t}'_r \dotplus \mathfrak{n}'_r, \qquad \textit{where } \mathfrak{t}'_r = \rho(\mathfrak{t}_0) + \mathfrak{t}_r,\ \mathfrak{n}'_r = \Delta \oplus \mathfrak{n}_r$$

and $\Delta = \{(\rho(a), -a),\ a \in \mathfrak{t}_0\}$ *is the "anti-diagonal" in* $\rho(\mathfrak{t}_0) \oplus \mathfrak{t}_0$.

Proof. By construction, $\mathfrak{t}'_r$ is an Abelian algebra of nontrivial semisimple derivations of $\mathfrak{r}'$, and $\mathfrak{n}'_r$ is the nilradical of

$$\mathfrak{r}'_s = \mathfrak{t}'_r \dotplus \mathfrak{r}' = (\rho(\mathfrak{t}_0) \oplus \mathfrak{t}_0) \dotplus \mathfrak{r}_s = (\rho(\mathfrak{t}_0) \oplus \Delta) \dotplus \mathfrak{r}'_s = \Delta \oplus (\rho(\mathfrak{t}_0) \dotplus \mathfrak{r}'_s).$$

□

3. Now let us consider an arbitrary Lie algebra $\mathfrak{g}$ with Levi decomposition $\mathfrak{g} = \mathfrak{l} \dotplus \mathfrak{r}$ into the radical $\mathfrak{r}$ and a maximal semisimple subalgebra $\mathfrak{l}$. The image of $\mathrm{ad}(\mathfrak{l})$ relative to the representation $\mathrm{ad} : \mathfrak{g} \to \mathrm{Der}(\mathfrak{r})$ is a semisimple subalgebra. Since maximal reductive subalgebras of $\mathrm{Der}(\mathfrak{r})$ are conjugate, one can select a maximal reductive subalgebra $\mathfrak{p} \oplus \mathfrak{t} \subset \mathrm{Der}(\mathfrak{r})$ such that $\mathrm{ad}(\mathfrak{l}) \subset \mathfrak{p}$, and hence the actions of $\mathfrak{l}$ and $\mathfrak{t}$ on $\mathfrak{r}$ commute. Define

$$\mathfrak{g}'_s = \mathfrak{l} \dotplus \mathfrak{r}'_s = (\mathfrak{l} \oplus \mathfrak{t}_r) \dotplus \mathfrak{r} = (\mathfrak{l} \oplus \mathfrak{t}_r) \dotplus \mathfrak{n}_r.$$

This Lie algebra $\mathfrak{g}'_s$ is called the *semisimple splitting* of the Lie algebra $\mathfrak{g}$.

It is not difficult to prove [**AB68**] that the split Lie algebra $\mathfrak{g}_s'$, just as $\mathfrak{r}_s'$, admits a Jordan decomposition for every element; that is, Lemma 5.1 holds true in the general case.

4. All our definitions and results can be applied almost literally to the class of connected simply connected Lie groups. First let R be a connected simply connected solvable Lie group and $\mathrm{Ad} : R \to \mathrm{Aut}(\mathfrak{r})$ the adjoint representation of R in the automorphism group $\mathrm{Aut}(\mathfrak{r})$ of its Lie algebra $\mathfrak{r}$. The real-algebraic group $\mathrm{Aut}(\mathfrak{r})$ splits as follows: $\mathrm{Aut}(\mathfrak{r}) = (PT) \cdot U$, where P is a maximal semisimple subgroup, T is an Abelian subgroup of semisimple automorphisms of $\mathfrak{r}$ that commutes with P (and the intersection $P \cap T$ is discrete), and U is the unipotent radical of $\mathrm{Aut}(\mathfrak{r})$. The group $\mathrm{Ad}(R)$ is a connected solvable normal subgroup of $\mathrm{Aut}(\mathfrak{r})$, and hence it belongs to the radical $T \cdot U$ of the algebraic group $\mathrm{Aut}(\mathfrak{r})$. Let $\theta : T \cdot U \to T$ be the epimorphism and let $T_R = \theta(\mathrm{Ad}(R))$. Then T_R is a connected Abelian group of semisimple automorphisms of the Lie algebra $\mathfrak{r}$, and hence of the Lie group R itself. The connected solvable Lie group $R_s = T_R \cdot R$ is called the *semisimple splitting* of the connected simply connected solvable Lie group R.

One can prove [**AB68**] the following result. If N_R is the nilradical of R_s (that is, the maximal connected normal nilpotent subgroup), then R_s is the semidirect product $T_R \cdot N_R$. Moreover, $R_s = RN_R$; that is, R and N_R generate R_s. This allows one to give an equivalent nonconstructive definition.

More specifically, let T_R be a connected Abelian group of nontrivial semisimple automorphisms of a connected simply connected solvable Lie group R. Define $R_s = T_R \cdot R$ and let N_R be the nilradical of R_s. If $R_s = T_R \cdot N_R$ and $R_s = RN_R$, then R_s is called the *semisimple splitting* of the connected simply connected solvable Lie group R, and N_R is called the *nilshadow* of R.

Denote by

$$p : R_s = T_R \cdot N_R \to T_R, \quad \pi : R_s = T_R \cdot N_R \to N_R$$

the projections along N_R and T_R. Then the restriction $p : R \to T_R$ is a group epimorphism, and the restriction $\pi : R \to N_R$ is a diffeomorphism. It is easy to see that the nilradical $N \subset R$ is of the form $N = (R \cap N_R)_0$. In fact, the nilradical $N \subset R$, being a characteristic subgroup of the normal subgroup $R \subset R_s$, is a connected nilpotent normal subgroup of R_s, and hence $N \subset (R \cap N_R)_0$. On the other hand, $(R \cap N_R)_0$ is a connected normal nilpotent subgroup of R, and hence $(R \cap N_R)_0 \subset N$.

One observes also that $[R_s, R_s] \subset [T_R \cdot R, T_R \cdot R] \subset R$, because R is normal in R_s, and T_R is Abelian. In particular,

$$[N_R, N_R] \subset (R \cap N_R)_0 = N \quad \text{and} \quad [T_R, N_R] \subset N.$$

On the other hand, if $a \in T_R$ and $\{a, N_R\} \subset [N_R, N_R]$, then $a = 1$. In fact, since T_R is reductive, it follows that N_R contains a T_R-submodule V complementary to $[N_R, N_R]$ (that is, $N_R = V[N_R, N_R]$ and $V \cap [N_R, N_R] = 1$). The module V is diffeomorphic to a vector space and generates N_R. Hence if V is trivial as an a-module, where $a \in T_R$, then N_R is also trivial as an a-module. But $T_R \subset \mathrm{Aut}(N_R)$, and hence $a = 1$.

5. Note that not only can the group T_R fail to be an algebraic subgroup of $\mathrm{Aut}(\mathfrak{r})$, but it can even fail to be closed in the usual (Euclidean) topology.

Example 5.5. Let $\mathrm{Aut}(V^4) = \mathrm{GL}(4,\mathbb{R})$ be the automorphism group of the real vector space $V^4 = V' \times V''$, $\dim V' = \dim V'' = 2$. It contains the torus $\mathbb{T}^2 = S' \times S''$, where S' is the group of orthogonal rotations of V' and S'' is the group of orthogonal rotations of V''. Choose an irrational number $\alpha \in \mathbb{R}$, and let $L_{\mathbb{R}} \subset \mathbb{T}^2$ be a dense one-parameter subgroup

$$L_t = \begin{pmatrix} \cos t & \sin t & 0 & 0 \\ -\sin t & \cos t & 0 & 0 \\ 0 & 0 & \cos\alpha t & \sin\alpha t \\ 0 & 0 & -\sin\alpha t & \cos\alpha t \end{pmatrix}.$$

Look at the semidirect product $R = L_{\mathbb{R}} \cdot V^4$. The Lie group R is a five-dimensional connected simply connected solvable group, called the *Mautner group*. In the theory of representations it is well known as one of "wild" Lie groups [**Kir**]. The automorphism group $\mathrm{Aut}(R)$ contains the 2-torus $\mathbb{T}^2$ of orthogonal rotations V^4 which keep both V' and V'' invariant (this is so because the actions of $\mathbb{T}^2$ and $L_{\mathbb{R}}$ on V^4 commute). It is easily seen that the semisimple splitting of the group R is as follows: $R_s = T_R \cdot R = T_R \cdot N_R$, where the subgroup T_R is isomorphic to $L_{\mathbb{R}} \subset \mathbb{T}^2$, $N_R = \Delta \times V^4$, and $\Delta = \{(L_t, L_{-t}),\ t \in \mathbb{R}\}$ is the "anti-diagonal" in $L_{\mathbb{R}} \times L_{\mathbb{R}}$. Here the subgroup T_R is not closed in $\mathrm{Aut}(R)$ even in the usual topology, because the subgroup $L_{\mathbb{R}}$ is dense in the torus $\mathbb{T}^2$.

Below we will see (as a corollary of refined the Mostow Theorem 6.2) that whenever a solvable Lie group R contains a quasilattice, the subgroup T_R is closed in the usual topology. It follows that a group R as in the example contains no lattices.

6. The following statements are immediate corollaries of Lemmas 5.2–5.4.

LEMMA 5.6. *If I is a connected (and hence simply connected) normal subgroup of a connected simply connected solvable Lie group R, then I_s is a connected normal subgroup of R_s and $I_s = T_I \cdot N_I$, where $T_I = p(I)$, $N_I = \pi(I)$*

LEMMA 5.7. *Let I be a connected normal subgroup of R, and let $R' = R/I$. Then $R'_s = T'_R \cdot N'_R$, where $T'_R = T_R/T(I)$, $N'_R = N_R/N_I$, and $T(I) = \{a \in T_R :\ \{a, R\} \subset I\}$.*

LEMMA 5.8. *Let $\rho : T_0 \to \mathrm{Aut}(\mathfrak{r}) \simeq \mathrm{Aut}(R)$ be a homomorphism of a connected simply connected Abelian group T_0 to the algebraic group $\mathrm{Aut}(\mathfrak{r})$, and let $\rho(T_0)$ consist of semisimple automorphisms of $\mathfrak{r}$ commuting with $T_R \subset \mathrm{Aut}(\mathfrak{r})$. Define $R' = T_0 \cdot R$. Then*

$$R'_s = T'_R \cdot R' = T'_R \cdot N'_R,$$

where $T'_R = \rho(T_0)T_R \subset \mathrm{Aut}(\mathfrak{r})$, $N'_R = \Delta \times N_R$, and $\Delta = \{(\rho(a), a^{-1}),\ a \in T_0\}$ is the "anti-diagonal" in $\rho(T_0) \times T_0$.

Remark. Given a homomorphism $\rho : T_0 \to \mathrm{Aut}(\mathfrak{r})$ of a torus T_0 to the algebraic group $\mathrm{Aut}(\mathfrak{r})$, one can take the semidirect product $R' = T_0 \cdot R$. Although the connected solvable group R' is not simply connected, nevertheless one can view the group $R'_s = T'_R \cdot R' = T'_R \cdot N'_R$, where T'_R and N'_R are as in Lemma 5.7, as its semisimple splitting. All properties of the semisimple splitting are satisfied. In particular, the restriction $p : R' \to T'_R$ is a group epimorphism, and the restriction $\pi : R' \to N'_R$ is a diffeomorphism.

7. To conclude, we define the semisimple splitting for an arbitrary connected simply connected Lie group G. If it is semisimple, we let $G_s = G$. In the general case, let $G = L \cdot R$ be a Levi decomposition, and let

$$\mathrm{Ad} : G \to \mathrm{Aut}(\mathfrak{r}) = \mathrm{Aut}(R) = (PT) \cdot U$$

be the adjoint representation of the group G on the Lie algebra $\mathfrak{r}$ of the radical R. Since maximal reductive subgroups of the algebraic group $\mathrm{Aut}(R)$ are conjugate, one can select a maximal semisimple subgroup $P \subset \mathrm{Aut}(R)$ and an Abelian reductive subgroup $T \subset \mathrm{Aut}(R)$ commuting with P in such a way that $\mathrm{Ad}(L) \subset P$ and hence the actions of L and T on R commute. So it makes sense to define

$$G_s = L \cdot R_s = (L \times T_R) \cdot R = (L \times T_R) \cdot N_R,$$

where $T_R \subset T \subset \mathrm{Aut}(R)$ and $R_s = T_R \cdot R$ is the semisimple splitting of the radical R. The connected group G_s is called the *semisimple splitting* of the connected simply connected Lie group G.

§6. Algebraic groups and homogeneous spaces of finite volume

1. In studying homogeneous spaces and flows, the theory of algebraic groups proves to be highly useful. The general idea of the "algebraic" approach to the study of a homogeneous space G/D consists in embedding the connected simply connected Lie group G into a certain real-algebraic group $\mathcal{A}(G)$ as a Zariski dense normal subgroup and considering the algebraic hull $\mathcal{A}(D) \subset \mathcal{A}(G)$. As a rule, such an embedding does exist, and if G/D is of finite volume, the algebraic hull $\mathcal{A}(D) \subset \mathcal{A}(G)$ is "almost equal" to $\mathcal{A}(G)$. This allows one to extract very valuable information on the structure of G/D.

Let R be a connected simply connected solvable Lie group, and let $R_s = T_R \cdot R = T_R \cdot N_R$ be its semisimple splitting. Since T_R acts on N_R by nontrivial automorphisms (via conjugations), one can view T_R as a subgroup of the algebraic group $\mathrm{Aut}(N_R)$. Let $\mathcal{A}(T_R)$ be the algebraic hull of the subgroup $T_R \subset \mathrm{Aut}(N_R)$. Then $\mathcal{A}(T_R)$ is an Abelian reductive group of automorphisms of N_R. Define $\mathcal{A}(R_s) = \mathcal{A}(T_R) \cdot N_R$. It turns out that the Lie group $\mathcal{A}(R_s)$ admits the structure of a real-algebraic solvable group, in which N_R is the unipotent radical, and $\mathcal{A}(T_R)$ is a maximal Abelian reductive subgroup. This directly follows from the Birkhoff embedding theorem (see [**Au**]):

THEOREM 6.1. *Given a connected simply connected nilpotent Lie group N, there exist a monomorphism $B : N \to \mathrm{U}(n, \mathbb{R})$ and an algebraic monomorphism $B : \mathrm{Aut}(N) \to \mathrm{GL}(n, \mathbb{R})$ such that*

$$B(\theta(u)) = B(\theta) B(u) B(\theta)^{-1} \quad \textit{for all } u \in N,\ \theta \in \mathrm{Aut}(N).$$

Since the subgroup T_R normalizes R and is Zariski dense in $\mathcal{A}(T_R)$ relative to the algebraic structure on $\mathcal{A}(R_s)$, we conclude that the subgroup R is normal in $\mathcal{A}(R_S)$ and Zariski dense in $\mathcal{A}(R_s)$, that is, $\mathcal{A}(R) = \mathcal{A}(R_s)$. Even if the original group R were an algebraic solvable group, it would no longer be algebraic relative to the structure on $\mathcal{A}(R_s)$ (clearly, this remark applies to nonnilpotent Lie groups R; otherwise $R = \mathcal{A}(R) = \mathcal{A}(R_s)$). In addition, $\mathcal{A}(R_s) = \mathcal{A}(T_R) \cdot R = \mathcal{A}(T_R) \cdot N_R$. Denote by

$$p : \mathcal{A}(R_s) = \mathcal{A}(T_R) \cdot N_R \to \mathcal{A}(T_R) \quad \text{and} \quad \pi : \mathcal{A}(R_s) = \mathcal{A}(T_R) \cdot N_R \to N_R$$

the projections along N_R and $\mathcal{A}(T_R)$. The basic theorems on the structure of compact solvable manifolds are the following results of Mostow in Auslander's modification [**Au**].

THEOREM 6.2. *If D is a quasilattice in a connected simply connected Lie group R, then $p(D)$ is a lattice in a closed (in the usual topology) group $T_R = p(R) \subset \operatorname{Aut}(N_R)$. In particular, the product ND of the nilradical and the quasilattice is closed.*

THEOREM 6.3. *If a solvable space R/D is compact, then the subgroup $\mathcal{A}(D)$ is a uniform normal subgroup of $\mathcal{A}(R)$. In particular, $N_R \subset \mathcal{A}(D)$, and there exists a torus $C \subset \mathcal{A}(T_R)$ such that $\mathcal{A}(R) = C\mathcal{A}(D)$.*

2. Now we proceed to consider a connected semisimple Lie group L. One knows that the adjoint group $\operatorname{Ad}(L)$ is a connected (in the usual topology) subgroup of finite index in its real-algebraic hull $\mathcal{A}(\operatorname{Ad}(L)) \subset \operatorname{Aut}(\mathfrak{l})$, where $\mathfrak{l}$ is the Lie algebra of L. Since we are not going to apply the theory of algebraic groups in its full power (we only need few typical properties of real-algebraic groups listed in the introduction), we can assume without loss of generality that the adjoint group $\operatorname{Ad}(L)$ is itself a real-algebraic group. More precisely, we will agree that a closed subgroup $H \subset \operatorname{Ad}(L)$ is "algebraic" if H is of finite index in the intersection $\mathcal{A}(H) \cap \operatorname{Ad}(L)$, where $\mathcal{A}(H)$ is the "true" algebraic hull of H in $\operatorname{Aut}(\mathfrak{l})$. Thus, a finite index subgroup of an "algebraic" group is itself "algebraic". It is also clear that any "algebraic" group has finitely many connected components.

One of the basic properties of an algebraic structure is that the intersection of two algebraic subgroups is still algebraic. If H_1, H_2 are two "algebraic" subgroups of $\operatorname{Ad}(L)$, then the intersection $H_1 \cap H_2$ is "algebraic", because

$$\mathcal{A}(H_1) \cap \mathcal{A}(H_2) \cap \operatorname{Ad}(L) = (\mathcal{A}(H_1) \cap \operatorname{Ad}(L)) \cap (\mathcal{A}(H_2) \cap \operatorname{Ad}(L)).$$

It is also easy to verify that product of two "algebraic" subgroups is locally closed in the usual topology. It is obvious as well that whenever $H \subset \operatorname{Ad}(L)$ normalizes a connected subgroup $C \subset \operatorname{Ad}(L)$, so does the "algebraic" hull $\mathcal{A}(H) \subset \operatorname{Ad}(L)$. These observations allow us to treat the structure just introduced on $\operatorname{Ad}(L)$ as a true algebraic structure.

Now let us try to introduce an algebraic structure on the group L itself. A subgroup $H \subset L$ will be called Ad-*algebraic* if the image $\operatorname{Ad}(H) \subset \operatorname{Ad}(L)$ is "algebraic". Clearly, such a definition also preserves all the basic properties of an algebraic structure whenever the center $Z = \operatorname{Ker}(\operatorname{Ad})$ is finite. What is not obvious is that the definition works for groups L with infinite center Z as well. For instance, according to such a definition, an infinite center itself becomes Ad-algebraic. Thus, we lose the property that an algebraic group has only finitely many connected components. Fortunately, this is the only property that we lose. Let us verify this.

LEMMA 6.4. *The intersection of two* Ad-*algebraic subgroups is* Ad-*algebraic.*

Proof. Let subgroups H_1, $H_2 \subset L$ be Ad-algebraic. Since the kernel $\operatorname{Ker}(\operatorname{Ad}) = Z$ is discrete, the identity components of the subgroups $\operatorname{Ad}(H_1) \cap \operatorname{Ad}(H_2)$ and $\operatorname{Ad}(H_1 \cap H_2)$ are the same. Since the intersection $\operatorname{Ad}(H_1) \cap \operatorname{Ad}(H_2)$, being algebraic, has finitely many connected components, it follows that $\operatorname{Ad}(H_1 \cap H_2)$ is of finite index in $\operatorname{Ad}(H_1) \cap \operatorname{Ad}(H_2)$. Hence $H_1 \cap H_2$ is Ad-algebraic. □

This result allows us to define the Ad-*algebraic hull* $\mathcal{A}(H)$ of a subgroup $H \subset L$ as the smallest Ad-algebraic subgroup of L containing H.

LEMMA 6.5. *If a subgroup $H \subset L$ normalizes a connected closed subgroup $C \subset L$, then so does the* Ad*-algebraic hull of H.*

Proof. By the hypothesis, $\mathrm{Ad}(H)$ normalizes the Lie subalgebra $\mathfrak{c} \subset \mathfrak{l}$ of $C \subset L$. Hence the algebraic hull $\mathcal{A}(\mathrm{Ad}(H))$ normalizes $\mathfrak{c}$. Since $\mathcal{A}(H) \subset \mathrm{Ad}^{-1}(\mathcal{A}(\mathrm{Ad}(H)))$ and $\mathrm{Ker}(\mathrm{Ad})$ is the center of L, we deduce that $\mathcal{A}(H)$ normalizes C. □

LEMMA 6.6. *The product of two* Ad*-algebraic subgroups is locally closed.*

Proof. If $H_1, H_2 \subset L$ are two Ad-algebraic subgroups, the inverse image ZH_1H_2 of a locally closed set $\mathrm{Ad}(H_1)\,\mathrm{Ad}(H_2)$ is also locally closed. Since Z is a discrete group, the set H_1H_2 is locally closed as well. □

Since any algebraic group has finitely many connected components, it follows that whenever an Abelian algebraic group A contains an algebraic subgroup $B \subset A$ with the quotient space A/B compact, there exists a torus $C \subset A$ such that $A = BC$ and the intersection $B \cap C$ is finite. We will make use of an analogue of the statement in the Ad-algebraic situation.

LEMMA 6.7. *Let B be an* Ad*-algebraic subgroup of an Abelian* Ad*-algebraic group A such that A/B is compact. Then there exists a connected* Ad*-algebraic subgroup $V \subset A$ such that $A = BV$ and $Z \cap V$ is a finite index subgroup of the discrete group $V \cap B$.*

Proof. In place of V one takes the identity component of the subgroup $\mathrm{Ad}^{-1}(C)$, where C is a subtorus in $\mathrm{Ad}(A)$ complementary to $\mathrm{Ad}(A)$. □

Thus, we were able to construct the analogue of an algebraic structure on a semisimple Lie group L. Suppose now that the quotient space L/D is of finite volume and $L = S \times K$ is the decomposition of L into noncompact and compact parts. Then, by the Borel Density Theorem E.4, given a finite-dimensional representation ρ of the group L, the algebraic hull of $\rho(D)$ contains $\rho(S)$. In particular, this applies to the adjoint representation Ad. Since $\mathrm{Ad}(L) = \mathrm{Ad}(S) \times \mathrm{Ad}(K)$ is precisely the decomposition of $\mathrm{Ad}(L)$ into noncompact and compact parts, the following hold.

LEMMA 6.8. *If a space L/D is of finite volume, then the noncompact part $S \subset L$ belongs to $\mathcal{A}(D)$ and, in particular, the space $L/\mathcal{A}(D)$ is compact.*

LEMMA 6.9. *Let D be a quasilattice in a semisimple Lie group $L = S \times K$, and let $L = \overline{SD}$. Then $L = \mathcal{A}(D)$, D is a lattice in S, and the product ZD is closed (Z being the center of L).*

Thus, so far we have managed to embed an arbitrary simply connected solvable Lie group G into the algebraic group $\mathcal{A}(G)$, and to simulate an algebraic structure on any semisimple Lie group. In both cases the algebraic hull $\mathcal{A}(D)$ of a subgroup $D \subset G$ having the quotient space G/D of finite volume turned out to be a uniform subgroup of the algebraic hull $\mathcal{A}(G)$ of our Lie group. Moreover, one can find a connected compact Lie group $C \subset \mathcal{A}(G)$ such that $\mathcal{A}(G) = C\mathcal{A}(D)$.

3. Let us try to introduce an algebraic structure on an arbitrary simply connected Lie group. This is now easy to do by combining the semisimple and solvable cases.

Let $G = L \cdot R$ be a Levi decomposition of a connected simply connected Lie group G, and let

$$G_s = (L \times T_R) \cdot R = (L \times T_R) \cdot N_R$$

be its semisimple splitting. Recall that T_R is an Abelian reductive subgroup of $\mathrm{Aut}(N_R)$ whose action on R (and hence on N_R) commutes with the action of L. Let $\mathcal{A}(T_R)$ be the algebraic hull of the subgroup $T_R \subset \mathrm{Aut}(N_R)$. Since the action of L on N_R commutes with that of T_R, it must commute with the action of $\mathcal{A}(T_R)$. Therefore we can introduce a Lie group

$$\mathcal{A}(G_s) = (L \times \mathcal{A}(T_R)) \cdot R = (L \times \mathcal{A}(T_R)) \cdot N_R.$$

As was mentioned, by the Birkhoff embedding theorem, the solvable Lie group $\mathcal{A}(R_s) = \mathcal{A}(T_R) \cdot N_R$ admits the structure of a real-algebraic group, in which N_R is the unipotent radical and $\mathcal{A}(T_R)$ is a maximal Abelian reductive subgroup. The problem in introducing an algebraic structure on $\mathcal{A}(G_s)$ lies in the semisimple group L. If L acted on R (or N_R) effectively, then, by the same Birkhoff theorem, one could introduce a true algebraic structure on $\mathcal{A}(G_s)$. This does not work in the general case.

Let $L = L_e \times L_t$ be the decomposition of L into the maximal connected normal subgroup L_e whose action on N_R is locally effective, and the maximal connected normal subgroup L_t commuting with N_R. Take the discrete subgroup $Z = L \cap Z(G)$. Clearly, Z is the maximal subgroup of L which is central in $\mathcal{A}(G_s)$, $Z \cap L_e$ is the kernel of the action of L_e on N_R, and $Z \cap L_t$ is the center of L_t. Denote by

$$\psi : \mathcal{A}(G_s) = (L \times \mathcal{A}(T_R)) \cdot N_R \to (L/Z \times \mathcal{A}(T_R)) \cdot N_R$$

the factorization by the central discrete subgroup $Z \subset \mathcal{A}(G_s)$. Then $\mathcal{A}(G_s)/Z$ is the product of the real-algebraic group $(L_e/Z \cap L_e \times \mathcal{A}(T_R)) \cdot N_R$ and the semisimple center-free Lie group $L_t/Z \cap L_t$ (which one can also view as a real-algebraic group). Similarly to the semisimple case, we introduce the following definition.

A subgroup $H \subset \mathcal{A}(G_s)$ will be called *ψ-algebraic* if the image $\psi(H) \subset \mathcal{A}(G_s)/Z$ is algebraic.

All the results we have proved for the Ad-algebraic structure can be directly transferred to the ψ-algebraic structure.

LEMMA 6.10. *The intersection of two ψ-algebraic subgroups is ψ-algebraic.*

Clearly, now we can define the *ψ-algebraic hull* $\mathcal{A}(H) \subset \mathcal{A}(G_s)$ of a subgroup $H \subset \mathcal{A}(G_s)$ as the smallest ψ-algebraic subgroup containing H.

LEMMA 6.11. *If a subgroup $H \subset \mathcal{A}(G_s)$ normalizes a connected closed subgroup $C \subset \mathcal{A}(G_s)$ then so does the ψ-algebraic hull $\mathcal{A}(H)$.*

LEMMA 6.12. *The product of two ψ-algebraic subgroups is locally closed.*

LEMMA 6.13. *Let B be a ψ-algebraic subgroup of an Abelian ψ-algebraic group A such that A/B is compact. Then there exists a connected ψ-algebraic subgroup $V \subset A$ such that $A = BV$ and $Z \cap V$ is a finite index subgroup in the discrete group $V \cap B$.*

4. Now we discuss two results [**St89**] on the structure of general spaces of finite volume.

THEOREM 6.14 [**St89**]. *If G/D is of finite volume, then $\mathcal{A}(D)$ is a uniform subgroup of $\mathcal{A}(G)$. Moreover, if C is the maximal compact connected normal subgroup of $L \times \mathcal{A}(T_R)$, then $\mathcal{A}(G) = C\mathcal{A}(D)$.*

Note that in the solvable case ($G = R$) the group $\mathcal{A}(G)$ is algebraic and we obtain the Mostow Theorem 6.3. As for the semisimple case ($G = L = S \times K$), the embedding $S \subset \mathcal{A}(D)$ follows immediately from the Borel density theorem.

The idea of the proof is as follows. Define $G' = (\overline{SRD})_0$. Then $G' = (S\times K')R'$, where $R' = T\cdot R$ and T is a torus in K which commutes with a connected semisimple subgroup $K' \subset K$, their intersection being finite. The subgroup $D \cap G'$ is of finite index in D, and without loss of generality we may assume that $D \subset G'$. Since G'/D is of finite volume and $G' = \overline{SR'D}$, by virtue of Theorem E.11 the space $R'/D \cap R'$ is compact. It follows that $\mathcal{A}(R'_s)/\mathcal{A}(D \cap R')$ is a compact space. Now, using the relation between $\mathcal{A}(R_s)$ and $\mathcal{A}(R'_s)$ (see Lemma 5.8), one can prove that the space $\mathcal{A}(R_s)/\mathcal{A}(D) \cap \mathcal{A}(R_s)$ is also compact, and hence $N_R \subset \mathcal{A}(D)$.

The space $S/K\overline{RD} \cap S$ is of finite volume and, according to the Borel density theorem,

$$S \subset \mathcal{A}(K\overline{RD}) = K\mathcal{A}(R)\mathcal{A}(D) = K\mathcal{A}(R_s)\mathcal{A}(D).$$

It follows that one can select a semisimple Levi subgroup $L \subset G$ such that $S \subset \mathcal{A}(D)$. But $N_R \subset \mathcal{A}(D)$, and the group SN_R is normal in $\mathcal{A}(G)$. Hence $SN_R \subset \mathcal{A}(D)$, regardless of the choice of $L \subset G$. Since the space $\mathcal{A}(R_s)/\mathcal{A}(D) \cap \mathcal{A}(R_s)$ is compact and $S \subset \mathcal{A}(D)$, the space $\mathcal{A}(G)/\mathcal{A}(D)$ is also compact. The second part of the theorem follows directly from the first part because $SN_R \subset \mathcal{A}(D)$. □

5. Now we turn to a result on spaces of finite volume which generalizes many structure theorems.

THEOREM 6.15 [**St89**]. *Let G/D be of finite volume and let*

$$T(D_0) = \{a \in \mathcal{A}(D) : \ \{a, \mathcal{A}(D)\} \subset D_0\}.$$

Then the product $T(D_0)N_RD$ is closed.

In the formulation we do not require that the subgroup $D \subset G$ be a quasi-lattice. Hence the conclusion holds for arbitrary spaces of finite volume. We mention some particular cases of the theorem:

1. If D is a lattice in a solvable Lie group R, then $T(D_0) = T(1) \subset N_R$, and we obtain the refined Mostow Theorem 6.2 saying that the product N_RD is closed.

2. If D is a lattice in a semisimple group S having no compact components, then $T(D_0) = T(1)$ is the center $Z \subset S$, and we conclude that the product ZD is closed.

3. If D is a lattice in a Lie group $G = K \cdot R$, where a compact semisimple group K acts on the radical R with finite kernel $Z \subset K$, then $T(D_0) = T(1) \subset ZN_R$. We again conclude that the group N_RD closed. Since $N = (N_R \cap R)_0$ is the nilradical and $(\overline{ND})_0 \subset (N_R \cap G)_0 = N$, it follows that the product ND is closed, and hence $N/D \cap N$ is a compact space. We have derived the Auslander Theorem 8.25 from [**Rag**] (note that both the formulation and the proof of the theorem in [**Rag**] need to be corrected; in this connection see the Appendix in [**St89**]).

4. If D is a lattice in a Lie group $G = S\cdot R$, where the semisimple Levi subgroup S has no compact components, then, according to Theorem E.10, the group RD is closed. It is easy to prove this by a double application of Theorem 6.15.

Note that the restrictions imposed on the Lie group G in cases 2 and 3 are essential. If one omits the restrictions, the right answer is given by Theorem 6.15.

We outline the idea of the proof of Theorem 6.15. First suppose that D is a lattice in G, and $\mathcal{A}(D) = \mathcal{A}(G)$. Then $T(D_0) = T(1)$ is the center of $\mathcal{A}(G)$, and hence $T(1)N_R = ZN_R$. The connected group $(\overline{ZD})_0$ normalizes D and hence belongs to $T(1)_0 \subset N_R$. Since the spaces $G/\overline{ZD}$ and $G/(\overline{ZD})_0D$ are of finite volume, it follows that the group $(\overline{ZD})_0D$ is of finite index in $\overline{ZD}$, and hence $(\overline{ZD})_0DN_R = DN_R$ is of finite index in $\overline{ZD}N_R$. The density condition $G = \overline{SRD}$ implies that the intersection $D \cap R$ is a lattice in R, and, by Mostow's Theorem E.3, the product DN_R is closed. It follows that the product $\overline{ZD}N_R$ is closed as well, and hence $T(D_0)N_RD = ZN_RD$ is also closed.

If G/D is of finite volume and $\mathcal{A}(D) = \mathcal{A}(G)$, the subgroup D_0 is normal in G. In this case the assertion follows from the previous case by applying Lemma 5.7 to the factor group G/D_0.

The general case reduces to the situation $\mathcal{A}(R) \subset \mathcal{A}(D)$ by the following trick. By Theorem 6.14, one can find a torus $T^* \subset \mathcal{A}(T_R)$ such that

$$\mathcal{A}(T_R) = T^*(\mathcal{A}(T_R) \cap \mathcal{A}(D))$$

and the intersection $T^* \cap \mathcal{A}(D)$ is finite. Define

$$G^* = (T^*G \cap \mathcal{A}(D))_0.$$

Then $T^*G^* = T^*G$, $T^* \cap G = 1$, and hence the projection $T^*G \to G$ along T^* induces a cover[11] $G^* \to G$. Since the group G is simply connected, the cover is one-to-one. By construction, $D^* = D \cap G^*$ is of finite index in D, and hence the space G^*/D^* is of finite volume. The nilradical of $\mathcal{A}(G^*)$ equals N_R, and it suffices to prove our theorem for the pair $D^* \subset G^* = L \cdot R^*$, where $\mathcal{A}(R^*) \subset \mathcal{A}(D^*)$.

Lastly, if $\mathcal{A}(R) \subset \mathcal{A}(D)$ we consider

$$G' = (\overline{SRD})_0 = (S \times K')(TR),$$

where a semisimple group $K' \subset K$ commutes with a torus $T \subset K$. Without loss of generality $K' \cap T = 1$. Let V be the universal covering group of the torus T, let $F = (S \times K')(VR)$, and let $C \subset D$ be the inverse image of $D \cap G'$ under the cover $F \to G'$ (note that $D \cap G'$ is of finite index in D). It is easy to see that the proof of the theorem for the pair $D \subset G$ reduces to considering the pair $C \subset F$. But $\mathcal{A}(C) = \mathcal{A}(F)$, and we find ourselves in a familiar situation. □

To conclude, we note that the property $\mathcal{A}(D) = \mathcal{A}(G)$ for a quasilattice $D \subset G$ (that is, its density in our "algebraic" structure) is equivalent to the Zariski density in the Ad-representation. Indeed, either condition ensures that $G = \overline{SRD}$, and hence, by Theorem E.11, the intersection $D \cap R$ is a quasilattice in the radical R. This allows us to reduce the problem to considering separately the semisimple case (where everything follows from the Borel density theorem) and the solvable case (where it suffices to recall the construction of the group $\mathcal{A}(R_s)$).

---

[11] Somewhat abusing the terminology, by a cover we mean any smooth fibration, rather than a fibration with discrete leaves.

§7. Solvable homogeneous flows

Here we finalize the study of ergodic properties of homogeneous flows on compact solvable manifolds. First we prove the ergodicity criterion and construct the ergodic decomposition for flows on class (I) homogeneous spaces. Then, along the lines of Auslander [**Au**], the ergodicity criterion for arbitrary solvable flows will be reduced to the class (I) case. Subsequently, following [**St87b**], we prove that flows on ergodic components are finitely covered by ergodic solvable flows.[12] Finally, we set forth an alternative approach of Brezin and Moore [**BM**] to checking solvable flows for ergodicity by considering maximal Euclidean factors.

1. From now on, $\mathcal{A}(R_s) = \mathcal{A}(T_R) \cdot R = \mathcal{A}(T_R) \cdot N_R$ is the real-algebraic group which contains a connected simply connected solvable group R as a Zariski dense normal subgroup, and $D \subset R$ is a quasilattice in R. As usual,

$$p : \mathcal{A}(R_s) = \mathcal{A}(T_R) \cdot N_R \to \mathcal{A}(T_R) \quad \text{and} \quad \pi : \mathcal{A}(R_s) = \mathcal{A}(T_R) \cdot N_R \to N_R$$

stand for the epimorphism along N_R onto a maximal Abelian algebraic reductive subgroup $\mathcal{A}(T_R)$ and the projection along $\mathcal{A}(T_R)$ onto the unipotent radical N_R. The maximal reductive subgroup $\mathcal{A}(T_R) \subset \mathcal{A}(R_s)$ will be selected to agree with a Jordan decomposition for a one-parameter subgroup $g_\mathbb{R} \subset R$. Hence, for all $t \in \mathbb{R}$,

$$g_t = a_t \times u_t, \qquad \text{where } a_\mathbb{R} \subset \mathcal{A}(T_R),\ u_\mathbb{R} \subset N_R,$$

and the subgroups $a_\mathbb{R}$ and $u_\mathbb{R}$ commute.

Let us formulate the ergodicity criterion for the flow $(R/D, g_\mathbb{R})$. We start by considering a class (I) group R. By the Mostow Theorem 6.2, the existence of a quasilattice $D \subset R$ implies that the subgroup T_R is closed (in the usual topology). Moreover, the projection $p(D)$ is a lattice in T_R. Since R is of class (I), the subgroup T_R is equal to $\mathcal{A}(T_R)$ and is a compact torus, whereas the discrete subgroup $p(D) \subset T_R$ is finite. Surely, for the original flow to be ergodic it is necessary that the quotient flow $(T_R/p(D), a_\mathbb{R})$ be ergodic, that is, $T_R = \overline{a_\mathbb{R}}$. Let us prove that under this condition the diffeomorphism $\pi : R \to N_R$ induces a smooth isomorphism of flows

$$\pi^* : (R/D \cap N_R, g_\mathbb{R}) \to (N_R/D \cap N_R, u_\mathbb{R}).$$

Indeed,

$$\begin{aligned}\pi(g_t h(D \cap N_R)) &= \pi(a_t u_t p(h) \pi(h)(D \cap N_R)) \\ &= \pi(a_t p(h) u_t \pi(h)(D \cap N_R)) = u_t \pi(h(D \cap N_R)),\end{aligned}$$

because the subgroups $u_\mathbb{R}$ and $T_R = \overline{a_\mathbb{R}}$ commute. It follows that under the condition $T_R = \overline{a_\mathbb{R}}$, the original flow is finitely covered by the nilflow $(N_R/D \cap N_R, u_\mathbb{R})$. Hence the original flow is ergodic if and only if we have $T_R = \overline{a_\mathbb{R}}$ and the nilflow $(N_R/D \cap N_R, u_\mathbb{R})$ is ergodic.

Moreover, following Auslander [**AG**] one can prove that the cover $\pi^* : N_R/D \cap N_R \to R/D$ is one-sheeted, that is, the quasilattice D is nilpotent and belongs to N_R whenever R/D carries an ergodic flow. We will give a simplified (as compared to the original) proof of the fact.

---

[12]Note that a complete study of homogeneous flows on *arbitrary* solvable manifolds G/D (of infinite volume or even without a G-invariant smooth measure) was made in [**St87b**].

Since the nilspace $N_R/(D \cap N_R)$ is compact, the product $(D \cap N_R)[N_R, N_R]$ is closed. Note that given any $d \in D$, the projection $p(d) \in T_R$ normalizes the product $(D \cap N_R)[N_R, N_R]$. On the other hand, since the flow $(R/D, g_{\mathbb{R}})$ is ergodic, the element $p(d) \in T_R$ commutes with the subgroup $u_{\mathbb{R}} \subset N_R$. The latter projects to a dense subgroup of the torus $N_R/(D \cap N_R)[N_R, N_R]$. It follows that the action of $p(d)$ on the torus $N_R/(D \cap N_R)[N_R, N_R]$ is trivial. But the group $p(D)$ is reductive and hence all its irreducible nontrivial submodules belong to $D_0[N_R, N_R]$. Since D_0 is normal in N_R, all the submodules belong to D_0 and generate a T_R-invariant subgroup $M \subset D_0$. But D is a quasilattice and hence $M = 1$, whence $p(D) = 1$ and $D \subset N_R$.

Hence the following holds.

THEOREM 7.1 [**AG**]. *An ergodic flow on a class (I) compact solvable manifold is smoothly isomorphic to a minimal nilflow.*

2. Suppose now that the flow $(R/D, g_{\mathbb{R}})$ is nonergodic (as Examples 4.1–4.3 show, for class (I) solvable groups this is an ordinary event). We want to construct the ergodic decomposition. Define

$$T_1 = \overline{a_{\mathbb{R}}} \subset T_R, \qquad R_1 = (T_1 N_R \cap R)_0, \qquad D_1 = D \cap R_1.$$

Then R_1 is a connected simply connected normal subgroup of R, the space R_1/D_1 is compact (though D_1 need not be a quasilattice in R_1), and $g_{\mathbb{R}} \subset R_1$. The compact subspace $R_1 D \subset R/D$ is invariant relative to our flow, and to construct the ergodic submanifold $E_1(g_{\mathbb{R}})$ it suffices to restrict our attention to the flow $(R_1/D_1, g_{\mathbb{R}})$.

According to Lemma 5.6, the semisimple splitting of the group R_1 is as follows:

$$(R_1)_s = T_1 \cdot R_1 = T_1 \cdot N_1, \qquad \text{where } N_1 = (T_1 R \cap N_R)_0.$$

Since $T_1 = \overline{a_{\mathbb{R}}}$, the diffeomorphism $\pi : R_1 \to N_1$ induces a smooth isomorphism of flows

$$\pi^* : (R_1/D_1 \cap N_1, g_{\mathbb{R}}) \to (N_1/D_1 \cap N_1, u_{\mathbb{R}}).$$

The nilflow on $N_1/D_1 \cap N_1$ may not be ergodic, and by Theorem 3.10 the ergodic submanifold $E_1(u_{\mathbb{R}}) \subset N_1/D_1 \cap N_1$ is homogeneous and is given by

$$E_1(u_{\mathbb{R}}) = \overline{u_{\mathbb{R}}(D_1 \cap N_1)} = F_1(D_1 \cap N_1),$$

where $F_1 = (\overline{u_{\mathbb{R}}(D_1 \cap N_1)})_0$ is a connected simply connected subgroup of N_1.

The connected closed subset

$$\widetilde{M}_1 = (\overline{g_{\mathbb{R}}(D_1 \cap N_1)})_0 \subset R_1$$

is the inverse image of the subgroup $F_1 \subset N_R$ under the diffeomorphism $\pi : R \to N_R$, and hence is a smooth $g_{\mathbb{R}}$-invariant submanifold (as is seen from Example 4.3, it can fail to be a subgroup because π is not a homomorphism). Define

$$E_1(g_{\mathbb{R}}) = \overline{g_{\mathbb{R}} D_1} = \widetilde{M}_1 D_1 \subset R_1/D_1.$$

Let us show that $E_1(g_{\mathbb{R}})$ is a smooth manifold as well. It suffices to prove that $E_1(g_{\mathbb{R}})$ is the quotient of the manifold $\widetilde{M}_1$ by orbits of the right free action of a subgroup $D' \subset D_1$. We will prove even more.

LEMMA 7.2 [**St84a**]. *Let G be a Lie group, let F be a subgroup of G, let Γ be a closed subgroup of G, and let $\widetilde{M} = (\overline{F\Gamma})_0 \subset G$. Then $\Gamma' = \widetilde{M} \cap \Gamma$ is a subgroup of Γ, and $M = \overline{F\Gamma} \subset G/\Gamma$ is the quotient of $\widetilde{M}$ by orbits of the right free action of the subgroup Γ'. In particular, if $\widetilde{M} \subset G$ is a smooth manifold, then so is $M \subset G/\Gamma$.*

Proof. Let us prove that, given $\gamma \in \Gamma$, one has $\widetilde{M}\gamma = \widetilde{M} \iff \gamma \in \Gamma'$. In fact, take $\gamma \in \Gamma' = (\overline{F\Gamma})_0 \cap \Gamma$. Then $(\overline{F\Gamma})_0\gamma$ is the connected component of the set $\overline{F\Gamma} \subset G$ that contains the point $\gamma \in \widetilde{M}$ and hence intersects $(\overline{F\Gamma})_0$. It follows that $\widetilde{M}\gamma = \widetilde{M}$. Conversely, if $\widetilde{M}\gamma = \widetilde{M}$, then since $1 \in \widetilde{M}$, we conclude that $\gamma \in \widetilde{M}$. Now it is clear that the set Γ' is invariant under multiplication and taking the inverse element. It follows that Γ' is a subgroup of Γ. □

Therefore $(E_1(g_{\mathbb{R}}), g_{\mathbb{R}})$ is a smooth flow which is finitely covered by a minimal ergodic nilflow $(F_1/D_1 \cap F_1, u_{\mathbb{R}})$. It follows that it is also minimal and ergodic relative to the smooth $g_{\mathbb{R}}$-invariant measure μ_1 obtained from a smooth F_1-invariant measure on $F_1/D_1 \cap F_1$. Since R_1/D_1 is a compact subspace of R/D, the ergodic submanifold $E_1(g_{\mathbb{R}})$ is given by $E_1(g_{\mathbb{R}}) = \overline{g_{\mathbb{R}}D} \subset R/D$.

Likewise, all the closed subsets $E_h(g_{\mathbb{R}}) = \overline{g_{\mathbb{R}}hD} \subset R/D$ are smooth ergodic submanifolds for our flow which are finitely covered by minimal nilflows

$$(F_h/D \cap F_h, h^{-1}u_{\mathbb{R}}h),$$

where $F_h = (\overline{h^{-1}u_{\mathbb{R}}h(D \cap N_R)})_0$ is a connected simply connected subgroup of N_R. Since all the flows on $E_h(g_{\mathbb{R}})$ are minimal, any two of the submanifolds either coincide or are disjoint. It follows that they form the ergodic partition E of our flow.

Recall that according to §2, the ergodic decomposition $\widetilde{E}$ of our flow consists of closed sets $\widetilde{E}_h(g_{\mathbb{R}}) = \overline{g_{\mathbb{R}}MhD} \subset R/D$, where $M \subset R$ is the Moore normal subgroup for $g_{\mathbb{R}}$. Since $M \subset N \subset N_R$, every flow $(\widetilde{E}_h(g_{\mathbb{R}}), g_{\mathbb{R}})$ is finitely covered by the nilflow $(F/D \cap F, h^{-1}u_{\mathbb{R}}h)$, where $F = (\overline{u_{\mathbb{R}}MD})_0$ is a normal subgroup of N_R. Note that $u_{\mathbb{R}}M$ is the smallest (normal in R) subgroup containing all the conjugate subgroups $hu_{\mathbb{R}}h^{-1}$, $h \in R$. Hence F contains F_h for every $h \in R$. Because E and $\widetilde{E}$ are ergodic decompositions of our flow, they coincide almost everywhere, and hence $F_h = F$ for almost all $h \in R$.

The same assertion can be derived if one observes that F contains countably many proper subgroups V_i containing D_0 and intersecting $D \cap F$ in a uniform subgroup. Then one uses Lemma 3.11, which ensures that $\{h \in R : \ hu_{\mathbb{R}}h^{-1} \subset V_i\}$ is a closed null set in R. It follows from Lemma 7.2 that all $\widetilde{E}_h(g_{\mathbb{R}})$ are smooth manifolds with smooth $g_{\mathbb{R}}M$-invariant measures obtained from an F-invariant measure on $F/D \cap F$. As is seen from Example 4.2 (in which E and $\widetilde{E}$ coincide everywhere), though the submanifolds $\widetilde{E}_h(g_{\mathbb{R}})$ are of the same dimension, the decomposition $\widetilde{E}$ can fail to be a smooth fibration.

Let us summarize our study of flows on class (I) solvable manifolds.

THEOREM 7.3. *Let R be a connected simply connected class (I) solvable Lie group, and let D be a quasilattice in R. Then every smooth flow $(E_h(g_{\mathbb{R}}), g_{\mathbb{R}})$ is finitely covered by an ergodic nilflow $(F_h/D \cap F_h, h^{-1}u_{\mathbb{R}}h)$, and every smooth flow $(\widetilde{E}_h(g_{\mathbb{R}}), g_{\mathbb{R}})$ is finitely covered by a nilflow $(F/D \cap F, h^{-1}u_{\mathbb{R}}h)$, where $F_h \subset F \subset N_R$*

for all $h \in R$, and $F_h = F$ for almost all $h \in R$. Explicitly,

$$E_h(g_{\mathbb{R}}) = \overline{g_{\mathbb{R}} hD} \subset R/D \quad \text{and} \quad \widetilde{E}_h(g_{\mathbb{R}}) = \overline{g_{\mathbb{R}} MhD} \subset R/D.$$

3. Now we proceed to the general case. Let A be the Auslander normal subgroup for a subgroup $g_{\mathbb{R}} \subset R$, and let C be the maximal connected subgroup of $(\overline{AD}_0)$ normal in R. Then $R' = R/C$ is a connected simply connected Lie group with $D' = \overline{AD}/C$ as a quasilattice. Let $\alpha : R \to R' = R/C$ be the epimorphism and let $g'_{\mathbb{R}} = \alpha(g_{\mathbb{R}}) \subset R'$. As follows from Corollary 2.3, the original flow $(R/D, g_{\mathbb{R}})$ is ergodic if and only if the quotient flow $(R'/D', g'_{\mathbb{R}})$ is (the latter being the maximal quotient flow of zero entropy). By construction, the subgroup $g'_{\mathbb{R}}$ is quasi-unipotent in R'. Hence in the semisimple splitting $R'_s = T'_R \cdot N'_R$ of the group R', the closure $T'_1 = \overline{p(g'_{\mathbb{R}})} \subset T'_R$ is a torus. It follows that the quotient flow is ergodic only if $T'_R = T'_1$. In particular, the Lie group R' must be of class (I), and the study of the quotient flow reduces to that of a nilflow on $N'_R/D' \cap N'_R$. This reduction is called Auslander's ergodicity criterion for a solvable flow.

If the flow $(R'/D', g'_{\mathbb{R}})$ is nonergodic, the group R' need not be of class (I). However, the normal subgroup $R'_1 = (T'_1 N'_R \cap R')_0$ is of class (I) anyway. In addition, it contains $g'_{\mathbb{R}}$, and the quotient space $R'_1/D'_1 = R'_1/D' \cap R'_1$ is compact. It follows that we can restrict our attention to the flow $(R'_1/D'_1, g'_{\mathbb{R}})$ on a class (I) solvable manifold and construct all the submanifolds $E'_1(hg'_{\mathbb{R}}h^{-1}) \subset \widetilde{E}'_1(hg'_{\mathbb{R}}h^{-1}) \subset R'_1/D'_1$, $h \in R'$. Defining as usual

$$E'_h(g'_{\mathbb{R}}) = hE'_1(h^{-1}g'_{\mathbb{R}}h) \quad \text{and} \quad \widetilde{E}'_h(g'_{\mathbb{R}}) = h\widetilde{E}'_1(h^{-1}g'_{\mathbb{R}}h)$$

for all $h \in R'$, we obtain the ergodic partitions E' and $\widetilde{E}'$ of the flow $(R'/D', g'_{\mathbb{R}})$.

Return to the original flow and denote by

$$\widehat{\alpha} : R/D \to R'/D' = R/\overline{AD}$$

the fibration of homogeneous spaces with the compact fiber $\overline{AD}/D$. Define

$$E_h(g_{\mathbb{R}}) = \widehat{\alpha}^{-1}(E'_{\alpha(h)}(g'_{\mathbb{R}})), \qquad \widetilde{E}_h(g_{\mathbb{R}}) = \widehat{\alpha}^{-1}(\widetilde{E}'_{\alpha(h)}(g'_{\mathbb{R}})).$$

It follows from the Mautner phenomenon that all the flows $(E_h(g_{\mathbb{R}}), g_{\mathbb{R}})$ are ergodic relative to smooth $g_{\mathbb{R}}A$-invariant measures μ_h, and, by construction, $E_h(g_{\mathbb{R}}) = \overline{g_{\mathbb{R}}AhD} \subset R/D$. Moreover, $\widetilde{E}$ is the ergodic decomposition of the original flow, since the partition $\widetilde{E}'$ of the space R'/D' into submanifolds

$$\widetilde{E}'_{\alpha(h)}(g'_{\mathbb{R}}) = \overline{g'_{\mathbb{R}}\alpha(M)\alpha(h)D'}$$

is the ergodic decomposition of the quotient flow (because $\alpha(M)$ is the Moore normal subgroup for $g'_{\mathbb{R}} = \alpha(g_{\mathbb{R}})$). We have proved the following result.

THEOREM 7.4. *The ergodic decomposition $\widetilde{E}$ of a solvable flow consists of smooth submanifolds $\widetilde{E}_h(g_{\mathbb{R}}) = \overline{g_{\mathbb{R}}MhD}$ of the same dimension and with smooth $g_{\mathbb{R}}M$-invariant measures $\widetilde{\mu}_h$. The ergodic partition E is a subpartition of $\widetilde{E}$ and coincides with the latter almost everywhere. Every flow $(E_h(g_{\mathbb{R}}), g_{\mathbb{R}})$ is ergodic relative to a smooth $g_{\mathbb{R}}A$-invariant measure μ_h, and $E_h(g_{\mathbb{R}}) = \overline{g_{\mathbb{R}}AhD}$.*

4. So, we have reduced the study of a solvable flow $(R/D, g_{\mathbb{R}})$ to the study of the family of smooth ergodic flows $(E_h(g_{\mathbb{R}}), g_{\mathbb{R}})$. Since an ergodic submanifold need not be a homogeneous subspace of R/D, such a reduction is not very good: we have lost the homogeneous structure of the flow. If the group R is of class (I), we can overcome this difficulty by constructing a finite-to-one cover of every ergodic flow $(E_h(g_{\mathbb{R}}), g_{\mathbb{R}})$ by a minimal homogeneous nilflow. In the general case it would be also nice to find a finite-to-one cover of the flow by an ergodic homogeneous flow. Fortunately, this is possible. Let us show how to construct such a cover for the flow $(E_1(g_{\mathbb{R}}), g_{\mathbb{R}})$ (the most general result will be obtained in §10, but in the solvable case the cover is arranged much more simply and deserves separate consideration).

Denote

$$T(C) = \{s \in \mathcal{A}(T_R) : \{s, N_R\} \subset C\}.$$

The subgroup $D' = \overline{AD}/C$ is a quasilattice in $R' = R/C$, and, by Lemma 5.7,

$$\mathcal{A}(R'_s) = \mathcal{A}(T'_R) \cdot R' = \mathcal{A}(T'_R) \cdot N'_R, \quad \text{where} \mathcal{A}(T'_R) = \mathcal{A}(T_R)/T(C), \ N'_R = N_R/C.$$

Since the product $N'_R D'$ is closed, the same applies to its inverse image $T(C)N_R D$. On the other hand, the image $a_{\mathbb{R}}T(C) \subset \mathcal{A}(T'_R)$ is relatively compact (because the subgroup $g'_{\mathbb{R}} = g_{\mathbb{R}}C/C \subset R'$ is quasi-unipotent). Hence if $B \subset \mathcal{A}(T_R)$ stands for the algebraic hull of $a_{\mathbb{R}} \subset T_R$, then the factor group $B/B \cap T(C)$ is compact. In turn, Theorem 6.3 ensures the the factor group $B/B \cap \mathcal{A}(D)$ is compact. Since the groups are algebraic, we can find a torus $T \subset B$ such that

$$B(T(C) \cap \mathcal{A}(D)) = T(T(C) \cap \mathcal{A}(D))$$

and the intersection $T \cap T(C) \cap \mathcal{A}(D)$ is finite. For all $t \in \mathbb{R}$, let $a_t = c_t \times b_t$, where $c_{\mathbb{R}} \subset T$, $b_{\mathbb{R}} \subset T(C)$. Define $f_t = b_t \times u_t$, $t \in \mathbb{R}$. Then $f_{\mathbb{R}} \subset \mathcal{A}(D)$. The homogeneous space TR/D is compact, and the Auslander subgroups for $g_{\mathbb{R}}$ and $f_{\mathbb{R}}$ are the same. Hence the ergodic submanifold $E_1(f_{\mathbb{R}}) \subset TR/D$ is given by $E_1(f_{\mathbb{R}}) = \overline{f_{\mathbb{R}}AD} \subset TR/D$. Let us prove that

1) the submanifold $E_1(f_{\mathbb{R}})$ is a homogeneous subspace of TR/D, and

2) the ergodic homogeneous flow $(E_1(f_{\mathbb{R}}), f_{\mathbb{R}})$ finitely covers the smooth flow $(E_1(g_{\mathbb{R}}), g_{\mathbb{R}})$.

In fact, since $f_t = b_t \times u_t$ for all $t \in \mathbb{R}$, where $b_{\mathbb{R}} \subset T(C)$ and $u_{\mathbb{R}} \subset N_R$, it follows that the image $f'_{\mathbb{R}} = f_{\mathbb{R}}C/C$ belongs to the nilradical N'_R of the group $\mathcal{A}(R'_s)$. Consequently, the closure $F'_1 = (\overline{f'_{\mathbb{R}}D'})_0$ is a subgroup of N'_R. This means that $E_1(f_{\mathbb{R}}) = F_1 D$, where $F_1 \subset TR \cap \mathcal{A}(D)$ is the inverse image of the subgroup $F'_1 \subset N'_R$. Let $\beta : T \cdot R \to R$ be the projection along the torus T. Then $\beta(f_{\mathbb{R}}) = g_{\mathbb{R}}$, and hence

$$\beta(F_1) = \beta(\overline{f_{\mathbb{R}}AD})_0 = (\overline{g_{\mathbb{R}}AD})_0 = \widetilde{M} \subset R.$$

Since F_1 fibers over the simply connected group F'_1 with the connected fiber C, the group F_1 is itself simply connected and hence $T \cap F_1 = 1$. It follows that the restriction $\beta : F_1 \to \widetilde{M}$ is a diffeomorphism. By Lemma 7.2, the intersection $\widehat{D} = D \cap \widetilde{M}$ is a subgroup and $E_1(g_{\mathbb{R}})$ is the orbit space of the manifold $\widetilde{M}$ relative to the right free action of $\widehat{D}$. Since β sends f_t-action on F_1 to g_t-action on $\widetilde{M}$, we conclude that β induces a cover of flows

$$\beta^* : (F_1/D \cap F_1, f_{\mathbb{R}}) \to (E_1(g_{\mathbb{R}}), g_{\mathbb{R}}).$$

Finally, since $\widehat{D} \subset T(D \cap F_1)$, then $D \cap F_1$ is of finite index in $\widehat{D}$, and hence the cover β^* is finitely-sheeted.

We have proved that the smooth ergodic flow $(E_1(g_{\mathbb{R}}), g_{\mathbb{R}})$ is finitely covered by an ergodic homogeneous flow on $F_1/D \cap F_1$. By construction, $f_{\mathbb{R}} \subset F_1 \subset \mathcal{A}(D)$, and hence the identity component D_0 is normal in F_1. It follows that the homogeneous space $F_1/D \cap F_1$ is isomorphic to the homogeneous space H_1/Γ_1, where $\Gamma_1 = D \cap F_1/D_0$ is a lattice in the solvable Lie group $H_1 = F_1/D_0$. In particular, in the case $E_1(g_{\mathbb{R}}) = R/D$ we obtain the following result due to Auslander.

THEOREM 7.5 [**Au**]. *An ergodic solvable flow is finitely covered by an ergodic homogeneous flow on a compact solvable manifold with a discrete isotropy subgroup.*

5. Here we present the theory of flows on so-called Euclidean solvable manifolds, which is due to Brezin and Moore [**BM**]. This allows us to prove a fairly effective ergodicity criterion for solvable flows. Our exposition is much more elementary; along the way we correct some inaccuracies from [**BM**].

One says that a solvable group R is *Euclidean* if its semisimple splitting is given by $R_s = T_R \cdot N_R$, where the group T_R is compact, and the nilshadow N_R is Abelian. Clearly, any Euclidean group is of class (I).

LEMMA 7.6. *A group R is Euclidean $\Longleftrightarrow$ $R = A \cdot N$, where the nilradical N is Abelian, and the action of A on N is reductive and all its orbits are compact.*

Proof. 2) $\Rightarrow$ 1) follows from Lemma 5.8. To prove 1) $\Rightarrow$ 2) it suffices to show that R splits. Let $V \subset N_R$ be a T_R-invariant complement to the nilradical $N \subset R$ in N_R. Then $A = T_R V \cap R$ is the desired group. □

Let $D \subset R$ be a quasilattice. We say that a compact manifold R/D is *Euclidean* if the group R is Euclidean.

Note that the group $R = O_{\mathbb{R}} \cdot \mathbb{R}^2$ in Examples 4.1–4.4 was Euclidean. In addition, a quasilattice $D \subset R$ need not be discrete. For instance, let $\mathbb{R}^2 = \mathbb{R}' \times \mathbb{R}''$ and let $D = O_{\pi\mathbb{Z}} \cdot (\mathbb{Z}' \times \mathbb{R}'')$, where $\mathbb{Z}'$ is a lattice in $\mathbb{R}'$. Then D has an Abelian subgroup of finite index $D_1 = O_{2\pi\mathbb{Z}} \times \mathbb{Z}' \times \mathbb{R}''$. The projection $\Pi : R \to N_R = \Delta \times \mathbb{R}^2$ induces a homeomorphism of R/D_1 and the torus N_R/D_1. The space R/D is two-sheetedly covered by the torus and has a non-Abelian fundamental group $D/D_0 = O_{\pi\mathbb{Z}} \cdot \mathbb{Z}'$; hence it is the Klein bottle.

LEMMA 7.7. *Let $D \subset R$ be a quasilattice in a connected simply connected solvable group R. Then the following conditions are equivalent*:

1) *The manifold R/D is Euclidean.*
2) *The fundamental group $\pi_1(R/D) = D/D_0$ is almost Abelian.*
3) *The space R/D is finitely covered by a torus.*

Proof. Take $p : R_s = T_R \cdot N_R \to T_R$. If the group R is Euclidean, then the projection $p(D)$ is finite, and the intersection $D \cap N_R$ is Abelian. Hence 1) $\Rightarrow$ 2). The projection $\Pi : T_R \cdot N_R \to N_R$ induces a homeomorphism

$$\Pi^* : R/D \cap N_R \to N_R/D \cap N_R,$$

and hence 1) $\Rightarrow$ 3). The implication 3) $\Rightarrow$ 2) is obvious. It now suffices to prove that 2) $\Rightarrow$ 1).

Without loss of generality we can assume that the group D/D_0 is Abelian, that is, $[D, D] \subset D_0$. By Mostow's Theorems 6.2–6.3, $N_R \subset \mathcal{A}(D)$ and $D_0 \subset N$. Hence $[N_R, N_R] \subset D_0$. Since D is a quasilattice, the nilshadow N_R is Abelian. But then $[p(D), D \cap N] \subset D_0$, and hence $[p(D), N] \subset D_0$. The reductivity and the inclusion $[T_R, N_R] \subset N$ imply that $[p(D), N_R] \subset D_0$. Let $V \subset N_R$ be a maximal subspace on which the action of the group $p(D)$ is trivial. Then V is invariant relative to T_R, and one can find a T_R-invariant complement W to V in N_R. Clearly, W is a normal subgroup of R_s, and $W \subset D_0$, whence W is trivial. It follows that $p(D)$ commutes with N_R, and hence the group $p(D)$ is also trivial. On the other hand, $p(D)$ is a lattice in T_R, and hence T_R is a torus. □

COROLLARY 7.8. *If D is a quasilattice in R and the manifold R/D is homeomorphic to a Euclidean manifold R'/D', then the group R is Euclidean.*

Proof. The fundamental groups of R/D and R'/D' are isomorphic. □

Denote by $U \subset N$ the Auslander normal subgroup for the Abelian group T_R. Suppose that $H \subset R$ is a connected normal subgroup such that the factor group R/H is Euclidean. Explicit expression for the semisimple splitting of a factor group (see Lemma 5.7) ensures that $U \subset N_H$ and $[N_R, N_R] \subset N_H$. Since $N_H = (H \cap N_R)_0$, we deduce that $U[N_R, N_R] \subset H$. On the other hand, the inclusion $U[N_R, N_R] \subset H$ implies that the factor group R/H is Euclidean. We have derived the following.

LEMMA 7.9. *The Euclidean manifold $R/\overline{U[N_R, N_R]D}$ is the maximal Euclidean quotient space of a compact manifold R/D.*

Now we turn to homogeneous flows on Euclidean manifolds. Pick a subgroup $g_{\mathbb{R}} \subset R \subset R_s = T_R \cdot N_R$, where the subgroup T_R is selected to commute with $g_{\mathbb{R}}$. Then $g_t = c_t \times u_t$, where $c_{\mathbb{R}} \subset T_R$, $u_{\mathbb{R}} \subset N_R$.

Let D be a quasilattice in R, and let the flow $(R/D, g_{\mathbb{R}})$ be ergodic. Then $D \subset N_R$, and the projection $\Pi : T_R \cdot N_R \to N_R$ induces an isomorphism of flows

$$\Pi^* : (R/D, g_{\mathbb{R}}) \to (N_R/D, u_{\mathbb{R}}).$$

This means that an ergodic flow on a Euclidean manifold is isomorphic to a rectilinear flow on the torus. In addition, ergodicity implies that $\overline{c_{\mathbb{R}}} = T_R$, which in turn implies that $u_{\mathbb{R}}$ belongs to the (connected) center $Z(R_s)$ of the group R_s. It follows that $g_{\mathbb{R}}$ has trivial Moore subgroup. Clearly, ergodicity of the action $(N_R/D \cap N_R, Z(R_s))$ is a necessary condition for R/D to carry an ergodic homogeneous flow. Conversely, if the action is ergodic, then there exists a one-parameter subgroup $u_{\mathbb{R}} \subset Z(R_s)$ which acts on $N_R/D \cap N_R$ ergodically. But then the subgroup $g_{\mathbb{R}} = (T_R \times u_{\mathbb{R}}) \cap R$ acts on $R/D \cap N_R$ ergodically (and, in particular, $D \subset N_R$, that is, R/D is homeomorphic to the torus N_R/D). We have obtained the following.

LEMMA 7.10. *Let $Z(R_s)$ be the center of the group R_s. Then $(\overline{Z(R_s)D})_0 = N_R$ if and only if the Euclidean manifold R/D carries an ergodic one-parameter flow.*

Remark. Theorem 5.2 in [**BM**] stated erroneously that the criterion for a Euclidean manifold to admit an ergodic flow consists in the following two conditions: 1) R/D is homeomorphic to a torus, and 2) $(\overline{Z(R)_0 D})_0 = N$. The "ergodic" Example 4.4, in which the center $Z(R)$ is discrete, shows that condition 2) is not necessary. On the other hand, condition 1) can be omitted. So, the two conditions in Theorem 5.2 in [**BM**] do not give a criterion; instead, a criterion is given by the condition in Lemma 7.10.

If the condition $(\overline{Z(R_s)_0 D})_0 = N_R$ is satisfied, the flow $(R/D, g_{\mathbb{R}})$ is isomorphic to a rectilinear flow on the torus $(N_R/D, u_{\mathbb{R}})$, the ergodicity criterion for the latter being well known.

Now let $(R/D, g_{\mathbb{R}})$ be a homogeneous flow on an arbitrary compact solvable manifold. We know that if $D \subset C \subset R$ and the quotient space R/C is Euclidean, then $D \subset L \subset C$, where G/L is the maximal Euclidean quotient space from Lemma 7.9. The following effective ergodicity criterion holds.

THEOREM 7.11 [**BM**]. *A flow* $(R/D, g_{\mathbb{R}})$ *is ergodic* $\iff$ *the flow* $(R/L, g_{\mathbb{R}})$ *on the maximal Euclidean quotient space is ergodic.*

Proof. It suffices to prove 2) $\Rightarrow$ 1). Let $M \subset N$ be the Moore normal subgroup for $g_{\mathbb{R}} \subset R$. Since any subgroup ergodic on a Euclidean manifold has trivial Moore subgroup, we deduce that $M \subset L$. Hence R/L is the maximal Euclidean quotient space of $R/\overline{MD}$.

Consequently, it suffices to consider the case when $g_{\mathbb{R}}$ has trivial Moore subgroup. Define as usual $p : T_R \cdot N_R \to T_R$, and let $g_t = c_t \times u_t$ be a Jordan decomposition, where $c_{\mathbb{R}} \subset T_R$ and $u_{\mathbb{R}} \subset N_R$. Since $T_R/p(D)$ is a Euclidean quotient space of R/D, we deduce that $T_R = \overline{c_{\mathbb{R}} p(D)}$. Since the Moore subgroup is trivial, the closure $\overline{c_{\mathbb{R}}}$ is a torus. Hence $T_R = \overline{c_{\mathbb{R}}}$; that is, the group R is of class (I), and the flow $(R/D \cap N_R, g_{\mathbb{R}})$ is isomorphic to the nilflow $(N_R/D \cap N_R, u_{\mathbb{R}})$. In addition, $L = [N_R, N_R]D$. The flows

$$(R/[N_R, N_R](D \cap N_R), g_{\mathbb{R}}) \quad \text{and} \quad (N_R/[N_R, N_R](D \cap N_R), u_{\mathbb{R}})$$

are isomorphic. But by Green's ergodicity criterion, if the latter flow is ergodic, then so is the nilflow $(N_R/D \cap N_R, u_{\mathbb{R}})$. Hence the same applies to the original flow. □

§8. Flows on semisimple homogeneous spaces

1. Here we study the Mautner phenomenon in the semisimple case. Our approach is based on Lemma 2.2 and does not involve spectral theory. Then we prove Moore's ergodicity criterion and construct the ergodic decomposition.

We start the study with unitary representations of the group $S = \mathrm{SL}(2, \mathbb{R})$. Define

$$A_t = \begin{pmatrix} e^{t/2} & 0 \\ 0 & e^{-t/2} \end{pmatrix}, \qquad U_t = \begin{pmatrix} 1 & t \\ 0 & 1 \end{pmatrix}, \qquad H_t = \begin{pmatrix} 1 & 0 \\ t & 1 \end{pmatrix}.$$

The Auslander subgroup for every element $A_s \in S$, $s \neq 0$, is equal to the group S, and hence $\mathrm{Fix}(A_s) = \mathrm{Fix}(S)$ (see §1 for notation) for any unitary representation $\rho : S \to \mathbb{U}(\mathbb{H})$.

Following [**Ma91b**], we will prove that $\mathrm{Fix}(U_s) = \mathrm{Fix}(S)$ for any $s \neq 0$. We use the relation

$$\begin{pmatrix} 2 & 0 \\ c & 1/2 \end{pmatrix} = \begin{pmatrix} 1 & 1/c \\ 0 & 1 \end{pmatrix} \begin{pmatrix} 1 & 0 \\ c & 1 \end{pmatrix} \begin{pmatrix} 1 & -1/2c \\ 0 & 1 \end{pmatrix},$$

or, equivalently,

$$f(c) = U_{1/c} H_c U_{-1/2c}, \qquad \text{where } f(c) = \begin{pmatrix} 2 & 0 \\ c & 1/2 \end{pmatrix}.$$

Given $c_n = \frac{1}{2ns}$, we obtain

$$f(c_n) \in U_{s\mathbb{Z}} H_{c_n} U_{s\mathbb{Z}}, \qquad \text{where } f(c_n) \to a = \begin{pmatrix} 2 & 0 \\ 0 & 1/2 \end{pmatrix}, \ n \to \infty.$$

Hence the element a belongs to the closure of the set $U_{s\mathbb{Z}} W U_{s\mathbb{Z}}$, where W is an arbitrary neighborhood of the element $1 \in \mathrm{SL}(2, \mathbb{R})$. According to Lemma 2.2,

$\mathrm{Fix}(U_s) \subset \mathrm{Fix}(a)$ and hence $\mathrm{Fix}(U_s) = \mathrm{Fix}(S)$. All the calculations are valid for any group S locally isomorphic to $\mathrm{SL}(2,\mathbb{R})$. Let us summarize.

LEMMA 8.1. *Given any continuous unitary representation of a connected group S locally isomorphic* $\mathrm{SL}(2,\mathbb{R})$, *one has* $\mathrm{Fix}(U_s) = \mathrm{Fix}(S)$ *for any* $s \neq 0$.

In particular, if $v \in \mathbb{H}$ is an eigenvector for $U_{\mathbb{R}}$, then $v \in \mathrm{Fix}(U_s)$ for some $s \in \mathbb{R}$, $s \neq 0$, and hence $v \in \mathrm{Fix}(S)$.

Every one-parameter subgroup of S either is conjugate to $A_{\mathbb{R}}$ or $U_{\mathbb{R}}$, or has its image in the Ad-representation compact (and its Moore subgroup trivial). Thus, we have completed the study of the Mautner phenomenon for groups locally isomorphic to $\mathrm{SL}(2,\mathbb{R})$.

2. Now let $\rho : L \to \mathbb{U}(\mathbb{H})$ be a unitary representation of a semisimple group L. Suppose that $v \in \mathbb{H}$ is an eigenvector for a subgroup $g_{\mathbb{R}} \subset G$. Then v is fixed under the Auslander normal subgroup $A = A(g_{\mathbb{R}})$, and we can replace $\mathbb{H}$ by the G-invariant closed subspace $\mathrm{Fix}(A)$.

Hence without loss of generality the subgroup $g_{\mathbb{R}} \subset L$ may be assumed to be quasi-unipotent. A Jordan decomposition for $g_{\mathbb{R}}$ is given by $g_t = k_t \times u_t$, where the reductive subgroup $k_{\mathbb{R}} \subset L$ commutes with the unipotent subgroup $u_{\mathbb{R}} \subset L$. There is a generalized Jacobson–Morozov lemma (see [**Ja**]):

LEMMA 8.2. *Let n be a nilpotent element in a semisimple Lie algebra $\mathfrak{l}$, commuting with a semisimple element $k \in \mathfrak{l}$. Then there exist a semisimple element $a \in \mathfrak{l}$ and a nilpotent element $m \in \mathfrak{l}$ such that $[a,n] = n$, $[a,m] = -m, [n,m] = a$, where a and m commute with k.*

Proof. The centralizer $Z_{\mathfrak{l}}(k)$ is reductive, and hence $Z_{\mathfrak{l}}(k) = \mathfrak{p} \oplus \mathfrak{h}$, where the subalgebra $\mathfrak{p}$ is semisimple and the subalgebra $\mathfrak{h}$ is Abelian and reductive. Clearly, $n \in \mathfrak{p}$, and we can apply the Jacobson–Morozov lemma to the nilpotent element n in the semisimple algebra $\mathfrak{p}$. □

Hence L has a subgroup S locally isomorphic to $\mathrm{SL}(2,\mathbb{R})$ that contains $u_{\mathbb{R}}$ and commutes with $k_{\mathbb{R}}$. Pick $v \in \mathrm{Fix}(g_s)$ for some $s \neq 0$. Computations made for $\mathrm{SL}(2,\mathbb{R})$ show that there exist a subgroup $h_{\mathbb{R}} \subset S$, an element $a \in S$, and a curve $f(t) \subset S$ such that

$$f(1/2ns) = u_{-2ns} h_{1/2ns} u_{ns}, \qquad \text{where } f(t) \to a \in S,\ t \to 0.$$

Suppose for simplicity that the center $Z(L)$ is finite. Then the subgroup $k_{\mathbb{R}} \subset L$ is relatively compact, and without loss of generality $k_{ns} \to k'$, $n \to \infty$, for some element $k' \in L$ commuting with a. It follows that $a \times k' \in \overline{g_{s\mathbb{Z}} W g_{s\mathbb{Z}}}$ for an arbitrary neighborhood $W = W(1) \subset L$, and hence $\mathrm{Fix}(g_s) \subset \mathrm{Fix}(a \times k')$. But

$$A(a \times k') = A(a) = M(u_{\mathbb{R}}) = M(g_{\mathbb{R}}),$$

and hence $\mathrm{Fix}(g_s) \subset \mathrm{Fix}(M)$, where $M = M(g_{\mathbb{R}})$ is the Moore normal subgroup for $g_{\mathbb{R}}$. We have examined the Mautner phenomenon in the semisimple case and proved a particular case of Theorem 2.1.

THEOREM 8.3 [**Mo66**]. *Let $\rho : L \to \mathbb{U}(\mathbb{H})$ be a unitary representation of a semisimple Lie group L, and $v \in \mathbb{H}$ be an eigenvector for a subgroup $g_{\mathbb{R}} \subset L$. Then $\rho(M)v \equiv v$, where $M \subset L$ is the Moore normal subgroup for $g_{\mathbb{R}}$.*

3. Following [**Mo66**], we will apply the Mautner phenomenon to finding the ergodicity criterion and studying the ergodic decomposition of a flow $(L/D, g_{\mathbb{R}})$, where D is a quasilattice in a connected simply connected semisimple group L. If $L = S \times K$ is the decomposition of L into noncompact and compact parts, we let $p : L \to K$ be the projection onto the compact part K.

Define $H = \overline{p(D)}$, $T = \overline{p(g_{\mathbb{R}})} \subset K$. Clearly, for the original flow to be ergodic it is necessary that the flow $(K/H, T)$ be ergodic. But if the action of the torus T on K/H is ergodic, it must be topologically transitive: there exists an element $k \in K$ such that $K = TkH$ and hence $K/H = k^{-1}Tk/k^{-1}Tk \cap H$. But the left-hand side of this equality is a space with finite fundamental group, whereas the right-hand side is a torus. It follows that $K = H$ and hence $L = \overline{SD}$. Now Borel's Theorem E.4 implies the following.

LEMMA 8.4 [**Mo66**]. *If a semisimple space of finite volume L/D admits an ergodic homogeneous flow, then D is Zariski dense in L.*

In particular, any quasilattice $D \subset L$ is a lattice.

The Dani normal subgroup $J \subset L$ in the semisimple case is equal to the Moore normal subgroup $M \subset L$. By Corollary 2.3, the original flow is ergodic if and only if the quotient flow $(L/\overline{MD}, g_{\mathbb{R}})$ is ergodic. In view of the condition $L = \overline{SD}$, the identity component $(\overline{MD})_0$ is a normal subgroup of L, and hence our quotient flow is isomorphic to the flow $(L'/D', g'_{\mathbb{R}})$, where

$$L' = L/(\overline{MD})_0, \qquad D' = \overline{MD}/(\overline{MD})_0, \qquad g'_{\mathbb{R}} = g_{\mathbb{R}}(\overline{MD})_0/(\overline{MD})_0.$$

By construction, the subgroup $g'_{\mathbb{R}} \subset L'$ is such that the closure $\overline{\mathrm{Ad}(g'_{\mathbb{R}})} \subset \mathrm{Ad}(L') = L'/Z'$ is compact. But because of the Zariski density, the intersection of the lattice $D' \subset L'$ with the center $Z' \subset L'$ is of finite index in Z'. Hence every orbit closure of the flow $(L'/D', g'_{\mathbb{R}})$ is a torus in L'/D'. But if the flow is ergodic, the space L'/D' must be a torus. Since the lattice $D' \subset L'$ cannot be a free Abelian group, we derive that $L' = 1$ and $L = \overline{MD}$.

Thus, we have proved that the weak mixing condition $L = \overline{MD}$ serves as an ergodicity criterion. To make it more effective and to avoid taking closures, we apply the decomposition of the lattice $D' = DK \cap S$ into the product of irreducible components. More specifically, there is a decomposition $S = \prod_i S_i$ into the product of normal subgroups such that $D'_i = D' \cap S_i$ is an irreducible lattice in S_i. Here the product $\prod_i D'_i$ is of finite index in D'. Let $p_i : L \to S_i$ be the corresponding projections, and let $g^i_{\mathbb{R}} = p_i(g_{\mathbb{R}})$. If every flow $(S_i/D'_i, g^i_{\mathbb{R}})$ is ergodic, then so is the original flow. Since every lattice $D'_i \subset S_i$ is irreducible, we have $\overline{M_i D'_i} = S_i \iff M_i \neq 1$, where M_i stands for the Moore normal subgroup for $g^i_{\mathbb{R}} \subset S_i$. The latter condition means that the subgroup $\mathrm{Ad}(g^i_{\mathbb{R}}) \subset \mathrm{Ad}(S_i)$ is closed and noncompact. We have deduced Moore's ergodicity criterion [**Mo66**] for homogeneous flows on semisimple spaces of finite volume. Let us extract an important particular case:

THEOREM 8.5. *Let Γ be an irreducible lattice in a semisimple group S having finite center and no compact components. Then a flow $(S/\Gamma, g_{\mathbb{R}})$ is ergodic $\iff$ the subgroup $g_{\mathbb{R}} \subset S$ is not bounded (that is, it is not contained in a compact subgroup of S).*

In contrast to solvable flows, the ergodic partitions E and $\widetilde{E}$ coincide in the semisimple case. In addition, the dimension of submanifolds $\widetilde{E}_h(g_{\mathbb{R}}) = E_h(g_{\mathbb{R}})$ may vary in $h \in G$.

Example 8.6. Let K be a compact semisimple group and let $\overline{g_{\mathbb{R}}} = D = T$, where T is a maximal torus in K. Then $\widetilde{E}_h(g_{\mathbb{R}}) = ThT \subset K/T$, and hence the dimension of the ergodic submanifold is equal to $\dim T$ on a dense open set of points $h \in K$. But on a closed null set of points $h \in K$, the dimension of the torus $ThT \subset K/T$ drops.

§9. The Mautner phenomenon, ergodicity, and mixing in the general case

1. Let G be a Lie group with connected subgroups H, F. Recall that F has the Mautner property with respect to H if $\mathrm{Fix}(H) \subset \mathrm{Fix}(F)$ for any continuous unitary representation of the group G. We will prove the Mautner phenomenon for arbitrary connected subgroups:

THEOREM 9.1 [**Mo80**]. *Let $F \subset G$ be a connected subgroup and let $M = M(F) \subset G$ be the Moore normal subgroup for F. Then M has the Mautner property with respect to F.*

First, following [**MT96**] we consider the Mautner phenomenon for a unipotent subgroup $U \subset G$. Clearly, in this case Theorem 9.1 is implied by the following assertion.

THEOREM 9.2. *Let U be a connected unipotent subgroup of a connected simply connected Lie group G. Then there exists a connected normal subgroup containing U and having the Mautner property with respect to U.*

First we need

LEMMA 9.3. *Let H be a connected Lie group with a connected simply connected unipotent normal subgroup $N \subset H$. Assume that a connected subgroup $V \subset N$ is not normal in H. Then $N - V$ contains a one-parameter subgroup having the Mautner property with respect to V.*

Proof. Pick a sequence $g_i \in H - N_H(V)$ convergent to $1 \in H$. The space N/V is diffeomorphic to a vector space, and every orbit $gVg^{-1}V \subset N/V$, $g \in H$, is closed in N/V. Let S be a sphere in N/V centered at the point $V \in N/V$. Since every orbit $g_iVg_i^{-1}V \subset N/V$ is closed and noncompact, it must intersect the sphere S in a point $s_i \in S$. Without loss of generality, $s_i \to s = aV \in S$. By Lemma 2.2, the element $a \in N - V$ has the Mautner property with respect to V. Since the element $a \in N$ can be made arbitrarily close to $1 \in N$, the maximal connected subgroup of N having the Mautner property with respect to V contains V as a proper subgroup. □

Proof of Theorem 9.2. Let $G = L \cdot R$ be a Levi decomposition and let $N \subset R$ be the nilradical of G. Without loss of generality we assume that U is the maximal connected unipotent subgroup of G having the Mautner property with respect to itself. Define $M = UR$. Since $[G, R] \subset N$, the unipotent subgroup UN is normal in M. It follows from Lemma 9.3 that the subgroup U is normal in M (otherwise $UN - U$ contains a one-parameter subgroup having the Mautner property with respect to U, which contradicts the maximality of U). In a similar fashion we prove that the intersection $U \cap N$ is a normal subgroup of G. Applying induction on the dimension of G, we can assume that $U \cap N = 1$ and hence $U \cap R = 1$.

Since U is normal in M, we derive that $M = U \times R$. Let $q : G \to L$ be the projection along R, and let V be an arbitrary one-parameter subgroup of U. By the Jacobson–Morozov lemma, $q(V)$ belongs to a connected subgroup $P \subset L$ locally isomorphic to $\mathrm{SL}(2, \mathbb{R})$. Define $H = P \cdot R$. Since the subgroup $Z_H(R)$ is normal in H and contains V, it follows that $q(Z_H(R))$ is normal in P and contains V. Consequently, $q(Z_H(R)) = P$, and hence $H = P \times R$. Let $B \subset P$ be a Borel subgroup of P containing $q(V)$. If $V \not\subset B$, we can apply Lemma 9.3 to the triple $B \times N, V \times N, V$ and find a one-parameter subgroup $V' \subset (V \times N) - V$ having the Mautner property with respect to V and hence to U. Since $U \cap N = 1$, it follows that $V' \not\subset U$, which contradicts the maximality of U. Consequently, $V \subset B$. Now applying Lemma 2.2, we find a reductive subgroup $a_{\mathbb{R}} \subset P$ having the Mautner property with respect to V and U. Let $A \subset G$ be the Auslander normal subgroup for $a_{\mathbb{R}}$. Then A has the Mautner property with respect to $a_{\mathbb{R}}, V$ and U. Now it suffices to observe that U is generated by one-parameter subgroups. □

Proof of Theorem 9.1. Since the connected subgroup $F \subset G$ is generated by one-parameter subgroups, it suffices to examine the case $F = g_{\mathbb{R}} \subset G$. First, as we noticed in §2, the problem easily reduces to the situation when the subgroup $g_{\mathbb{R}} \subset G$ is quasi-unipotent. Then the algebraic hull $\mathrm{Ad}(g_{\mathbb{R}}) \subset \mathrm{Aut}(\mathfrak{g})$ is the direct product of a torus $C \subset \mathrm{Aut}(\mathfrak{g})$ and a unipotent subgroup. Let $G' = C \cdot G$, and let $g_t = c_t \times u_t$ be a Jordan decomposition for g_t, where $c_{\mathbb{R}} \subset C$ and the subgroup $u_{\mathbb{R}} \subset G'$ is unipotent. Denote by $M = M(g_{\mathbb{R}}) \subset G$ the Moore normal subgroup for $g_{\mathbb{R}}$. Since C commutes with $g_{\mathbb{R}}$, the subgroup M is normal in G' and hence is equal to the Moore normal subgroup for $u_{\mathbb{R}} \subset G'$.

Let $\rho : G \to \mathbb{U}(\mathbb{H})$ be a continuous unitary representation. It induces a unitary representation of G' as follows. Let m be the Haar probability measure on the torus C, and let $\mathbb{H}' = L^2(C, \mathbb{H})$ be the Hilbert space of square-integrable functions $f : C \to \mathbb{H}$ equipped with the scalar product

$$(f, f') = \int_C (f(c), f'(c)) dm_c.$$

One can define a representation $\rho' : G' \to \mathbb{U}(\mathbb{H}')$ in the following way:

$$(\rho'(a)f)(c) = f(a^{-1}c), \qquad a, c \in C,$$
$$(\rho'(g)f)(c) = \rho(cgc^{-1})(f(c)), \qquad g \in G,\ c \in C.$$

Now let $v \in \mathbb{H}$ and $\rho(g_{\mathbb{R}})(v) \equiv v$. Take $f \in \mathbb{H}' : \ f \equiv v$. Since C commutes with $g_{\mathbb{R}}$, it follows that $\rho'(g_t)f \equiv v$ for all $t \in \mathbb{R}$, that is, $\rho'(g_{\mathbb{R}})(f) \equiv f$, and hence $\rho'(u_{\mathbb{R}})(f) \equiv f$. But then $\rho'(M)(f) \equiv f$, and therefore $\rho(M)v \equiv v$. Thus, $\mathrm{Fix}(g_{\mathbb{R}}) \subset \mathrm{Fix}(M)$. □

2. We return to considering a one-parameter flow $(G/D, g_{\mathbb{R}})$ on a space of finite volume.

As was already mentioned, the ergodicity criteria for nilpotent, solvable and semisimple Lie groups G were found by Green [**G**], Auslander [**Au**] and Moore [**Mo66**] respectively. In [**Da77**] Dani proved the ergodicity criterion for a general Lie group G with a lattice D. Brezin and Moore [**BM**], refining the result of Dani, proved the ergodicity criterion for homogeneous flows on the so-called admissible spaces of finite volume. The space G/D is said to be *admissible* if there exists a closed connected solvable subgroup $A \subset G$ such that A contains the radical R of the Lie group G, the space $A/D \cap A$ is compact and D normalizes A.

Clearly, if D is a lattice in G, the space G/D is admissible (one can take $A = (\overline{RD})_0$; then the compactness of $A/D \cap A$ follows from the Auslander Theorem E.8). In their paper Brezin and Moore conjectured that all spaces of finite volume are admissible. This was independently proved by many authors [**Wu, St86a, Wi87, Z87**]. Thus, the ergodicity criterion of Brezin and Moore was made complete.

The ergodicity criterion for homogeneous flows consists in reduction of the problem to the solvable and semisimple cases, which are well understood thanks to Auslander and Moore. Following [**St87a**], we outline such a reduction, which does not involve admissibility and is carried out more simply than in [**Da77**]. (Note that Dani [**Da77**] used a weaker condition of admissibility than that of [**BM**]; it is not clear whether all finite volume spaces are admissible in his sense.)

Let $G = L \cdot R$ be a Levi decomposition of a connected simply connected Lie group G, and let $q : G = L \cdot R \to L$ be the corresponding projection.

Suppose that the space $R/D \cap R$ is compact. Then the projection $q(D)$ is a closed subgroup of L, and G/D fibers over the semisimple space of finite volume $L/q(D)$ with leaves isomorphic to the compact solvable manifold $R/D \cap R$. In a sense, this allows one to reduce the study of the homogeneous space G/D to the study of semisimple and solvable homogeneous spaces.

Unfortunately, in the general case the subgroup D can fail to have a "good" intersection with the radical R. As usual, the obstacle is in the compact part of a Levi subgroup L. Let $L = S \times K$ be the decomposition into noncompact and compact parts. If $G = \overline{SRD}$, then, by Theorem E.11, the space $R/D \cap R$ is compact and we have no problems. An important role in the general ergodicity criterion is played by the following.

LEMMA 9.4 [**St87a, Wi87**]. *If a homogeneous space of finite volume G/D carries an ergodic homogeneous flow, then $G = \overline{SRD}$, and hence the space $R/D \cap R$ is compact.*

Proof. Let $(G/D, g_{\mathbb{R}})$ be an ergodic flow. Then so is the flow $(L/\overline{q(D)}, q(g_{\mathbb{R}}))$ on the semisimple homogeneous space of finite volume $L/\overline{q(D)}$. As was shown in §8, this is possible only if $L = \overline{Sq(D)}$, that is, if $G = \overline{SRD}$. □

Remark. Actually, we have proved even more: if the semisimple quotient space $L/\overline{q(D)}$ carries an ergodic homogeneous flow, then $G = \overline{SRD}$, and the space $R/D \cap R$ is compact.

Now it is not difficult to formulate and prove the general ergodicity criterion (first obtained by Dani [**Da77**] in the case when D is a lattice in G). Let G_∞ be the smallest connected subgroup normal in G and containing L (we use such a notation because the Lie subalgebra $\mathfrak{g}_\infty$ is the limit of the sequence of ideals $[\mathfrak{g}, \mathfrak{g}]$, $[[\mathfrak{g}, \mathfrak{g}], [\mathfrak{g}, \mathfrak{g}]], \dots$). It is clear that G_∞ is generated by the set

$$\operatorname{Int}_G L = \{gLg^{-1},\ g \in G\}$$

and the factor group G/G_∞ is solvable (actually, this is the maximal solvable factor group), being isomorphic to $R/R \cap G_\infty$.

THEOREM 9.5 [**Da77, St87a**]. *A flow $(G/D, g_{\mathbb{R}})$ on a homogeneous space of finite volume is ergodic if and only if the semisimple flow $(G/\overline{RD}, g_{\mathbb{R}})$ and the solvable flow $(G/\overline{G_\infty D}, g_{\mathbb{R}})$ are ergodic.*

Proof. It suffices to establish the ergodicity of the original flow provided that the semisimple and solvable quotient flows are ergodic. According to the Mautner phenomenon, it suffices to prove that the quotient flow $(G/\overline{MD}, g_{\mathbb{R}})$ is ergodic, where $M = M(g_{\mathbb{R}})$ is the Moore normal subgroup. But this follows from ergodicity of the solvable quotient flow once we prove that $G_\infty \subset \overline{MD}$.

Since $q(M) \subset L$ is the Moore normal subgroup for $q(g_{\mathbb{R}})$, it follows from the ergodicity criterion of the semisimple flow

$$(L/\overline{q(D)}, q(g_{\mathbb{R}})) \simeq (G/\overline{RD}, g_{\mathbb{R}})$$

that $L = \overline{q(M)q(D)}$, and hence $G = \overline{RMD}$. On the other hand, as we know, if $L/\overline{q(D)}$ carries an ergodic homogeneous flow, then the space $R/D \cap R$ is compact. It follows that the space $R/\overline{MD} \cap R$ is also compact. Then the product $R\overline{MD}$ is closed, and, since $G = R\overline{MD}$, there exists a Levi subgroup $L \subset G$ such that $L \subset (\overline{MD})_0$. Without loss of generality we may assume that D is a quasilattice in G. Then by Theorem E.11, $D \cap R$ is a quasilattice in the radical R, and by Mostow's Theorem E.3, the nilspace $N/D \cap N$ is compact. But since $D \cap N$ normalizes $(\overline{MD})_0$, then N normalizes $(\overline{MD})_0$ as well, and hence $\mathrm{Int}_N L = \mathrm{Int}_G L \subset (\overline{MD})_0$. Finally, we note that G_∞ is generated by the set $\mathrm{Int}_G L$, and hence $G_\infty \subset (\overline{MD})_0$. □

It is not difficult to prove (see [**BM**]) that if D is a quasilattice in G and the space G/D is homeomorphic to a solvable manifold, then the group G is solvable. This implies that the maximal solvable quotient space of G/D is unique and is none other than $G/\overline{G_\infty D}$. Now Lemma 7.9 ensures that the space G/D has a unique maximal Euclidean factor G/L. Explicitly,

$$L = \overline{G_\infty U[N_R, N_R]D},$$

where $U \subset N$ is the smallest normal subgroup such that the factor group R/U is of class (I) and N_R is the nilshadow of the radical R.

COROLLARY 9.6 [**BM**]. *A flow $(G/D, g_{\mathbb{R}})$ on a homogeneous space of finite volume is ergodic if and only if the semisimple quotient flow $(G/\overline{RD}, g_{\mathbb{R}})$ and the Euclidean quotient flow $(G/L, g_{\mathbb{R}})$ are ergodic.*

Recall that effective ergodicity criteria in the Euclidean and semisimple cases were obtained in §§7 and 8.

3. Now we prove the weak mixing criterion for a homogeneous flow $(G/D, g_{\mathbb{R}})$. For general Lie groups G it was found by Dani [**Da77**] under the assumption that D is a lattice in G. We extend this result to all homogeneous spaces G/D of finite volume.

THEOREM 9.7. *The weak mixing criterion for a flow $(G/D, g_{\mathbb{R}})$ on a homogeneous space G/D of finite volume is the condition $G = \overline{JD}$, where $J = J(g_{\mathbb{R}})$ is the Dani normal subgroup.*

Proof. The sufficiency of the condition $G = \overline{JD}$ for the mixing property follows from the Mautner phenomenon and was established in Corollary 2.4.

Suppose now that our flow is weak mixing, and hence the rotation group of the circle is not a quotient flow of it. Let H be the maximal connected subgroup of $\overline{JD}$ normal in G, and let

$$G' = G/H, \qquad D' = \overline{JD}/H \subset G', \qquad g'_{\mathbb{R}} = g_{\mathbb{R}}H/H \subset G'.$$

Then the flow $(G'/D', g'_{\mathbb{R}})$ is isomorphic to the flow $(G/\overline{JD}, g_{\mathbb{R}})$ and hence is a weak mixing quotient flow of the original flow. Let $q' : G' = L' \cdot R' \to L'$ be the projection of G' onto a semisimple Levi subgroup L'. Since the flow $(G'/D', g'_{\mathbb{R}})$ is ergodic, the projection $q'(D')$ of the quasilattice $D' \subset G'$ is a lattice in L'. But since the Dani normal subgroups for $g'_{\mathbb{R}}$ and $q(g'_{\mathbb{R}})$ are trivial, the flow $(L'/q'(D'), q(g'_{\mathbb{R}})$ is ergodic only if $L' = 1$, that is, if the group G' is solvable. But a nontrivial solvable homogeneous flow always covers the rotation group of the circle, and hence $G' = 1$ and $G = \overline{JD}$. ☐

In what follows, we need the criterion for G/D to carry a mixing flow.

THEOREM 9.8 [**Da77, St95c**]. *If $G = \overline{G_\infty D}$, then every ergodic homogeneous flow on G/D is mixing. Otherwise G/D carries no mixing homogeneous flows at all. The condition $G = \overline{G_\infty D}$ is satisfied if and only if G/D does not fiber homogeneously over the circle.*

Proof. Let $G = \overline{G_\infty D}$, let a flow $(G/D, g_{\mathbb{R}})$ be ergodic, and let H be the maximal connected subgroup of $\overline{JD}$ normal in G. As in the proof of the preceding theorem, we derive that $L \subset H$ and hence $G_\infty \subset H$. But then $H = G$ and, clearly, weak mixing takes place.

Conversely, if the solvable space $G/\overline{G_\infty \Gamma}$ is nontrivial, then G/Γ carries no mixing homogeneous flows. This is so because a compact solvable space fibers homogeneously over the circle, and no homogeneous flow on the circle can be mixing.☐

Remark. In particular, the properties of topological mixing and weak mixing for homogeneous flows are equivalent, and, as will be shown in §22, imply measure-theoretic mixing of all orders.

Let us prove a result due to Witte on the structure of homogeneous spaces admitting a mixing homogeneous flow.

THEOREM 9.9 [**Wi87**]. *If G/D carries a mixing homogeneous flow, then the quasilattice $D \subset G$ is a lattice which is Zariski dense in the* Ad*-representation of the group G. In addition, the radical of the group G is nilpotent.*

Proof. Let $G = L \cdot R$. Consider the ψ-algebraic structure on the group $\mathcal{A}(G) = (L \times \mathcal{A}(T_R)) \cdot N_R$, and denote by $q : \mathcal{A}(G) \to L$ the projection. Since $q(D) \subset L$ is a Zariski dense lattice in L, it follows that $L \subset \mathcal{A}(D)$, and hence $\mathcal{A}(D) \cap G = L \cdot (\mathcal{A}(D) \cap R)$. Then $G/\mathcal{A}(D) \cap G \simeq R/\mathcal{A}(D) \cap R$, and since G/D does not fiber over the circle, we have $G \subset \mathcal{A}(D)$. Consequently, D is Zariski dense in the Ad-representation of the group G, and hence is a lattice. By Theorem E.12, the product LND is closed (recall that Witte in essence proved the theorem under the condition $G = \overline{SRD}$, which follows from the Zariski density). Since G/LND is a torus, we derive that $G = LN$, that is, $N = R$. ☐

4. Now we apply the Mautner phenomenon to the study of the ergodic decomposition of a *unipotent* flow $(G/D, U)$. The Moore normal subgroup $M = M(U) \subset G$ for a unipotent subgroup $U \subset G$ is such that the product UM is normal in

G. According to the Mautner phenomenon, the flow $(G/D, U)$ is ergodic $\iff$ $G = \overline{UMD}$. As was shown in §2, the ergodic decomposition $\widetilde{E}$ of the space G/D is the partition into closed subsets $\widetilde{E}_g(U) = \overline{UMgD} \subset G/D$. Clearly, for a unipotent U, all $\widetilde{E}_g(U)$ are homogeneous subspaces of finite volume: $\widetilde{E}_g(U) = F(g)gD$, where $F(g) = g(\overline{UMD})g^{-1} \subset G$.

If the homogeneous flow $\widetilde{E}_g(U)$ is not ergodic relative to an $F(g)$-invariant measure, one can repeat the procedure by taking the Moore normal subgroup $M' = M'(U) \subset F(g)$ (clearly, $J \subset M' \subset M$, where $J = J(U) \subset G$ is the Dani normal subgroup for U) and considering the homogeneous subspace $\overline{UM'gD} \subset \widetilde{E}_g(U)$. This can be repeated until we get an ergodic U-flow on a homogeneous subspace of finite volume containing the point $gD \in G/D$. Clearly, the subspace obtained is none other than the ergodic component $E_g(U) = \overline{UJgD}$. One could prove directly that all the subsets $E_g(U)$ are homogeneous, but we have preferred a simpler inductive way.

So, any unipotent flow $(G/D, U)$ gives rise to the partition into homogeneous subspaces of finite volume $E_g(U) = \overline{UJgD} \subset G/D$, where all the flows $(E_g(U), U)$ are ergodic. Thus, while studying a unipotent flow it suffices to assume its ergodicity.

Reduction to the ergodic case for nonunipotent flows is more technically involved and we present it in the following section.

§10. The ergodic decomposition

1. Let $(G/D, g_{\mathbb{R}})$ be a nonergodic flow on a space G/D of finite volume. We already know that closed subsets $E_h(g_{\mathbb{R}}) = \overline{g_{\mathbb{R}}JhD}$, where J is the Dani normal subgroup for $g_{\mathbb{R}}$, form the ergodic partition E of the space G/D. Simultaneously, closed subsets $\widetilde{E}_h(g_{\mathbb{R}}) = \overline{g_{\mathbb{R}}MhD}$, where M is the Moore normal subgroup for $g_{\mathbb{R}}$, are components of the ergodic decomposition $\widetilde{E}$ of our flow, which differs from E on a null subset of G/D.

As the solvable case shows, every $E_h(g_{\mathbb{R}})$ is a smooth submanifold carrying a smooth ergodic $g_{\mathbb{R}}J$-invariant measure μ_h, and the partition E is topological (that is, manifolds $E_h(g_{\mathbb{R}})$ and $E_{h'}(g_{\mathbb{R}})$ are either disjoint, or the same). Moreover, every smooth ergodic flow $(E_h(g_{\mathbb{R}}), g_{\mathbb{R}})$ is finitely covered by an ergodic homogeneous flow. The same refers to $\widetilde{E}$, with the only difference that a smooth $g_{\mathbb{R}}M$-invariant measure $\widetilde{\mu}_h$ on the submanifold $\widetilde{E}_h(g_{\mathbb{R}})$ is ergodic only for almost all $h \in G$.

The same situation persists in the general case. However the proof gets more intricate. But if we only want to construct a compact (not necessarily finite) cover[13] of flows on manifolds $E_h(g_{\mathbb{R}})$ and $\widetilde{E}_h(g_{\mathbb{R}})$ by homogeneous flows, no special problems arise. Take, for instance, the maximal torus C in the algebraic hull of the subgroup $\mathrm{Ad}(g_{\mathbb{R}}) \subset \mathrm{Aut}\, G$. Then inside $G' = C \cdot G$ the subgroup $g_{\mathbb{R}}$ admits a decomposition $g_t = c_t \times h_t$, $t \in \mathbb{R}$, such that the subgroup $c_{\mathbb{R}}$ is dense in C, and the subgroup $h_{\mathbb{R}} \subset G'$ has only purely real eigenvalues (relative to the adjoint representation). Consequently, the image of $h_{\mathbb{R}}$ in G'/M is a central subgroup, and that in G'/J belongs to the nilradical. It follows that $F = (\overline{h_{\mathbb{R}}MD})_0$ and $F_1 = (\overline{h_{\mathbb{R}}JD})_0$ are subgroups of G'. The projection $\alpha : G' = C \cdot G \to G$ along C induces a compact cover of the flow $(G/D, g_{\mathbb{R}})$ by the flow $(G'/D, h_{\mathbb{R}})$ (since C commutes with $h_{\mathbb{R}}$).

[13] As was already mentioned in §6, we use the term "cover" in a more general setting than is accepted in topology.

Here the component $E_1(g_{\mathbb{R}})$ is compactly covered by the homogeneous space $F_1/D\cap F_1$, and the component $\widetilde{E}_1(g_{\mathbb{R}})$ by the space $F/D\cap F$. Using Lemma 7.20, it is not difficult to prove that $E_1(g_{\mathbb{R}})$ and $\widetilde{E}_1(g_{\mathbb{R}})$ are smooth manifolds.

To construct a finite-to-one cover of the ergodic flow $(E_h(g_{\mathbb{R}}), g_{\mathbb{R}})$, we involve the ψ-algebraic group $\mathcal{A}(G) = (L\times\mathcal{A}(T_R))\cdot N_R$. The strategy is to find a ψ-algebraic subgroup $B(D) \subset A(D)$ such that

(1) the product $B(D)D$ is closed,

(2) the group $B(D)/(\overline{JD})_0$ is nilpotent, and

(3) the group $\mathcal{A}(hg_{\mathbb{R}}h^{-1})/B(D)\cap\mathcal{A}(hg_{\mathbb{R}}h^{-1})$ is compact for every $h \in G$.

First we assume for simplicity that the group $\mathcal{A}(G)$ is algebraic. Then there exists a torus $C_1 \subset \mathcal{A}(g_{\mathbb{R}})$ such that $\mathcal{A}(g_{\mathbb{R}})B(D) = C_1B(D)$ and the intersection $C_1 \cap B(D)$ is finite. Let C_1^* be a copy of C_1, and let $\beta : C_1^* \times B(D) \to C_1B(D)$ be a finite-to-one cover given as

$$\beta(c^*, b) = cb, \qquad c \in C_1,\ b \in B(D),$$

with the fiber

$$\Delta_{C_1\cap B(D)} = \{c^* \times c^{-1},\ c \in C_1 \cap B(D)\}.$$

Let $f_{\mathbb{R}}(1) \subset C_1^* \times B(D)$ be a subgroup such that $\alpha(f_t(1)) = g_t$. The closure $F_1 = (\overline{f_{\mathbb{R}}(1)JD})_0$ is a subgroup of $C_1^* \times B(D)$, and we have constructed a finite-to-one cover $F_1/D\cap F_1 \to E_1(g_{\mathbb{R}})$.

Similarly one constructs each finite-to-one cover

$$(F_h/D\cap F_h, f_{\mathbb{R}}(h)) \to (E_h(g_{\mathbb{R}}), g_{\mathbb{R}}).$$

It makes sense to find how the groups F_h and $f_{\mathbb{R}}(h)$ vary in $h \in G$. Our main result is as follows.

THEOREM 10.1 [**St89**]. *Let G be a connected simply connected Lie group, and let $(G/D, g_{\mathbb{R}})$ be a homogeneous flow on a space G/D of finite volume. Then G embeds into a connected Lie group G^* as a normal subgroup such that G^*/G is a torus and G^* contains a connected subgroup F and a one-parameter subgroup $f_{\mathbb{R}}$ with the following properties:*

1) *For all $h \in G$, there exists a compact cover of flows*

$$(F/D\cap F, h^{-1}f_{\mathbb{R}}h) \to (\widetilde{E}_h(g_{\mathbb{R}}), g_{\mathbb{R}}),$$

and it is finite-to-one for almost all $h \in G$.

2) *For all $h \in G$, there exists a finite-to-one cover of flows*

$$(F_h/D\cap F_h, f_{\mathbb{R}}(h)) \to (E_h(g_{\mathbb{R}}), g_{\mathbb{R}}).$$

where F_h is a connected subgroup of F. Moreover, $F_h = F$ and $f_{\mathbb{R}}(h) = h^{-1}f_{\mathbb{R}}h$ for almost all $h \in G$.

3) *The subgroup D_0 is normal in F.*

Thus, the study of nonergodic homogeneous flows reduces to that of a family of ergodic homogeneous flows on quotient spaces of Lie groups by lattices, because $F_h/D\cap F_h = (F_h/D_0)/(D\cap F_h/D_0)$. In particular, we obtain the following result generalizing Auslander's Theorem 7.5 for solvable flows.

COROLLARY 10.2 [**St89**]. *An ergodic flow* $(G/D, g_{\mathbb{R}})$ *on a homogeneous space of finite volume is finitely covered by an ergodic flow of the form* $(H/\Gamma, f_{\mathbb{R}})$, *where* Γ *is a lattice in a Lie group* H.

Moreover, we will show that the decomposition $\widetilde{E}$ is a smooth fibration almost everywhere.

THEOREM 10.3 [**St89**]. *There exists a closed* $\widetilde{E}$*-invariant null subset* Q *of* G/D *such that the restriction of* $\widetilde{E}$ *onto* $G/D - Q$ *is a smooth fibration.*

2. We proceed to the proof of Theorems 10.1 and 10.3. As usual, we use the notation

$$q : \mathcal{A}(G) = (L \times \mathcal{A}(T_R)) \cdot N_R \to L \times \mathcal{A}(T_R) \quad \text{and} \quad Z = L \cap Z(G).$$

First let us find the required ψ-algebraic group $B(D)$. If $J = 1$, the subgroup $q(g_{\mathbb{R}})$ is quasi-unipotent and reductive, and hence $q(\mathcal{A}(g_{\mathbb{R}}))/q(\mathcal{A}(g_{\mathbb{R}})) \cap Z$ is a compact torus. Consequently, in this case the quotient space

$$\mathcal{A}(hg_{\mathbb{R}}h^{-1})/\mathcal{A}(hg_{\mathbb{R}}h^{-1}) \cap ZN_R$$

is compact for every $h \in G$. In the general case the following holds.

LEMMA 10.4. *Define* $T'(J) = \{a \in \mathcal{A}(G) : \ \{a, \mathcal{A}(G)\} \subset J\}$. *Then the space* $\mathcal{A}(hg_{\mathbb{R}}h^{-1})/\mathcal{A}(hg_{\mathbb{R}}h^{-1}) \cap T'(J)N_R$ *is compact for every* $h \in G$.

Proof. Apply Lemma 5.7 and reduce the situation to the case $J = 1$. □

LEMMA 10.5. *Define* $B(D) = T((\overline{JD})_0)N_R$, *where*

$$T((\overline{JD})_0) = \{a \in \mathcal{A}(D) : \ \{a, \mathcal{A}(D)\} \subset (\overline{JD})_0\}.$$

Then the ψ*-algebraic subgroup* $B(D) \subset \mathcal{A}(D)$ *satisfies the properties* (1)–(3).

Proof. Applying Theorem 6.15 to the subgroups $\overline{JD} \subset G$, we prove that the group $B(D)\overline{JD}$ is closed. But $\overline{JD} = (\overline{JD})_0 D$ and $(\overline{JD})_0 \subset B(D)$. It follows that $B(D)\overline{JD} = B(D)D$, and the first property is established. Let us check the second property (for $h = 1$). By definition, $T'(J) \cap \mathcal{A}(D) \subset T((\overline{JD})_0)$. According to Lemma 10.4 and Theorem 6.14, the spaces $\mathcal{A}(g_{\mathbb{R}})/\mathcal{A}(g_{\mathbb{R}}) \cap T'(J)N_R$ and $\mathcal{A}(G)/\mathcal{A}(D)$ are compact. Hence, since all the groups involved are ψ-algebraic, we prove that the space $\mathcal{A}(g_{\mathbb{R}})/\mathcal{A}(g_{\mathbb{R}}) \cap T'(J)N_R \cap \mathcal{A}(D)$ is compact, and so is $\mathcal{A}(g_{\mathbb{R}})/\mathcal{A}(g_{\mathbb{R}}) \cap B(D)$. The third property is obviously satisfied. □

Now we introduce a normal ψ-algebraic subgroup

$$B = \bigcap_{h \in G} hB(D)h^{-1} \subset \mathcal{A}(G).$$

Again, since all groups $hB(D)h^{-1}$ are ψ-algebraic, the factor group

$$\mathcal{A}(g_{\mathbb{R}})/\mathcal{A}(g_{\mathbb{R}}) \cap B$$

is compact. Using Lemma 3.11, one easily proves the following.

LEMMA 10.6. *There exists a closed null subset* $Q_1 \subset G$ *such that*

$$\mathcal{A}(g_{\mathbb{R}}) \cap hB(D)h^{-1} = \mathcal{A}(g_{\mathbb{R}}) \cap B \quad \text{for all } h \in G - Q_1.$$

COROLLARY 10.7. *There exists a connected subgroup $V \subset \mathcal{A}(g_{\mathbb{R}})$ such that*

$$h\mathcal{A}(g_{\mathbb{R}})h^{-1}B(D) = hVh^{-1}B(D),$$

where $Z \cap V$ is a finite index subgroup of the discrete subgroup $hVh^{-1} \cap B$ for every $h \in G$, and $hVh^{-1} \cap B(D) = hVh^{-1} \cap B$ for every $h \in G - Q_1^{-1}$.

Proof. The connected ψ-algebraic subgroup $V \subset \mathcal{A}(g_{\mathbb{R}})$ from Lemma 6.13 suits us perfectly. Recall that $\mathcal{A}(g_{\mathbb{R}})B = VB$, the intersection $V \cap B$ is discrete, and the factor group $Z \cap V \backslash V$ is a torus. Note that the subgroup $B \subset \mathcal{A}(G)$ is normal and the subgroups $V, B \subset \mathcal{A}(G)$ are ψ-algebraic. It now suffices to apply Lemma 10.6. □

We proceed with the construction of a cover of the flow $(\widetilde{E}_1(g_{\mathbb{R}}), g_{\mathbb{R}})$ by a homogeneous flow. Note that the Moore normal subgroup M belongs to B, because $M \subset JN_R \subset B(D)$. Define

$$\widetilde{\mathcal{E}}(g_{\mathbb{R}}) = (\overline{g_{\mathbb{R}}MD})_0 \subset G.$$

It is easy to see that $\widetilde{\mathcal{E}}(g_{\mathbb{R}}) \subset \mathcal{A}(g_{\mathbb{R}})B(D)$, because $g_{\mathbb{R}} \subset \mathcal{A}(g_{\mathbb{R}})$, $M \subset B \subset B(D)$, the subgroup $B(D)D$ is closed, and $\mathcal{A}(g_{\mathbb{R}})/\mathcal{A}(g_{\mathbb{R}}) \cap B(D)$ is a torus.

Pick the connected subgroup $V \subset \mathcal{A}(g_{\mathbb{R}})$ from Corollary 10.7 and a copy V^* of it. All the events to follow will take place in the new Lie group

$$\Delta_{Z\cap V}\backslash(V^* \times \mathcal{A}(G)),$$

where $\Delta_{Z\cap V}$ is the "anti-diagonal" in $(Z \cap V)^* \times (Z \cap V)$. Consider a smooth fibration

$$\alpha : \Delta_{Z\cap V}\backslash(V^* \times \mathcal{A}(G)) = (\Delta_{Z\cap V}\backslash\Delta_V) \cdot \mathcal{A}(G) \to \mathcal{A}(G)$$

with the compact fiber $\Delta_{Z\cap V}\backslash\Delta_V = \widehat{\Delta}_V$, where Δ_V is the "anti-diagonal" in $V^* \times V$. The fibration α is well defined by the formula

$$\alpha(\Delta_{Z\cap V}(v^*, h)) = vh \quad \text{for all } v \in V,\ h \in \mathcal{A}(G),$$

because $\Delta_V \cap \mathcal{A}(G) = 1$.

The restriction $\alpha : \Delta_{Z\cap V}\backslash(V^* \times B) \to VB$ has the finite fiber $\Delta_{Z\cap V}\backslash\Delta_{V\cap B} = \widehat{\Delta}_{V\cap B}$, which commutes with the one-parameter subgroup $g_{\mathbb{R}} \subset \mathcal{A}(g_{\mathbb{R}}) \subset VB$. Hence there exists a subgroup $f_{\mathbb{R}} \subset \Delta_{Z\cap V}\backslash V^* \times B$ such that $\alpha(f_t) = g_t$ for all $t \in \mathbb{R}$. Moreover, the cover

$$\alpha : \Delta_{Z\cap V}\backslash(V^* \times B(D)) \to VB(D)$$

is equivariant relative to the right action of $B(D)$ and sends the left action of $f_{\mathbb{R}}$ to the left action of $g_{\mathbb{R}}$. Thus, we have obtained a cover of flows

$$(\Delta_{Z\cap V}\backslash(V^* \times B(D))/D \cap B(D), f_{\mathbb{R}}) \to (\mathcal{A}(g_{\mathbb{R}})B(D)/D \cap B(D), g_{\mathbb{R}})$$

with the compact fiber $\Delta_{Z\cap V}\backslash(\Delta_V \cap (V^* \times B(D))) = \Delta_{Z\cap V}\backslash\Delta_{V\cap B(D)}$.

According to Lemma 10.6, the factor group $B(D)/(\overline{MD})_0$ is nilpotent. On the other hand, by construction of M, the image of $g_{\mathbb{R}}$ in G/M is reductive. It follows that the subgroup

$$f_{\mathbb{R}}(\overline{MD})_0/(\overline{MD})_0$$

is central in the group $\Delta_{Z\cap V}\backslash(V^* \times B(D))/(\overline{MD})_0$. Hence the closed connected set

$$F = (\overline{f_{\mathbb{R}}MD})_0 \subset \Delta_{Z\cap V}\backslash(V^* \times B(D))$$

is a Lie group. Note that $\alpha(F) = \widetilde{\mathcal{E}}(g_{\mathbb{R}})$, and we have constructed a composite cover of flows

$$(F/D \cap F, f_{\mathbb{R}}) \to (\widetilde{\mathcal{E}}(g_{\mathbb{R}})/D \cap F, g_{\mathbb{R}}) \to (\widetilde{E}_1(g_{\mathbb{R}}), g_{\mathbb{R}})$$

with the fiber

$$(\widehat{\Delta}_V \cap F) \times (D \cap \widetilde{\mathcal{E}}(g_{\mathbb{R}})/D \cap F).$$

Here, according to Lemma 7.2, the intersection $D \cap \widetilde{\mathcal{E}}(g_{\mathbb{R}})$ is a group which contains $D \cap F$ as a finite index subgroup.

In a completely similar fashion, by considering a fibration

$$\alpha_h : \Delta_{Z\cap V}\backslash(V^* \times \mathcal{A}(G)) \to \mathcal{A}(G)$$

with the fiber $h\widehat{\Delta}_{V\cap B(D)}h^{-1}$, we obtain a cover of flows

$$(F/D \cap F, hf_{\mathbb{R}}h^{-1}) \to (\widetilde{E}_1(hg_{\mathbb{R}}h^{-1}), hg_{\mathbb{R}}h^{-1})$$

with the compact fiber

$$(h\widehat{\Delta}_V h^{-1} \cap F) \times (D \cap \widetilde{\mathcal{E}}(hg_{\mathbb{R}}h^{-1})/D \cap F).$$

The only comment to be made here is that $F = \overline{(hf_{\mathbb{R}}h^{-1}MD)}_0$ for all $h \in G$. To show this, note that $\{f_{\mathbb{R}}, \mathcal{A}(G)\} \subset B$, because $f_{\mathbb{R}} \subset (\Delta_{Z\cap V})\backslash V^* \times B$ and B is a normal subgroup of $\mathcal{A}(G)$. In addition, $\{f_{\mathbb{R}}, B\} \subset (\overline{MD})_0$, as we observed above. Since the image of $f_{\mathbb{R}}$ in $\mathcal{A}(G)/M$ is reductive, we conclude that $\{f_{\mathbb{R}}, \mathcal{A}(G)\} \subset (\overline{MD})_0$, and the equality in question becomes obvious.

We have shown that all

$$\widetilde{E}_h(g_{\mathbb{R}}) \simeq E_1(h^{-1}g_{\mathbb{R}}h) \simeq (h^{-1}\widehat{\Delta}_V h \cap F)\backslash F/D \cap \widetilde{\mathcal{E}}(h^{-1}g_{\mathbb{R}}h)$$

are smooth submanifolds. The F-invariant probability measure on $F/D \cap F$ gives rise to a smooth $g_{\mathbb{R}}M$-invariant measure $\widetilde{\mu}_h$ on $\widetilde{E}_h(g_{\mathbb{R}})$. Since the group $F/(\overline{MD})_0$ is Abelian, the quotient flow on $F/\overline{MD} \cap F$ is minimal. It follows that under the factorization $G/D \to G/\overline{MD}$, every flow $(\widetilde{E}_h(g_{\mathbb{R}}), g_{\mathbb{R}})$ covers a minimal flow. Hence any two submanifolds of the family $\widetilde{E}_h(g_{\mathbb{R}}), h \in G$, are either disjoint or the same. Consequently, the family forms a topological partition $\widetilde{E}$ of the space G/D.

To prove that the partition $\widetilde{E}$ is smooth almost everywhere, one must understand how the compact fiber of the cover $F/D \cap F \to \widetilde{E}_1(hg_{\mathbb{R}}h^{-1})$ varies in $h \in G$. To do this, note that

$$h\Delta_V h^{-1} \cap (V^* \times B(D)) = \Delta_{hVh^{-1}\cap B(D)} = \Delta_{hVh^{-1}\cap B} = h\Delta_{V\cap B}h^{-1}$$

for all $h \in G - Q_1^{-1}$, and hence

$$h\widehat{\Delta}_V h^{-1} \cap F = h(\widehat{\Delta}_{V\cap B} \cap h^{-1}Fh)h^{-1}$$

also for all $h \in G - Q_1^{-1}$. Using Lemma 3.11, it is easy to prove the following result.

LEMMA 10.8. *Define* $F^* = \bigcap_{h\in G} hFh^{-1}$. *Then there exists a closed null subset* $Q_2 \subset G$ *such that*

$$\widehat{\Delta}_{V\cap B} \cap hFh^{-1} = \widehat{\Delta}_{V\cap B} \cap F^* \quad \textit{for all } h \in G - Q_2.$$

By similar arguments one can prove the following.

LEMMA 10.9. *Define*

$$D(g_{\mathbb{R}}, h) = D \cap \widetilde{\mathcal{E}}(hg_{\mathbb{R}}h^{-1}) \quad \text{and} \quad D(g_{\mathbb{R}}) = \bigcap_{h\in G} D(g_{\mathbb{R}}, h).$$

Then there exists a closed null subset $Q_3 \subset G$ *such that* $D(g_{\mathbb{R}}, h) = D(g_{\mathbb{R}})$ *for all* $h \in G - Q_3$.

COROLLARY 10.10. *For all* $h \in G - (Q_1^{-1} \cup Q_2^{-1} \cup Q_3)$, *the fiber of the cover* $F/D \cap F \to \widetilde{E}_1(hg_{\mathbb{R}}h^{-1})$ *is finite and is equal to*

$$h(\widehat{\Delta}_{V\cap B} \cap F^*)h^{-1} \times (D(g_{\mathbb{R}})/D \cap F).$$

Thus, we have proved Theorem 10.3 and the part of Theorem 10.1 which concerns the structure of the ergodic decomposition $\widetilde{E}$. In fact, define $G^* = \Delta_{Z\cap V}\backslash(V^* \times G)$. Then G is a normal subgroup of G^*, $F = (\overline{f_{\mathbb{R}}MD})_0 \subset G^*$, and G^*/G is a torus because $G^*/G \simeq Z \cap V\backslash V$. In addition, the identity component D_0 is normal in F because $F \subset \Delta_{Z\cap V}\backslash(V^* \times \mathcal{A}(D))$. A smooth $g_{\mathbb{R}}$-invariant measure $\widetilde{\mu}_g$ on every $\widetilde{E}_h(g_{\mathbb{R}})$ is obtained from the same smooth F-invariant measure on $F/D \cap F$.

3. Now we turn to the ergodic partition E. As usual, it suffices to consider the ergodic flow $(E_1(hg_{\mathbb{R}}h^{-1}), hg_{\mathbb{R}}h^{-1})$. Fix $h \in G$, and let V_h be a connected ψ-algebraic subgroup of $\mathcal{A}(g_{\mathbb{R}})$ such that

$$h\mathcal{A}(g_{\mathbb{R}})h^{-1}B(D) = hV_hh^{-1}B(D)$$

and $Z \cap V_h$ is a finite index subgroup of the discrete group $hV_hh^{-1} \cap B(D)$. Lemma 10.6 ensures that $V_h \subset V$, $h \in G$, and $V_h = V$ for almost all $h \in G$. Taking a copy V_h^* of the connected Abelian group V_h and the "anti-diagonal" $\Delta_{Z\cap V_h}$ in $(Z \cap V_h)^* \times (Z \cap V_h)$, we obtain a topological fibration

$$\beta_h : \Delta_{Z\cap V_h}\backslash(V_h^* \times \mathcal{A}(G)) \to \mathcal{A}(G)$$

with the compact fiber $\Delta_{Z\cap V_h}\backslash h\Delta_{V_h}h^{-1}$. The fiber of β_h commutes with the subgroup $hg_{\mathbb{R}}h^{-1}$, and hence there exists a one-parameter subgroup

$$f_{\mathbb{R}}(h) \subset \Delta_{Z\cap V_h}\backslash(V_h^* \times B(D))$$

such that $\beta_h(f_t(h)) = hg_th^{-1}$ for all $t \in \mathbb{R}$. Since $V_h = V$ for almost all $h \in G$, it follows that $f_{\mathbb{R}}(h) = hf_{\mathbb{R}}h^{-1}$ for almost all $h \in G$.

Define

$$\mathcal{E}(hg_{\mathbb{R}}h^{-1}) = (\overline{hg_{\mathbb{R}}h^{-1}JD})_0 \subset G, \qquad F_h = (\overline{f_{\mathbb{R}}(h)JD})_0 \subset \Delta_{Z\cap V_h}\backslash(V_h^* \times B(D)).$$

In contrast to $\mathcal{E}(hg_{\mathbb{R}}h^{-1})$, the connected set F_h is a subgroup. The explanation is that by Lemma 10.5, the factor group $V_h^* \times B(D)/(\overline{JD})_0$ is nilpotent, and orbit closures of nilflows are always homogeneous. Since $\beta_h(F_h) = \mathcal{E}(hg_{\mathbb{R}}h^{-1})$, we have constructed a cover of flows

$$\beta_h : (F_h/D \cap F_h, f_{\mathbb{R}}(h)) \to (\mathcal{E}(hg_{\mathbb{R}}h^{-1})/D \cap F_h, hg_{\mathbb{R}}h^{-1}).$$

Since the restriction $\beta_h : \Delta_{Z\cap V_h}\backslash(V_h^* \times B(D)) \to hV_hh^{-1}B(D)$ has the finite fiber $\Delta_{Z\cap V_h}\backslash h\Delta_{V_h\cap B(D)}h^{-1}$, the map $\beta_h : F_h/D \cap F_h \to E_1(hg_{\mathbb{R}}h^{-1})$ is a finite-to-one cover with the fiber

$$(\Delta_{Z\cap V_h}\backslash h\Delta_{V_h}h^{-1} \cap F_h) \times (D \cap \mathcal{E}(hg_{\mathbb{R}}h^{-1})/D \cap F_h).$$

It follows that all $E_h(g_{\mathbb{R}})$ are submanifolds. For different $h \in G$, they either coincide or are disjoint (because the quotient flow of any ergodic flow $(E_h(g_{\mathbb{R}}), g_{\mathbb{R}})$ under the fibration $G/D \to G/\overline{JD}$) is minimal), and hence they form a topological partition E of the space G/D.

Note that

$$F_h \subset F = (\overline{hf_{\mathbb{R}}h^{-1}MD})_0 \subset \Delta_{Z\cap V}\backslash(V^* \times B(D)),$$

because $V_h^* \subset V^*$, $\Delta_{Z\cap V} \cap (V_h^* \times B(D)) = \Delta_{Z\cap V_h}$, and hence the restrictions of β_h and α_h onto $\Delta_{Z\cap V_h}\backslash(V_h^* \times B(D))$ coincide.

All we have to prove now is that $F_h = F$ for almost all $h \in G$. The space $F/\overline{JD} \cap F$ is compact, since the factor group $F/(\overline{JD})_0$ is nilpotent and $F/D \cap F$ is a space of finite volume. The Lie group F contains at most countably many proper subgroups $F(i) \subset F$ such that $(\overline{JD})_0 \subset F(i)$ and the space $F(i)/D \cap F(i)$ is compact. Every

$$Q^i = \{h \in G: \ hf_{\mathbb{R}}h^{-1} \subset F(i)\}$$

is a closed null subset of G. In fact, otherwise $Q^i = G$ by Lemma 3.11, and in this case $M \subset F(i)$ because the Moore normal subgroup of G^* for $f_{\mathbb{R}}$ is equal to M. But then

$$F(i) \supset (\overline{hf_{\mathbb{R}}h^{-1}MD})_0 = F,$$

which is a contradiction because $F(i)$ is a proper subgroup of F. Hence if

$$Q = \{h \in G: \ (\overline{hf_{\mathbb{R}}h^{-1}JD})_0 \neq F\},$$

then $Q \subset \bigcup_i Q^i$, and therefore Q is a null subset of G.

On the other hand, $f_{\mathbb{R}}(h) = hf_{\mathbb{R}}h^{-1}$ for almost all $h \in G$. This means that $F_h = F$ for almost all $h \in G$ as well. Note that the Dani normal subgroup of F_h for $f_{\mathbb{R}}(h)$ is equal to J. Hence every flow $(F_h/D \cap F_h, f_{\mathbb{R}}(h))$ is ergodic by the Mautner phenomenon (because $F_h = (\overline{f_{\mathbb{R}}(h)JD})_0 \subset G^*$). This yields Theorems 10.1 and 10.3. □

4. One could prove more directly (see [**St83**]) that all the components $E_h(g_{\mathbb{R}}) = \overline{g_{\mathbb{R}}JhD} \subset G/D$ are smooth submanifolds equipped with a smooth ergodic measure μ_h. It follows from Lemma B.4 that almost all orbits of the restriction of the flow onto $E_h(g_{\mathbb{R}})$ are dense. On the other hand, the G-invariant measure ν on G/D is glued from the measures $\{\mu_h, \ hD \in G/D\}$. This implies

COROLLARY 10.11 [**St83**]. *If $(G/D, g_{\mathbb{R}})$ is a homogeneous flow on a space of finite volume, then for almost all $hD \in G/D$, the closure $\overline{g_{\mathbb{R}}hD} \subset G/D$ is a smooth submanifold (moreover, it coincides with $E_h(g_{\mathbb{R}})$).*

§11. Spectral theory of homogeneous flows

1. To start with, we give some preliminaries from the spectral theory of operators on a complex separable Hilbert space $\mathbb{H}$ (which may by finite-dimensional); see also [**AGH, Kir, GGP**]. Any self-adjoint operator $A : \mathbb{H} \to \mathbb{H}$ is completely reducible and all its eigenvalues are real. For $\dim \mathbb{H} < \infty$ this means that, given a basis of eigenvectors, the operator A is written by a diagonal matrix with real entries. This provides the decomposition of $\mathbb{H}$ into irreducible components. In the case $\dim \mathbb{H} = \infty$, it also makes sense to find the decomposition into irreducible components (recall that all of them are one-dimensional by the Schur lemma). However, the family of the components may be uncountable, and one has to be careful. More specifically, one has to introduce the concept of the direct integral of Hilbert spaces.

Let μ be a Lebesgue–Stieltjes measure on $\mathbb{R}$, and let a Hilbert space $\mathbb{H}_\lambda$ (possibly, zero-dimensional) be assigned for every $\lambda \in \mathbb{R}$. One says that a Hilbert space $\mathbb{H}$ is the *direct integral* of spaces $\mathbb{H}_\lambda$ by measure μ (this is written as $\mathbb{H} = \int_{\mathbb{R}} \mathbb{H}_\lambda d\mu_\lambda$), if $\mathbb{H}$ is the space of measurable functions f on $\mathbb{R}$ such that for every $\lambda \in \mathbb{R}$, the value f_λ is an element of $\mathbb{H}_\lambda$, and the norm $\|f\|$ is finite relative to the scalar product

$$(f, f') = \int_{\mathbb{R}} (f_\lambda, f'_\lambda)_\lambda d\mu_\lambda,$$

where $(\cdot, \cdot)_\lambda$ is the scalar product in $\mathbb{H}_\lambda$. If the measure μ is discrete, then in essence $\mathbb{H}$ is a finite or countable orthogonal sum of its subspaces.

An example of such a decomposition of Hilbert space can be obtained via the Fourier transform. More precisely, let the measure μ be supported at integer points $n \in \mathbb{Z}$, each of them having measure 1. Take the Hilbert space $\mathbb{H} = L^2(I)$ of functions on the segment $I = [0, 2\pi]$ which are square-integrable relative to the usual Lebesgue measure dx. Then every function $f \in L^2(I)$ decomposes into the (convergent in mean square) series

$$(*) \qquad f(x) = \sum_{n\in\mathbb{Z}} f_n(x), \qquad \text{where } f_n(x) = c_n e^{inx} = \frac{e^{inx}}{2\pi} \int_0^{2\pi} e^{-ins} f(s) ds$$

and c_n is the n-th Fourier coefficient of the function f. Here $\mathbb{H}_\lambda \simeq \mathbb{C}$ for all $\lambda \in \mathbb{Z}$. A continuous sum of Hilbert spaces can be obtained in a similar fashion. Let $\mathbb{H} = L^2(\mathbb{R})$ and let μ be the usual Lebesgue measure on $\mathbb{R}$. Then every function $f \in L^2(\mathbb{R})$ decomposes into the integral

$$(**) \qquad f(x) = \int_{\mathbb{R}} f_\lambda(x) d\lambda, \qquad \text{where } f_\lambda(x) = \frac{e^{i\lambda x}}{2\pi} \int_{\mathbb{R}} e^{-i\lambda s} f(s) ds.$$

Now we can formulate the theorem on spectral decomposition for self-adjoint operators.

THEOREM 11.1. *Given any self-adjoint operator A on $\mathbb{H}$, there exists a decomposition $\mathbb{H} = \int_{\mathbb{R}} \mathbb{H}_\lambda d\mu_\lambda$ into the direct integral of Hilbert spaces by a Lebesgue-Stieltjes measure μ on $\mathbb{R}$ such that if $f = \int_{\mathbb{R}} f_\lambda d\mu_\lambda$, then*

$$Af = \int_{\mathbb{R}} \lambda f_\lambda d\mu_\lambda, \qquad f \in \mathbb{H}.$$

Here the operator A defines the equivalence class of the measure μ uniquely.

The type of the measure μ on $\mathbb{R}$ (that is, its equivalence class) is called the *maximal spectral type* of A. One says that A has an *absolutely continuous* (*Lebesgue*) spectrum if the measure μ is absolutely continuous (equivalent) with respect to the Lebesgue measure on $\mathbb{R}$. If the measure is discrete or singular, the same refers to the spectrum. The function $n(\lambda) = \dim \mathbb{H}_\lambda$ is called the *multiplicity function.* If $n(\lambda) < n_0 < \infty$ for μ-almost all λ, the spectrum is said to be of *finite multiplicity.* If $n(\lambda) = n_0$ for μ-almost all λ, the spectrum is said to be *homogeneous of multiplicity* n_0 (if $n_0 = \infty$ one talks about a *homogeneous spectrum of countable multiplicity*).

In general, the measure μ decomposes into the sum of a discrete measure, a singular measure, and an absolutely continuous measure. The (measure-theoretic) support of the discrete component is denoted by $\Lambda_d(A)$ and is called the *discrete component of the spectrum*. Clearly, $\Lambda_d(A)$ consists of $\lambda \in \mathbb{R}$ such that there exists a nonzero element $f \in \mathbb{H}$ with $Af = \lambda f$.

Using the maximal spectral type measure μ, one can easily define the so-called *projection measure* P. This "measure" is defined on the σ-algebra $\mathcal{B}(\mathbb{R})$ of Borel subsets of $\mathbb{R}$ and takes values in the space $\mathrm{End}(\mathbb{H})$ of endomorphisms of the Hilbert space $\mathbb{H}$. By definition,

$$P(B)h = \int_B h_\lambda d\mu_\lambda \quad \text{for every Borel } B \subset \mathbb{R}.$$

Note that P only depends on the equivalence class of the measure μ, and is uniquely determined by the operator A. Here are the main properties of the operators $P(B)$:

1) $P(B)$ is self-adjoint for every Borel $B \subset \mathbb{R}$.
2) $P(B_1 \cap B_2) = P(B_1)P(B_2)$ for any Borel $B_1, B_2 \subset \mathbb{R}$.
3) $P(B_1 \cup B_2) = P(B_1) + P(B_2)$ for any disjoint Borel $B_1, B_2 \subset \mathbb{R}$.
4) If $B_n \to B$ (that is, if $B = \bigcap_n \bigcup_{k \geq n} B_k = \bigcup_n \bigcap_{k \geq n} B_k$), then $P(B_n) \to P(B)$ in the weak topology.

In terms of the projection measure it is convenient to formulate the following criterion: a self-adjoint operator A has Lebesgue spectrum $\iff$ the operator $P(B)$ is trivial only if the Borel set B is of zero Lebesgue measure.

Let $\mathbb{H} = \int_{\mathbb{R}} \mathbb{H}_\alpha d\nu_\alpha$ and $A = \int_{\mathbb{R}} A_\alpha d\nu_\alpha$. Suppose that, for ν-almost all $\alpha \in \mathbb{R}$, the spectrum of A_α is Lebesgue. Then A has Lebesgue spectrum as well. This is immediate from the following remark: the projection measure $P(B)$ of the operator A is expressed via the projection measures $P_\alpha(B)$ of the operators A_α by the formula $P(B) = \int_{\mathbb{R}} P_\alpha(B) d\nu_\alpha$.

Note that if a Hilbert space $\mathbb{H}$ decomposes into the orthogonal sum of two invariant Hilbert subspaces $\mathbb{H} = \mathbb{H}_1 \oplus \mathbb{H}_2$, then $\mu = \mu_1 + \mu_2$, where μ_i is the maximal spectral type measure of $A|_{\mathbb{H}_i}$, $i = 1, 2$. Hence, whenever an absolutely continuous spectrum contains a Lebesgue component, the spectrum is Lebesgue itself.

2. Now let $U : \mathbb{H} \to \mathbb{H}$ be a unitary operator. Such an operator is also completely reducible. All of its eigenvalues belong to the unit circle $\mathbb{C}_1 = \{z \in \mathbb{C} : |z| = 1\}$ and hence are of the form $e^{i\lambda}$, $\lambda \in I = [0, 2\pi]$. In this case the theorem on spectral decomposition is as follows:

THEOREM 11.2. *Given any unitary operator U on $\mathbb{H}$, there exists a decomposition $\mathbb{H} = \int_I \mathbb{H}_\lambda d\mu_\lambda$ into the direct integral of Hilbert spaces by a Lebesgue-Stieltjes measure μ on $I = [0, 2\pi]$ such that $Uf = \int_I e^{i\lambda} f_\lambda d\mu_\lambda$, $f \in \mathbb{H}$.*

Let the maximal spectral type measure μ of a unitary operator U on $\mathbb{H}$ be Lebesgue, that is, $d\mu_\lambda / d\lambda = s(\lambda)$ for some positive function $s \in L^1(I)$. Pick an element $h = \int_I h_\lambda d\mu_\lambda$ and define

$$h' = \int_I \frac{h_\lambda}{|h_\lambda| \sqrt{s(\lambda)}} d\mu_\lambda$$

(if $h_\lambda = 0$, we set $\frac{h_\lambda}{|h_\lambda|} = 0$). Then $U^n h' \perp h'$, $n \in \mathbb{N}$ and hence the sequence $\{U^n h',\ n \in \mathbb{Z}\}$ is orthogonal.

Conversely, if for some vector $h \in \mathbb{H}$, the sequence $\{U^n h,\ n \in \mathbb{Z}\}$ is orthogonal, then the measure $d\nu_\lambda = |h_\lambda|^2 d\mu_\lambda$ satisfies $\int_I e^{in\lambda} d\nu_\lambda = 0$ for all $n \in \mathbb{N}$. Hence, up to a constant multiplier, the measure ν is the usual Lebesgue measure (because the functionals $f \to \int_I f d\nu_\lambda$ and $f \to \int_I f d\lambda$ coincide up to a constant multiplier on the space of continuous functions on I such that $f(0) = f(1)$). Thus, the restriction of μ onto the *cyclic subspace* $C(h) \subset \mathbb{H}$ (that is, the smallest closed subspace containing all of $U^n h,\ n \in \mathbb{Z}$) is a Lebesgue measure. If $C(h) = \mathbb{H}$, the spectrum is Lebesgue of multiplicity 1. If $\mathbb{H} = \bigoplus_{i\in\mathbb{N}} C(h_i)$, where $U^n h_i \perp h_i,\ i, n \in \mathbb{N}$, the spectrum is Lebesgue of countable multiplicity.

The classical example of a Lebesgue spectrum of countable multiplicity is an ergodic toral automorphism $A \in \operatorname{Aut}(\mathbb{T}^n)$ (more precisely, the associated unitary operator $a : L^2(\mathbb{T}^n) \to L^2(\mathbb{T}^n)$ defined by $af(\overline{x}) = f(A\overline{x}),\ \overline{x} \in \mathbb{T}^n$). As we said in §4, the space $L^2(\mathbb{T}^n)$ has the orthogonal basis of the functions

$$u_I(\overline{x}) = e^{2\pi i (I,x)}, \qquad \text{where } x \in \mathbb{R}^n,\ \overline{x} = x \ (\operatorname{mod} \mathbb{Z}^n) \in \mathbb{T}^n \text{ and } I \in \mathbb{Z}^n.$$

Then $a^n u_I = u_{(A^\intercal)^n I}$, and hence $a^n u_I \perp u_I$ for all $n \neq 0,\ I \neq 0$. Clearly, $\mathbb{Z}^n - \{0\}$ breaks into countably many infinite $A^\intercal$-orbits, and hence the operator a on $L^2(\mathbb{T}^n) \ominus \mathbb{C}$ has Lebesgue spectrum of countable multiplicity.

Finally, let us consider a continuous group of unitary operators $U_t : \mathbb{H} \to \mathbb{H}$. By the Stone theorem, there exists a self-adjoint operator $A : \mathbb{H} \to \mathbb{H}$ (in general, unbounded) such that $U_t = e^{itA}$ (here A is called the *generator* of the group U_t). This formula can be understood in the sense that

$$iAh = \frac{dU_t h}{dt}\Big|_{t=0}$$

for all $h \in \mathbb{H}$ for which the right-hand side is defined. By the spectrum of U_t one understands the spectrum of the generator A. It makes sense to give an analogue of Theorem 11.1 for the group U_t.

THEOREM 11.3. *Given a group U_t of unitary operators on $\mathbb{H}$, there exists a decomposition $\mathbb{H} = \int_{\mathbb{R}} \mathbb{H}_\lambda d\mu_\lambda$ into the direct integral of Hilbert spaces by a Lebesgue-Stieltjes measure μ on $\mathbb{R}$ such that $U_t f = \int_{\mathbb{R}} e^{it\lambda} f_\lambda d\mu_\lambda,\ f \in \mathbb{H}$.*

Let us give the simplest examples of discrete and Lebesgue spectra. If $U_t f = e^{it\lambda} f$ for all $f \in \mathbb{H}$, then the measure μ is supported at the point λ, and $\mathbb{H} = \mathbb{H}_\lambda$. Now take the group of translation operators on $L^2(\mathbb{R})$:

$$U_t f(x) = f(x + t), \qquad t, x \in \mathbb{R}.$$

In the decomposition (**), the operator U_t is written as follows:

$$(U_t f)(x) = \int_{\mathbb{R}} e^{it\lambda} f_\lambda(x) d\lambda.$$

Hence the maximal spectral type measure of U_t is the usual Lebesgue measure, and U_t has a homogeneous Lebesgue spectrum of multiplicity 1. Note that $U_t = e^{itA}$, where $iA = \frac{d}{dt}$ is the differentiation operator on $L^2(\mathbb{R})$.

More generally, let $\psi : \mathbb{R} \to \mathbb{R}$ be a locally integrable function, and let

$$Bf = \psi f + \frac{1}{i}\frac{df}{dt}.$$

Then the self-adjoint operator B is unitarily equivalent to the operator $\frac{1}{i}\frac{d}{dt}$ (via multiplication by the function $a(t) = e^{-i\int_0^t \psi(s)ds}$). Hence B has Lebesgue spectrum as well.

3. Let $(X, \mu, g_{\mathbb{R}})$ be a continuous flow on a probability space. By the spectrum of the flow one understands the spectrum of the group of unitary operators $U_t f(x) = f(g_{-t}x)$ on the Hilbert space $L^2(X,\mu) \ominus \mathbb{C}$ of square-integrable functions with zero average. Clearly, the flow is ergodic iff the point $\lambda = 0$ does not occur in the discrete component of the spectrum (that is, there are no invariant functions in $L^2(X,\mu) \ominus \mathbb{C}$), and it is weak mixing iff there is no discrete component at all. It is also known that the mixing property follows from the absolute continuity of the spectrum.

Let us proceed to the study of the spectrum of a homogeneous flow $(G/\Gamma, g_{\mathbb{R}})$ with respect to a G-invariant finite measure. Here the situation is simplified by the fact that the unitary representation of $g_{\mathbb{R}}$ in $L^2(G/\Gamma)$ is contained in the *regular* representation of G given by the formula

$$\rho(g)f(h\Gamma) = f(g^{-1}h\Gamma), \qquad g, h \in G, \quad f \in L^2(G/\Gamma).$$

Any unitary representation is completely reducible, and we have the theorem on the decomposition into irreducible components.

THEOREM 11.4 (Kolmogorov–Mautner; see [**AGH**]). *Given any unitary representation ρ of a locally compact second countable group G in a separable Hilbert space $\mathbb{H}$, there exists a Lebesgue-Stieltjes measure μ on $\mathbb{R}$ such that $\mathbb{H}$ is the direct integral $\mathbb{H} = \int_{\mathbb{R}} \mathbb{H}_x d\mu_x$ of Hilbert spaces $\mathbb{H}_x$ with unitary representations ρ_x of the group G on $\mathbb{H}_x$, where $\rho(g)h = \int_{\mathbb{R}} \rho_x(g)h_x d\mu_x$, and, for μ-almost all $x \in \mathbb{R}$, the representation ρ_x is irreducible.*

Surely, to compute the spectrum one needs to decompose the regular representation of G on $L^2(G/\Gamma)$ into irreducible components. It is worth noting that the direct integral into irreducible representations degenerates to an infinite discrete sum whenever the space G/Γ is compact [**GGP**].

We start with an ergodic rectilinear flow $(\mathbb{T}^n, \mathbb{R}v)$ on the torus, acting by translations along the vector $v = (v_1, \dots, v_n) \in \mathbb{R}^n$. We take the basis in $L^2(\mathbb{T}^n)$ consisting of the functions

$$u_I(\overline{x}) = e^{2\pi i(I,x)}, \qquad \text{where } I \in \mathbb{Z}^n,\ x \in \mathbb{R}^n,\ \overline{x} = x \ (\operatorname{mod} \mathbb{Z}^n).$$

Any function $f \in L^2(\mathbb{T}^n)$ decomposes into the series

$$f(\overline{x}) = \sum_{I \in \mathbb{Z}^n} c_I e^{2\pi i(I,x)}, \qquad \text{where } c_I = \frac{1}{(2\pi)^n} \int_{\mathbb{T}^n} f(\overline{y}) e^{-2\pi i(I,y)} d\overline{y}.$$

Then

$$f(\overline{x} + tv) = \sum_{I \in \mathbb{Z}^n} e^{2\pi i(I,tv)} c_I e^{2\pi i(I,x)}.$$

Now it is clear that the basis consists of eigenfunctions, and that the spectrum of the flow is discrete and is the additive subgroup $\Lambda_d = (\mathbb{Z}^n, v)$ (note that since the vector v has rationally independent coordinates, the formula $I \to \lambda = (I, v)$ determines a one-to-one correspondence $\mathbb{Z}^n \to \Lambda_d$, and hence the basis in $L^2(\mathbb{T}^n)$ can be enumerated by eigenvalues $\lambda \in \Lambda_d$).

4. Let us turn to the nilpotent case. An irreducible unitary representation of a nilpotent Lie group N is either one-dimensional (that is, it is a multiplicative character $\rho : N \to \mathbb{C}_1$), or infinite-dimensional.

To give an example of an infinite-dimensional representation, we have to define cocycles. Let M be a manifold equipped with a smooth action of a group N. One says that a C^∞-map $\beta : N \times M \to \mathbb{C}_1$ is a *multiplicative C^∞-cocycle* relative to the action N on M if

$$\beta(g_1 g_2, x) = \beta(g_1, g_2 x)\beta(g_2, x), \qquad g_1, g_2 \in N,\ x \in M.$$

Note that $\beta(1, x) \equiv 1$.

In our situation, a homogeneous space N/H equipped with an N-invariant measure m will play the role of the manifold M. We need the following result.

THEOREM 11.5 (Takenouchi; see [**AGH**]). *Let U be an irreducible unitary infinite-dimensional representation of a connected nilpotent Lie group N. Then there exist a connected subgroup $H \subset N$ and a multiplicative C^∞-cocycle $\beta : N \times N/H \to \mathbb{C}_1$ such that U is equivalent to the representation in the space $L^2(N/H, m)$ given by*

$$(U_g f)(x) = \beta(g, x) f(gx), \qquad g \in N,\ x \in N/H.$$

As a rule, the theorem is formulated in a shorter way: every irreducible representation of N is induced (in the sense of Mackey) from a one-dimensional representation of a connected subgroup $H \subset N$ (see [**AGH**] or [**Kir**] for the definition of the induced representation in a more general setup).

LEMMA 11.6. *Let U be an irreducible unitary infinite-dimensional representation of a connected nilpotent Lie group N. Suppose that $g_\mathbb{R} \subset N$ is not a subgroup of the kernel of U. Then the restriction of U onto $g_\mathbb{R}$ has a Lebesgue spectrum.*

Proof. We will use the explicit form of U given by Theorem 11.5. Clearly, the kernel of U is a normal subgroup of the form

$$\operatorname{Ker}(U) = \{g \in H :\ ngn^{-1} \in H,\ n \in N\}.$$

All $g_\mathbb{R}$-orbits in N/H are closed, and hence, by Glimm's theorem, the space X of the partition of N/H into $g_\mathbb{R}$-orbits is a Lebesgue one. In other words,

$$L^2(N/H, m) = \int_X \mathbb{H}_x d\mu_x,$$

where the measure μ on X is the image of the measure m (and the measure space (X, μ) is isomorphic to the space $\mathbb{R}$ equipped with a Lebesgue-Stieltjes measure). Moreover, the measure m is glued from $g_\mathbb{R}$-invariant measures on orbits.

Since $g_\mathbb{R} \not\subset \operatorname{Ker}(U)$, the set of $g_\mathbb{R}$-fixed points in N/H is a proper analytic submanifold (see Lemma 3.11), and all the $g_\mathbb{R}$-orbits outside the submanifold are noncompact. Since the measure m on N/H is smooth, then μ-almost every point $x \in X$ corresponds to a noncompact $g_\mathbb{R}$-orbit. For a noncompact orbit, the corresponding space $\mathbb{H}_x$ is isomorphic to $L^2(\mathbb{R}, l_x)$, and up to a constant multiplier the measure l_x is the usual Lebesgue measure. Moreover, the representation of g_t on $\mathbb{H}_x = L^2(\mathbb{R}, l_x)$ is given by

$$U^x_{g_t} \phi(s) = \beta_x(g_t, s)\phi(s + t)$$

with a smooth multiplicative cocycle $\beta_x : g_{\mathbb{R}} \times \mathbb{R} \to \mathbb{C}_1$ (we applied a diffeomorphism between a noncompact $g_{\mathbb{R}}$-orbit and the real line $\mathbb{R}$ which sends the g_t-translation to the t-translation).

Note that $U^x_{g_t} = e^{itA_x}$, where the self-adjoint generator A_x acts as follows:

$$A_x f = \psi_x f + \frac{1}{i}\frac{df}{dt}.$$

Here ψ_x is a C^∞-smooth real function; explicitly,

$$\psi_x(s) = \left.\frac{d\beta_x(g_t, s)}{dt}\right|_{t=0}.$$

Hence the spectrum of $g_{\mathbb{R}}$ on H_x is Lebesgue (see §2). Since almost all $g_{\mathbb{R}}$-orbits are noncompact, the restriction of U onto $g_{\mathbb{R}}$ also has Lebesgue spectrum. □

Let us find the spectrum of an ergodic nilflow $(N/\Gamma, g_{\mathbb{R}})$. By $q : N/\Gamma \to T = N/[N,N]\Gamma$ we denote the fibration over the torus T. The space $\mathbb{H} = L^2(N/\Gamma)$ decomposes into the direct sum $\mathbb{H} = \mathbb{H}' \oplus \mathbb{H}''$, where $\mathbb{H}'$ consists of functions $f \in L^2(N/\Gamma)$ that are constant along the fibers of q, and $\mathbb{H}''$ is the orthogonal complement to $\mathbb{H}'$.

THEOREM 11.7 [**AGH, Ste69**]. *The spectrum of an ergodic nilflow on H' is discrete, and one on H'' is Lebesgue of infinite multiplicity.*

Proof. Clearly, $\mathbb{H}'$ can be identified with $L^2(T)$. Hence all irreducible subrepresentations on $\mathbb{H}'$ are one-dimensional, and the spectrum of $g_{\mathbb{R}}$ on $\mathbb{H}'$ is discrete.

On the other hand, since every one-dimensional representation is trivial on $[N,N]$, the space $\mathbb{H}''$ decomposes into the discrete sum of N-irreducible infinite-dimensional components (the discreteness follows from the compactness of N/Γ). Because of ergodicity, the space H'' contains no $g_{\mathbb{R}}$-invariant functions, and hence the action of $g_{\mathbb{R}}$ on each irreducible component is nontrivial. It follows that the spectrum of $g_{\mathbb{R}}$ on H'' is Lebesgue.[14]

Now we compute the multiplicity of the Lebesgue spectrum following [**Ste69**]. If there are infinitely many N-irreducible components in the space H'', the spectrum is clearly countable. Let $z_{\mathbb{R}}$ be a one-parameter subgroup of the center $Z(N)$. Since on every N-irreducible space the action of z_t commutes with that of N, it is a multiplication by a multiplicative character. If the number of components were finite, the self-adjoint operator $\frac{dU(z_t)}{dt}|_{t=0}$ would be bounded on $H'' = L^2(N/\Gamma) \ominus L^2(T)$. On the other hand, this is the operator of differentiation along $z_{\mathbb{R}}$, and hence it cannot be bounded. This contradiction shows that the spectrum of $g_{\mathbb{R}}$ on $\mathbb{H}''$ is Lebesgue of countable multiplicity. □

The fact that the spectrum of a nilflow is the sum of a discrete spectrum and a Lebesgue spectrum of countable multiplicity can be illustrated by the following

[14] Our proof that the spectrum is Lebesgue differs essentially from the original proof of Green [**AGH**]. We applied the general form of irreducible representations of the group N given by Takenouchi's Theorem 11.5. Instead, Green utilized a particular case of the construction (see Corollary 3.2 in [**AGH**]). However, the description used in [**AGH**] of all irreducible unitary representations of N is incomplete: for instance, it implies that *every* such representation restricted onto $[N,N]$ is the multiplication by a multiplicative character, which is far from true. An alternative proof which does not involve the classification of unitary representations of nilpotent groups was suggested by Parry [**P70**].

simple example. Let

$$N = \mathrm{U}(3,\mathbb{R}), \qquad \Gamma = N_{\mathbb{Z}} = \mathrm{U}(3,\mathbb{Z}), \qquad a_t = \begin{pmatrix} 1 & tA & tB + \frac{t^2}{2}AC \\ 0 & 1 & tC \\ 0 & 0 & 1 \end{pmatrix}.$$

Then for $C \neq 0$ the flow $(N/\Gamma, a_t)$ is the standard suspension over an affine diffeomorphism α of the torus $\mathbb{T}^2 = M/M_{\mathbb{Z}}$, where the Abelian normal subgroup $M \subset N$ consists of matrices of the form $\begin{pmatrix} 1 & a & b \\ 0 & 1 & 0 \\ 0 & 0 & 1 \end{pmatrix}$. It is not difficult to prove that

$$\alpha(\overline{x}) = \sigma(\overline{x}) + v, \quad \text{where } \sigma = \begin{pmatrix} 1 & 0 \\ -1 & 1 \end{pmatrix}, \; v = \begin{pmatrix} A/C \\ B/C - A/2C \end{pmatrix}, \; \overline{x} \in \mathbb{T}^2.$$

Take the orthogonal basis

$$\{u_I(\overline{x}) = e^{2\pi i (I,x)}, \; I \in \mathbb{Z}^2, \; x \in \mathbb{R}^2\}$$

in $\mathbb{H} = L^2(\mathbb{T}^2)$. Then $\alpha(u_I) = u_I(v)u_{\sigma^\intercal(I)}$. We see that the Hilbert space $\mathbb{H}$ decomposes into the orthogonal sum $\mathbb{H} = \mathbb{H}' \oplus \mathbb{H}''$, where $\mathbb{H}'$ is spanned by basis functions u_I, $I \in \mathbb{Z} \times \{0\}$, that are constant in the second coordinate, and $\mathbb{H}''$ is spanned by the remaining functions. Clearly, the spectrum of α in $\mathbb{H}'$ is discrete. It is easy to verify that $(\sigma^\intercal)^k(I) \neq I$, and hence $\alpha^k(u_I) \perp u_I$ if $u_I \in \mathbb{H}''$, $k \in \mathbb{N}$. It follows that the spectrum of α in $\mathbb{H}''$ is Lebesgue. Since $\mathbb{Z}^2 - (\mathbb{Z} \times \{0\})$ breaks into countably many $\sigma^\intercal$-orbits, the spectrum of α in $\mathbb{H}''$ is of countable multiplicity.

Since the original flow $(N/\Gamma, a_t)$ is the standard suspension over $\alpha : \mathbb{T}^2 \to \mathbb{T}^2$, its spectrum is also the sum of a discrete spectrum and a Lebesgue spectrum of countable multiplicity (we omit the details).

5. Now we turn to the general case. To start with, we show how to reduce the spectral study of a homogeneous flow $(G/\Gamma, g_{\mathbb{R}})$ to that of the maximal quotient flow of zero entropy $(G/\overline{A\Gamma}, g_{\mathbb{R}})$, where as usual $A \subset G$ stands for the Auslander normal subgroup for $g_{\mathbb{R}}$. Surprisingly, the problem reduces to finding irreducible unitary representations of two elementary solvable groups. More specifically, let S_1 be the two-dimensional group of affine transformations of the real line: $S_1 = \mathbb{R} \cdot N$, where $\mathbb{R}$ acts on the nilradical $N \simeq \mathbb{R}$ by dilations (that is, S_1 consists of pairs (a, b), $a, b \in \mathbb{R}$, with the product $(a, b)(a', b') = (a + a', e^a b' + b)$. The group can also be realized as the group of real matrices $\begin{pmatrix} e^a & b \\ 0 & 1 \end{pmatrix}$. The group N is the unique proper normal subgroup of S_1. An obvious series of unitary representations of S_1 arises as follows: these are one-dimensional unitary representations of the factor group $\mathbb{R} = S_1/N$: $\rho_\lambda(a, b) = e^{i\lambda a}$ for $\lambda \in \mathbb{R}$. There are two more infinite-dimensional representations:

THEOREM 11.8 (Gelfand–Naimark; see [**AGH**]). *Any irreducible unitary representation of the group S_1 which is not one-dimensional is equivalent to one of the two representations*

$$\rho^{\pm}(a, b)f(x) = \exp(\pm i e^x b) f(x + a)$$

on the space $L^2(\mathbb{R})$.

Every one-parameter subgroup of S_1 either is N or is conjugate to $\mathbb{R}$. Thus, essentially there are two such subgroups: $\{g_t = (t, 0)\}$ and $\{u_s = (0, s)\}$. Take a unitary representation ρ of S_1 on $\mathbb{H}$ and let μ be the maximal spectral type measure for the restriction of our representation onto $u_{\mathbb{R}}$. Since μ is a Lebesgue-Stieltjes measure, the projection measure $P : \mathcal{B}(\mathbb{R}) \to \mathrm{End}(\mathbb{H})$ is the weak limit of linear combinations of the operators $\rho(u_s)$. Hence

$$\rho(a_t)P(B) = P(e^t B)\rho(a_t) \quad \text{for every Borel } B \subset \mathbb{R}.$$

If $\mu(B) = 0$ for some Borel $B \subset \mathbb{R}$, then $P(B) \equiv 0$. Since $|\rho(a_t) - I| \to 0$, it follows that $P(e^t B) \equiv 0$ for small enough t, and hence $P(e^t B) \equiv 0$ for all $t \in \mathbb{R}$. Consequently, the measure μ is sent to an equivalent measure under the map $s \to e^t s$, $s \in \mathbb{R}$. This means that μ is a linear combination of three measures – the Lebesgue measure on $(-\infty, 0)$, the atomic measure δ_0, and the Lebesgue measure on $(0, +\infty)$.

Let us consider the spectrum of the subgroup $g_{\mathbb{R}}$. Theorem 11.8 implies

LEMMA 11.9 (Mautner; see [**AGH**]). *Let ρ be a unitary representation of the group S_1 on a Hilbert space $\mathbb{H}$ and let* $\mathrm{Fix}(N) \subset \mathbb{H}$ *be the closed subspace of elements in $\mathbb{H}$ fixed under N. Then the restriction of $g_{\mathbb{R}}$ to* $\mathrm{Fix}(N)^{\perp} = \mathbb{H} \ominus \mathrm{Fix}(N)$ *has Lebesgue spectrum.*

Proof. Decompose ρ into the direct integral of irreducible representations of the group S_1: $\mathbb{H} = \int_{\mathbb{R}} \mathbb{H}_\lambda d\mu_\lambda$. Let $\mathbb{H}'$ be the sum of all one-dimensional representations, and $\mathbb{H}''$ the sum of all infinite-dimensional. Clearly, $\mathbb{H}' \subset \mathrm{Fix}(N)$. Take $f \in \mathbb{H}''$, $f \neq 0$, and let $\lambda \in \mathbb{R}$ correspond to an infinite-dimensional representation. Then

$$\rho_\lambda(0, b) f_\lambda(x) = e^{\pm i e^x b} f_\lambda(x),$$

and hence $\mathrm{Fix}(N) = \mathbb{H}'$. Finally,

$$\rho_\lambda(t, 0) f_\lambda(x) = f_\lambda(x + t),$$

and hence the spectrum of $g_{\mathbb{R}}$ on $\mathbb{H}'' = \mathrm{Fix}(N)^{\perp}$ is Lebesgue. □

Now consider the three-dimensional solvable Lie group $S_2 = \mathbb{R} \cdot N$, where the action of $\mathbb{R}$ on $N \simeq \mathbb{C}$ is a composite of a dilation and an orthogonal rotation. In other words, S_2 consists of pairs (t, w), $t \in \mathbb{R}$, $w \in \mathbb{C}$, with the product

$$(t, w)(t', w') = (t + t', e^{(1-i\alpha)t} w' + w)$$

for some $\alpha \in (0, 2\pi)$. The subgroup $N = \mathbb{C}$ is the unique proper normal subgroup of S_2, and obvious irreducible unitary representations of S_2 arise as one-dimensional representations of the factor group $\mathbb{R} = S_2/N$. Similarly to Theorem 11.8, one proves

THEOREM 11.10 [**AGH**]. *Any irreducible unitary representation of the group S_2 which is not one-dimensional is equivalent to a representation of the form*

$$\rho^r(t, w) f(x) = \exp(ir \operatorname{Re}(e^{(1-i\alpha)x} w)) f(x + t)$$

on the space $L^2(\mathbb{R})$, where $r \in [1, 2\pi/\alpha)$.

Every one-parameter subgroup of S_2 either belongs to $N = \mathbb{C}$ or is conjugate to $\mathbb{R}$. Pick $g_t = (t, 0)$. Similarly to Lemma 11.9, Theorem 11.10 implies

LEMMA 11.11 [**AGH**]. *Let U be a unitary representation of the group S_2 on a Hilbert space $\mathbb{H}$, and let $\operatorname{Fix}(N) \subset \mathbb{H}$ be the closed subspace of elements in $\mathbb{H}$ fixed under N. Then the spectrum of $g_{\mathbb{R}}$ on $\operatorname{Fix}(N)^{\perp}$ is Lebesgue.*

Now we are ready to prove the following:

THEOREM 11.12 [**AG**]. *Consider a unitary representation of the Lie group G on a Hilbert space $\mathbb{H}$, and let $A \subset G$ be the Auslander normal subgroup for a one-parameter subgroup $g_{\mathbb{R}} \subset G$. Then the restriction of $g_{\mathbb{R}}$ to the subspace $\operatorname{Fix}(A)^{\perp} \subset \mathbb{H}$ has Lebesgue spectrum.*

Proof. The Lie algebra $\mathfrak{a} \subset \mathfrak{g}$ of the normal subgroup A is generated by subalgebras $\mathfrak{g}^{\lambda}$, $|\lambda| \neq 1$, where λ is an eigenvalue of the operator Ad_{g_1} on $\mathfrak{g}_{\mathbb{C}}$. Denote by $G^{\lambda} \subset G$ the connected subgroup corresponding to $\mathfrak{g}_{\lambda}$.

Suppose that λ is a real number. Clearly, G^{λ} is a nilpotent subgroup and has a filtration invariant under $g_{\mathbb{R}}$:

$$G^{\lambda} = H_0 \supset H_1 \supset \cdots \supset H_{n+1} = 1,$$

in which $\dim(H_i/H_{i+1}) = 1$ and the operator Ad_{g_1} on H_i/H_{i+1} is the multiplication by λ (that is, $g_{\mathbb{R}}$ and H_i/H_{i+1} generate the group S_1). It follows that the spectrum of $g_{\mathbb{R}}$ on $\operatorname{Fix}(H_n)^{\perp}$ is Lebesgue (because $g_{\mathbb{R}}$ and H_n generate S_1). Note that

$$\operatorname{Fix}(H_{n-1})^{\perp} = \operatorname{Fix}(H_n)^{\perp} \oplus (\operatorname{Fix}(H_n) \cap \operatorname{Fix}(H_{n-1})^{\perp}).$$

The action of the group generated by $g_{\mathbb{R}}$ and H_{n-1} on the second space reduces to the action of the group generated by $g_{\mathbb{R}}$ and H_{n-1}/H_n, that is, again to the action of S_1. Hence the spectrum of $g_{\mathbb{R}}$ on the second space is also Lebesgue. Now by induction one easily proves that the spectrum of $g_{\mathbb{R}}$ on $\operatorname{Fix}(G^{\lambda})^{\perp}$ is Lebesgue.

If λ is not a real number, then every group H_i/H_{i+1} together with $g_{\mathbb{R}}$ generates a group isomorphic to S_2. Quite similarly to the first case, we show that the spectrum of $g_{\mathbb{R}}$ on $\operatorname{Fix}(G^{\lambda})^{\perp}$ is Lebesgue. Since A is generated by all G^{λ}, then $\operatorname{Fix}(A)^{\perp}$ is the closed linear hull of all $\operatorname{Fix}(G^{\lambda})^{\perp}$. □

Note that the theorem implies the well known fact that every eigenvector in $\mathbb{H}$ for $g_{\mathbb{R}}$ is A-invariant (which is a particular case of the Mautner phenomenon).

Now let us decompose the Hilbert space $\mathbb{H} = L^2(G/\Gamma)$ into the direct sum $\mathbb{H} = \operatorname{Fix}(A) \oplus \operatorname{Fix}(A)^{\perp}$. Here $\operatorname{Fix}(A)$ is the smallest closed subspace of functions in $L^2(G/\Gamma)$ invariant under A, and hence this is the space of functions constant along the fibers of the fibration $G/\Gamma \to G/\overline{A\Gamma}$, that is, constant along the Pinsker partition π for the flow $(G/\Gamma, g_{\mathbb{R}})$. We already know that the spectrum of $g_{\mathbb{R}}$ on $\operatorname{Fix}(A)^{\perp}$ is Lebesgue. In our situation, when $\mathbb{H} = L^2(G/\Gamma)$, it is also of countable multiplicity, as is seen from the following result due to Sinai:

THEOREM 11.13 [**Si66, P69**]. *Let $(X, \mu, g_{\mathbb{R}})$ be a continuous flow on a probability space, and let π be the Pinsker partition of X. Then the restriction of the flow to $\operatorname{Fix}(\pi)^{\perp} \subset L^2(X,\mu)$ has Lebesgue spectrum of countable multiplicity.*

This result generalizes a well known theorem due to Sinai saying that a K-flow has Lebesgue spectrum of countable multiplicity, and is proved by entropy considerations (not involving the theory of group representations).

6. Thus, we have reduced the spectral theory of homogeneous flows to the case of zero entropy. Now we can claim that the spectrum of an ergodic flow on a solvable manifold is in general the sum of a discrete spectrum and a Lebesgue spectrum of countable multiplicity. This is so because an ergodic solvable flow of zero entropy is isomorphic to a nilflow.

Following [**Mo66**] and [**Ste73**], we will study the case when G is a semisimple group and $(G/\Gamma, g_{\mathbb{R}})$ is an ergodic flow of zero entropy, that is, the subgroup $g_{\mathbb{R}} \subset G$ is quasi-unipotent. Without loss of generality we assume that the center $Z(G)$ is finite. By the Jacobson–Morozov lemma, there exist subgroups $a_{\mathbb{R}}$, $c_{\mathbb{R}}$, $u_{\mathbb{R}} \subset G$ such that $g_t = c_t \times u_t$ is a Jordan decomposition into semisimple and unipotent parts, $a_t u_s a_{-t} = u_{e^t s}$ (that is, $a_{\mathbb{R}}$ is diagonal for $u_{\mathbb{R}}$) and $a_{\mathbb{R}}$ commutes with $c_{\mathbb{R}}$. Note that the unipotent flow $(G/\Gamma, u_{\mathbb{R}})$ is ergodic and $(G/\Gamma, a_{\mathbb{R}})$ is a K-flow. As regards $(G/\Gamma, c_{\mathbb{R}})$, it is uniformly continuous and its ergodic decomposition is formed by orbits of the torus $T = \overline{c_{\mathbb{R}}}$.

The subgroups $a_{\mathbb{R}}$ and $u_{\mathbb{R}}$ generate a group isomorphic to S_1. Almost all irreducible components in the decomposition of the representation of S_1 on $L^2(G/\Gamma)$ are infinite-dimensional, because $u_{\mathbb{R}}$ induces an ergodic flow. We know that the restriction of an infinite-dimensional representation of the group S_1 to $a_{\mathbb{R}}$ has Lebesgue spectrum, and the restriction to $u_{\mathbb{R}}$ has Lebesgue spectrum on $(-\infty, 0)$ or on $(0, +\infty)$. Let $n^{\mp}$ be the multiplicity of the corresponding "semi-Lebesgue" measure in the spectrum of $(G/\Gamma, u_{\mathbb{R}})$. Since $a_{\mathbb{R}}$ induces a K-flow with Lebesgue spectrum of countable multiplicity, we have $n^- + n^+ = \infty$. The complex conjugation operator on $L^2(G/\Gamma)$ commutes with any unitary operator and permutes the "semi-Lebesgue" components, whence $n^- = n^+ = \infty$. Thus, $(G/\Gamma, u_{\mathbb{R}})$ also has Lebesgue spectrum of countable multiplicity.

The flow $(G/\Gamma, c_{\mathbb{R}})$ has discrete spectrum, and hence the space $\mathbb{H} = L^2(G/\Gamma)$ decomposes into the direct sum of subspaces $\mathbb{H} = \bigoplus_i \mathbb{H}_i$, each of which corresponds to an eigenvalue $\lambda_i \in \mathbb{R}$. Since the subgroups $c_{\mathbb{R}}$ and $u_{\mathbb{R}}$ commute, all the spaces $\mathbb{H}_i$ are invariant under $u_{\mathbb{R}}$. Let μ_i be the maximal spectral type measure for the restriction of $u_{\mathbb{R}}$ onto $\mathbb{H}_i$. Since μ is a Lebesgue measure, every measure μ_i is absolutely continuous. The shift $\widetilde{\mu}_i$ of the measure μ_i by λ_i is also absolutely continuous, and is the maximal spectral type of the restriction of $g_{\mathbb{R}}$ on $\mathbb{H}_i$. Thus, the maximal spectral type measure $\widetilde{\mu}$ for $g_{\mathbb{R}}$ on $\mathbb{H}$ is absolutely continuous (we cannot ensure at once that it is Lebesgue). On the other hand, the flow $(T\backslash G/\Gamma, a_{\mathbb{R}})$ is a K-flow, and the actions of $u_{\mathbb{R}}$ and $g_{\mathbb{R}}$ on $T\backslash G/\Gamma$ coincide. It follows that the flow $(T\backslash G/\Gamma, g_{\mathbb{R}})$ has Lebesgue spectrum of countable multiplicity. Since $L^2(T\backslash G/\Gamma)$ is a closed subspace of $\mathbb{H} = L^2(G/\Gamma)$, we have obtained a Lebesgue component of countable multiplicity of the absolutely continuous measure $\widetilde{\mu}$. It follows that the flow $(G/\Gamma, g_{\mathbb{R}})$ has Lebesgue spectrum of countable multiplicity.

7. We have completed the study of the spectrum of an ergodic homogeneous flow in the solvable and semisimple cases. To reduce the general case to these cases, we have to utilize the following generalization of the Mautner phenomenon

THEOREM 11.14 [**Mo80**]. *Let ρ be a unitary representation of a Lie group G on a Hilbert space $\mathbb{H}$, and let $M \subset G$ be the Moore normal subgroup for a one-parameter subgroup $g_{\mathbb{R}} \subset G$. Then the restriction of $g_{\mathbb{R}}$ to the subspace* $\mathrm{Fix}(M)^{\perp} \subset \mathbb{H}$ *has absolutely continuous spectrum.*

We are not able to reproduce the proof here. We only note that the subgroup M is constructed in three steps: $A \subset J \subset M$, and hence the proof breaks into three parts. First, the spectrum of $g_{\mathbb{R}}$ on $\mathrm{Fix}(A)^{\perp}$ is Lebesgue (we have already proved this). Second, suppose that $A = 1$, that is, the subgroup $g_{\mathbb{R}}$ is quasi-unipotent. One can show that it suffices to handle the case when $g_{\mathbb{R}} \subset G$ is a unipotent subgroup. Dani in [**Da77**] proved that the spectrum of $g_{\mathbb{R}}$ on $\mathrm{Fix}(J)^{\perp}$ is absolutely continuous (this is the most technically involved part of the proof, which allowed Dani to find the ergodicity criterion for homogeneous flows in the general case). Finally, let the subgroup $g_{\mathbb{R}}$ be unipotent and $J = 1$; then $g_{\mathbb{R}}$ belongs to the nilradical $N \subset G$. Now one can play the same game as in the nilpotent case to prove that the spectrum of $g_{\mathbb{R}}$ on $\mathrm{Fix}(M)^{\perp}$ is Lebesgue.

Now we can prove that in the general case the spectrum of an ergodic homogeneous flow $(G/\Gamma, g_{\mathbb{R}})$ is the sum of a discrete spectrum and a Lebesgue spectrum of countable multiplicity (where one of the components may be missing). The space $\mathrm{Fix}(M)$ is isomorphic to $L^2(G/\overline{M\Gamma})$, and since the flow $(G/\overline{M\Gamma}, g_{\mathbb{R}})$ is isomorphic to a rectilinear flow on the torus, the spectrum of $g_{\mathbb{R}}$ on $\mathrm{Fix}(M)$ is discrete. It suffices to find a Lebesgue component of countable multiplicity on $\mathrm{Fix}(M)^{\perp}$. Consider the projection $q : G = L \cdot R \to L$ and the quotient flow $(L/q(R\Gamma), g_{\mathbb{R}})$ (recall that Γ is assumed to be a quasilattice, and, in view of ergodicity, the product $R\Gamma$ is closed). If $L \neq 1$, the quotient flow has Lebesgue spectrum of countable multiplicity, and we are done. Otherwise, the group G is solvable, and this case has been also examined.

It suffices to find conditions under which discrete and Lebesgue components occur in the spectrum. It follows from Theorem 11.14 that the Lebesgue component appears whenever $M \neq 1$. The condition $M = 1$ combined with ergodicity means that the group G is Euclidean. As for the discrete component, it appears whenever $G \neq \overline{M\Gamma}$, that is, whenever the mixing condition fails to hold. Now we can summarize our results as follows.

THEOREM 11.15 [**BM**]. *The spectrum of an ergodic homogeneous flow is the sum of a discrete spectrum and a Lebesgue soectrum of countable multiplicity. The discrete component is trivial if and only if the flow is mixing. The Lebesgue component is trivial if and only if the flow is uniformly continuous.*

CHAPTER 2

Dynamics of Unipotent Flows

As was already mentioned, the best-known representative of unipotent flows is the *horocycle flow* on (the unit tangent bundle of) a surface of constant curvature -1. Already in the thirties Hedlund discovered that, for a surface of finite area, a horocycle orbit is either dense or periodic, and the latter exist only on a noncompact surface. Under the compactness condition, the horocycle flow is minimal and, moreover, uniquely ergodic (Furstenberg [**Fu72**]).[1] A noncompact surface carries no ergodic measures other than the smooth volume measure and those supported on periodic horocycles (Dani and Smillie [**DS**]; a short proof is given in [**R92**]).

The results of Hedlund and Furstenberg were refined by Bowen [**Bo76**], Veech [**V75,77**], and Ellis and Perrizo [**EP**]. They proved that any *horospherical flow* on a compact homogeneous space G/Γ is minimal and uniquely ergodic whenever it is ergodic (see §13). Subsequently Dani [**Da86a**] proved that all orbit closures of a horospherical flow on a semisimple space G/Γ of finite volume are homogeneous subspaces. If one does not require the group G to be be semisimple, but instead assumes that the space G/Γ is compact, then all orbits of a horospherical flow look alike: their closures are isomorphic and coincide with the corresponding ergodic homogeneous subspaces (Starkov, [**St91**]), see §13.

On the other hand, any ergodic nilflow is minimal (Auslander, [**AGH**]) and uniquely ergodic (Furstenberg [**Fu61**]); we have already proved this in §3. Moreover, we have already established that all orbit closures and ergodic measures for a unipotent flow on a solvable manifold of finite volume are algebraic[2] (most likely, the results for nilflows were already known in the sixties to Parry, although he did not formulate them explicitly).

The results enable one to conjecture that all orbit closures and *locally finite* ergodic measure of a unipotent flow are algebraic. The first conjecture is known as the *Raghunathan topological conjecture*, and the second will be called the *Dani measure conjecture* (sometimes it is referred to as the Raghunathan measure conjecture). The interest in such statements rose considerably when in the mid-seventies Raghunathan discovered a remarkable connection between unipotent flows and solution of Diophantine inequalities for quadratic forms. More specifically, if one manages to prove that compact orbit closures of the subgroup

$$u_t = \begin{pmatrix} 1 & t & t^2/2 \\ 0 & 1 & t \\ 0 & 0 & 1 \end{pmatrix}$$

---

[1] For a more modern proof of the unique ergodicity, which is valid for the uniformly parametrized horocycle flow on a surface of variable negative curvature, see Ratner's paper [**R92**].

[2] It is of interest that orbit closures of such flows are homogeneous, and finite ergodic measures are algebraic even if one omits the finite volume condition (see [**St87b**] and [**R90b**]).

on $\mathrm{SL}(3,\mathbb{R})/\mathrm{SL}(3,\mathbb{Z})$ are homogeneous subspaces, then a well-known conjecture of Oppenheim and Davenport in number theory follows (see §28). The main difficulty here is that $u_{\mathbb{R}}$ *is not a horospherical subgroup* of G (though it is contained in one). Exactly in this connection Dani formulated in [**Da81**] the conjecture on the algebraicity of orbit closures and locally finite ergodic measures for a unipotent flow $(G/\Gamma, U)$ on a finite volume space.[3] The most general formulation for an arbitrary subgroup U generated by unipotent elements is due to Margulis [**Ma88,91b**].

The main events in the study of unipotent flows were as follows. In the late eightis Margulis published a proof of the Oppenheim–Davenport conjecture in [**Ma87,89**], giving a thorough treatment of compact orbit closures of the flow $(\mathrm{SL}(3,\mathbb{R})/\mathrm{SL}(3,\mathbb{Z}), u_{\mathbb{R}})$. Then the result was sharpened in joint papers of Dani and Margulis [**DM89,90a,b**]. In particular, in [**DM90b**] the Raghunathan conjecture was proved for any lattice $\Gamma \subset \mathrm{SL}(3,\mathbb{R})$ and any regular unipotent subgroup $u_{\mathbb{R}} \subset G$ (this means that the matrix $u_t - I$ is of rank 2). Thus, for the first time one managed to study orbits of a nonhorospherical flow on a semisimple homogeneous space. The method of [**DM90b**] involved finding orbits of larger subgroups inside $\overline{u_{\mathbb{R}}x}$ by studying $u_{\mathbb{R}}$-orbits near $u_{\mathbb{R}}$-minimal subsets $Y \subset \overline{u_{\mathbb{R}}x}$.

This approach seemed likely to lead to the proof of the Raghunathan conjecture in the general case. However, success was reached much faster than might have been expected. Namely, in the early nineties, in a series of papers [**R90b,90c,91a**] Ratner proved the measure conjecture, and in [**R91b**] also the topological conjecture (simultaneously Shah [**Sh91**] proved it for regular unipotent subgroups of a semisimple group of $\mathbb{R}$-rank 1). The measure theorem turned out to be central; in particular, it is crucial for the proof of the topological theorem. It is important to emphasize that the measure theorem was proved in a more general setting than it was conjectured: for *arbitrary* discrete subgroups Γ and *finite* ergodic measures (as will be shown in §12, a locally finite ergodic measure for a unipotent action on a finite volume space is necessarily finite). A new way to prove the measure conjecture (by using some ideas due to Ratner, as well as those of earlier papers of Dani and Margulis) was suggested by Margulis and Tomanov [**MT94,96**]. Earlier, Dani and Margulis [**DM93**] gave a new proof to the topological conjecture.

A few words about the structure of the chapter. Most likely, the first important tool for the proof of the Raghunathan conjecture was obtained before it was formulated in [**Da81**]; this is a very nontrivial theorem of Margulis [**Ma71**] that *no unipotent trajectory in* $\mathrm{SL}(n,\mathbb{R})/\mathrm{SL}(n,\mathbb{Z})$ *exits to infinity.* Later this result was sharpened by Dani [**Da84,86b**], who eventually proved that any unipotent trajectory comes back to a compact subset, and the time it spends there has a positive density. We consider these questions in §12.

On the other hand, in the early eighties Ratner [**R82a**] discovered remarkable (and surprising even for experts) results on *rigidity of the horocycle flow*: a measure-theoretic isomorphism of two horocycle flows coincides almost everywhere with an affine map (that is, has an algebraic origin). The main new ingredient for this is now known as the *R-property* of the horocycle flow, which describes the divergence of its close trajectories (later, precisely this property was intensively used by Ratner for the proof of the Raghunathan and Dani conjectures). We describe these results in

[3] Originally, the conjectures were formulated only for reductive G and one-parameter U. In addition, the formulation of the topological conjecture in [**Da81**] did not require that the closure of a unipotent trajectory be a homogeneous space of *finite volume*.

greater detail. Before that, in §14 we briefly present preliminaries from hyperbolic geometry and consider the dynamics of geodesic and horocycle flows.

The theorem on the measure rigidity of the horocycle flow is stated in §15; also we discuss the *algebraicity* of its *quotients and joinings* along the lines of [**R82a,b,83**]. In §18 we show how the results are generalized to the class of arbitrary unipotent flows (here the main contribution is due to Witte, who established in [**Wi85**] the measure-theoretic rigidity of unipotent ergodic flows, and in [**Wi94**] studied the structure of their quotient flows). Note that all the properties for nilflows were obtained earlier by Parry [**P71,73**]). Also we show that these results are based on the Dani measure conjecture (and hence they are naturally involved in our discussion).

The proof of the measure conjecture is still too complicated to reproduce within the framework of this book. To give an idea of what is involved, we use two approaches. In §16 we present Ratner's proof for $G = \mathrm{SL}(2, \mathbb{R})$ and an arbitrary discrete subgroup $\Gamma \subset G$ (not using the fact that unipotent subgroups of $\mathrm{SL}(2, \mathbb{R})$ are horospherical), and in §17 we show how this approach works for arbitrary unipotent flows. Simultaneously, in §17 we outline the proof of the measure theorem given in [**MT94**], using the case $G = \mathrm{SL}(2, \mathbb{C})$ as an example.

As for the deduction of the Raghunathan topological conjecture from the measure conjecture, in §19 we have chosen the way suggested by Dani and Margulis [**DM93**]. For one-parameter flows the topological theorem can be considerably sharpened: a unipotent trajectory not only has homogeneous closure, but also is *uniformly distributed* in it. This was conjectured in [**Da82**] for $\mathrm{SL}(n, \mathbb{R})$ and in [**Ma88**] in the general case, and proved in [**R91b**]; another proof was given in [**DM93**]. Essentially, the result concerns convergence in the weak* topology in the space of probability measures on G/Γ. Generalizations of this theorem due to Dani and Margulis [**DM93**], Ratner [**R94**], and Eskin, Mozes and Shah [**MS, Sh94, EMS96,97**] are given in §20. For instance, we show that a family of growing pieces of trajectories is uniformly distributed provided that among all limit points there is at least one generic point (not contained in an invariant subspace of smaller dimension). Here one can vary not only trajectories, but unipotent subgroups as well [**DM93**]. The space of ergodic probability measures for a unipotent subgroup turns out to be closed [**MS**]; similar results on the algebraicity of the limit measure for a family of translates of an ergodic algebraic measure are obtained in [**EMS96,97**]. The uniform distribution property (and the algebraicity of closures) is valid not only for unipotent trajectories, but also for polynomial curves in $\mathrm{SL}(n, \mathbb{R})$ [**Sh94**].

Recall (see §9, Subsection 4) that ergodic components $E_g(U)$ of a unipotent flow on a space with finite G-invariant measure are homogeneous subspaces of finite volume. Clearly, every U-ergodic measure on G/Γ is supported on a certain subspace $E_g(U)$. Hence while studying orbits and ergodic measures of unipotent flows, one can always require that the flow be ergodic relative to the G-invariant measure.

§12. Homecoming of unipotent trajectories

1. One of the basic features of unipotent flows is the absence of divergent semiorbits.

THEOREM 12.1 (Margulis [**Ma71**]). *Let Γ be a lattice in G and let a subgroup $u_{\mathbb{R}} \subset G$ be unipotent. Then for every point $x \in G/\Gamma$ there exist a compact set $K \subset G/\Gamma$ and a sequence $t_n \to +\infty$ such that $u_{t_n}x \in K$.*

Actually, Margulis proved this very nontrivial result for unipotent flows on $\mathrm{SL}(n,\mathbb{R})/\mathrm{SL}(n,\mathbb{Z})$ (and used it to prove the arithmeticity Theorem E.7). For semisimple Lie groups of $\mathbb{R}$-rank 1 it was proved by Dani [**Da84**]. The case of an arbitrary semisimple Lie group G reduces to these two cases by the arithmeticity theorem (see [**Da86b**]). In the general case one can argue like this: if $G = (S\times K)\cdot R$ is a Levi decomposition, then by Theorem E.10 the intersection $\Gamma \cap KR$ is a uniform lattice in KR, and hence everything reduces to the case when the group G is semisimple and has no compact components.

An essential refining of the Margulis theorem is the following result, which states that a unipotent trajectory not only comes back to a compact subset infinitely often, but also spends most of its life therein.

THEOREM 12.2 (Dani [**Da86b**]). *Given a unipotent flow $(G/\Gamma, u_{\mathbb{R}})$, a point $x \in G/\Gamma$ and $\varepsilon > 0$, there exists a compact subset $K \subset G/\Gamma$ such that*

$$l(t \in [0,T] : \ u_t x \in K) \geq (1-\varepsilon)T \quad \textit{for all } T \geq 0.$$

Here l stands for the Lebesgue measure on $\mathbb{R}$. To prove the theorem, Dani develops ideas of Margulis from [**Ma71**]. We illustrate the proof of Theorems 12.1 and 12.2 on the simplest example: $G = \mathrm{SL}(2,\mathbb{R})$, $\Gamma = \mathrm{SL}(2,\mathbb{Z})$. Define

$$D_s = \{y \in \mathbb{R}^2 : \ \|y\| < s\}, \qquad K_s = \{g\Gamma \in G/\Gamma : \ g\mathbb{Z}^2 \cap D_s = 0\}.$$

By Mahler's criterion, K_s is compact in G/Γ. A vector $e \in \mathbb{Z}^2$ is called *primitive* if $\alpha e \in \mathbb{Z}^2 \iff \alpha \in \mathbb{Z}$. Let $\{e_i\}$ be a maximal family of pairwise noncollinear primitive vectors (that is, $\mathbb{Z}e_i \cap \mathbb{Z}e_j = 0$, $i \neq j$, and $\bigcup_i \mathbb{Z}e_i = \mathbb{Z}^2$). Fix an element $g \in G$ such that $x = g\Gamma$, and choose a unipotent subgroup $\{u_t\} \subset G$ and $\varepsilon > 0$. If $u_{\mathbb{R}} g e_i \equiv g e_i$ for some i, then the subgroup $g^{-1}u_{\mathbb{R}} g \subset G$ is $\mathbb{Q}$-defined, and the orbit $u_{\mathbb{R}}x$ is periodic. Hence we can assume that $u_{\mathbb{R}}$ keeps no nonzero vector of the lattice $g\mathbb{Z}^2 \subset \mathbb{R}^2$ fixed.

Since $G = \mathrm{SL}(2,\mathbb{R})$ preserves the Lebesgue measure on $\mathbb{R}^2$, the lattice $g\mathbb{Z}^2$ for any $g \in G$ has a fundamental parallelogram of area equal to 1. We set

$$A_i(T) = \{t \in [0,T] : \ u_t g e_i \in D_\varepsilon\}, \qquad B_i(T) = \{t \in [0,T] : \ u_t g e_i \in D_1\}$$

and

$$a_i(T) = l(A_i(T)), \qquad b_i(T) = l(B_i(T)).$$

Since the point $u_t g e_i$ moves along a line with constant speed, it follows that $a_i(T) \leq 2\varepsilon b_i(T)$ for large enough T. Hence

$$\limsup_{T\to\infty} \frac{\sum_i a_i(T)}{\sum_i b_i(T)} \leq 2\varepsilon.$$

The key point is that

$$B_i(T) \cap B_j(T) = \varnothing, \qquad i \neq j$$

(otherwise e_i and e_j would generate a lattice with the area of a fundamental parallelogram less than 1). It follows that $\sum_i b_i(T) \leq T$. Now since

$$\bigcup_i A_i(T) = \{t \in [0,T] : \ u_t x \notin K_\varepsilon\},$$

we derive

$$l(t \in [0,T]: \ u_t x \in K_\varepsilon) \geq (1-4\varepsilon)T \quad \text{for } T \geq T_0.$$

It now suffices to take $K = K_{\varepsilon/4} \cup u_{[0,T_0]}x$. □

We say a few words about the proof of the theorem for $\mathrm{SL}(n,\mathbb{R})/\mathrm{SL}(n,\mathbb{Z})$ with $n > 2$. One of the differences from the case $n = 2$ is that the point $u_t y$, $t \in \mathbb{R}$, moves according to a polynomial law. Hence it is more difficult to estimate the ratio of the time spent by the unipotent trajectory $u_{\mathbb{R}}y$ in the ball D_ε to the time spent in D_1. Denote this fraction by $\delta = \delta(\varepsilon, y, u_{\mathbb{R}})$. In essence, it was proved in [**Ma71**] that $\delta \to 0$ as $\varepsilon \to 0$, where the convergence is uniform in vectors $y \in \mathbb{R}^n$ and in unipotent subgroups $u_{\mathbb{R}} \subset \mathrm{SL}(n,\mathbb{R})$ (provided y is not a fixed point for $u_{\mathbb{R}}$). To prove this, note that $\|u_t y\|^2$ is a polynomial in t of degree at most $2n^2$. Given an interval $B \subset \mathbb{R}$ and a real polynomial f on $\mathbb{R}$, we define $\|f\|_B = \sup\{|f(t)|, \ t \in B\}$. We use the following result.

LEMMA 12.3 [**R92**], [**DM93**]. *For every $k \in \mathbb{N}$ there exists a function $\delta_k(\varepsilon) \to 0$ as $\varepsilon \to 0$ such that for any interval $B \subset \mathbb{R}$ and any nonconstant polynomial f of degree at most k with $\|f\|_B = 1$ we have*

$$\frac{l(t \in B: \ |f(t)| < \varepsilon)}{l(B)} < \delta_k(\varepsilon).$$

Proof (see [**KM98**]). Let $\sigma = l(t \in B: \ |f(t)| < \varepsilon)$. Then there exist points $t_1, \dots, t_{k+1} \in B$ such that

$$|f(t_i)| < \varepsilon, \ 1 \leq i \leq k+1, \qquad \text{and} \qquad |t_i - t_j| \geq \sigma/2k, \ 1 \leq i < j \leq k+1.$$

The Lagrange interpolation formula implies that

$$|f(t)| = \left| \sum_{i=1}^{k+1} f(t_i) \frac{\prod_{j \neq i}(t - t_j)}{\prod_{j \neq i}(t_i - t_j)} \right| \leq (k+1)\varepsilon \frac{l(B)^k}{(\sigma/2k)^k},$$

whence $\sigma/l(B) < \delta_k(\varepsilon) = C(k)\varepsilon^{1/k}$. □

The use of this lemma[4] enables one to control how the point $\{u_t g e_i\}$ moves inside the unit ball D_1 in the proof of Theorem 12.2. However, a much more serious difficulty, arising for $n > 2$ and resolved in [**Ma71**], is that the sets $B_i(T)$ need not be disjoint. Therefore one has to consider the action of u_t not only on vectors but also on subgroups in $g\mathbb{Z}^n$, and to apply a rather nontrivial inductive argument; see also [**DM90b**], [**KM98**] for details.

A deeper analysis allows one to prove a "global" homecoming result.

THEOREM 12.4 (Dani [**Da86b**]). *Given $\varepsilon > 0$, there exists a compact subset $K \subset G/\Gamma$ such that for every unipotent subgroup $u_{\mathbb{R}} \subset G$ and any point $x \in G/\Gamma$ either there exists $T_0 > 0$ such that*

$$l(t \in [0,T]: \ u_t x \in K) \geq (1-\varepsilon)T \quad \textit{for all } T \geq T_0,$$

or the orbit $u_{\mathbb{R}}x$ belongs to a proper subspace of G/Γ of finite volume.

[4] An elementary proof of Theorems 12.2 and 12.4 for *any* lattice $\Gamma \subset \mathrm{SL}(2,\mathbb{R})$ with the help of Lemma 12.3 was given in [**R92**].

It is easily seen that the compact set $K_{\varepsilon/4}$ constructed in the proof of Theorem 12.2 for $\Gamma = \mathrm{SL}(2,\mathbb{Z})$ satisfies the conclusion of Theorem 12.4. Also it is clear that Theorem 12.4 strengthens the "individual" Theorem 12.2 (induction on dimension!).

2. Let μ be a locally finite ergodic measure for a unipotent flow. Then, for each point $x \in G/\Gamma$, the compact set K from Theorem 12.2 can be chosen such that $\mu(K) > 0$. Now taking an everywhere positive continuous function $f \in L^1(G/\Gamma, \mu)$ and applying the Birkhoff–Khintchine ergodic theorem, one gets a positive constant $f^* \in L^1(G/\Gamma, \mu)$ as an average (almost everywhere). This implies the following result.

THEOREM 12.5 (Dani [**Da84**]). *A locally finite ergodic measure for a unipotent flow on a space of finite volume is necessarily finite.*

This can be extended to a larger class of subgroups. The finiteness of ergodic measures for all connected unipotent subgroups $H \subset G$ was proved by Dani [**Da84**]. There is a stronger result:

THEOREM 12.6 (Margulis [**Ma88**]). *If $H = S_H N_H$, where N_H is unipotent in G and normal in H, and S_H is a semisimple group with no compact components, then any locally finite H-ergodic measure on G/Γ is finite.*

Note that this result implies the first part of Theorem E.6 of Borel and Harish-Chandra for semisimple groups. Indeed, let $H \subset \mathrm{SL}(n,\mathbb{R})$ be a semisimple $\mathbb{Q}$-defined group with no compact components. Then the product of H and $\mathrm{SL}(n,\mathbb{Z})$ is closed, and hence the Haar measure on H induces an H-ergodic locally finite measure on $H/\mathrm{SL}(n,\mathbb{Z}) \cap H$. It has to be finite, and hence H intersects $\mathrm{SL}(n,\mathbb{Z})$ in a lattice.

Finally we note that every subgroup $H \subset G$ generated by one-parameter subgroups unipotent in G satisfies the conditions of Theorem 12.6 (see, for instance, [**Sh91**]). It follows that any locally finite H-ergodic measure on G/Γ is finite.

3. Another corollary of homecoming of unipotent trajectories is that all minimal sets are compact.

THEOREM 12.7 (Margulis [**Ma91a**]). *Any minimal set for a unipotent flow $(G/\Gamma, U)$ on a space of finite volume is compact.*

We consider the case where U is a one-parameter subgroup, $U = u_{\mathbb{R}}$ (in this case the result was first proved in [**DM89**]). Clearly, Theorem 12.7 is a direct corollary of Theorem 12.1 and of the following simple observation (see [**Ma91a**] or [**KW**]).

PROPOSITION 12.8. *Let $(X, \phi_{\mathbb{R}})$ be a continuous minimal flow on a noncompact topological space. Then there exists a point $x \in X$ with its positive semiorbit exiting to infinity.*

Proof. Choose $x_0 \in X$, and let $S = S(x_0)$ be a compact disc centered at x_0 and transversal to our flow. Take an open subset $U \subset S$, and define a continuous function $f(x)$ on S as the first positive moment for the trajectory $\phi_t x$ to enter U (where $f(x) = \infty$ if $\phi_{(0,+\infty)}x \cap U = \varnothing$). Suppose that the function f takes only finite values on S. Then, due to continuity, $T = \sup_{x\in S} f(x) < \infty$, whence $\phi_{[0,+\infty)}S = \phi_{[0,T]}S$. Hence the limit set of the positive semiorbit of any point in S belongs to the compact set $\phi_{[0,T]}S$, and, due to minimality, the space X itself is compact (and is equal to $\phi_{[0,T]}S$). This contradiction shows that there exists a point $x \in S$ for which $\phi_{(0,+\infty)}x \cap U = \varnothing$. Consequently, the limit set for the positive semiorbit of the point x does not intersect the open set $\phi_{(-\infty,0)}U$. By minimality, the limit set must be empty, that is, $\phi_t x \to \infty, t \to +\infty$. □

4. Since minimal sets of a unipotent flow are compact, we deduce that no unipotent flow on a noncompact homogeneous space can be minimal. This rather useful result can also be deduced without using Zorn's lemma to ensure the existence of minimal sets.

A compact orbit presents the simplest example of a minimal set. Surely, compact orbits do not always exist. One of the few cases when they do exist is given by the following result.

THEOREM 12.9 (Garland and Raghunathan [**GR**]). *A maximal horospherical flow on a finite volume noncompact space of a simple group of rank* 1 *has a compact orbit.*

Following [**Da85**], we will prove a related result in the general situation.

THEOREM 12.10. *Any unipotent flow on a homogeneous space of finite volume has a compact invariant subspace.*

Proof. It suffices to find a proper invariant subspace of finite volume for a unipotent flow on a noncompact space (then one applies induction on dimension). The problem obviously reduces to considering an action $(G/\Gamma, U)$, where Γ is an irreducible nonuniform lattice in a semisimple group G with no compact components. On the other hand, U can be assumed to be a maximal horospherical subgroup (since every unipotent subgroup of G belongs to such a subgroup).

Let $\operatorname{rank}_{\mathbb{R}} G > 1$. By arithmeticity, we can assume without loss of generality (adding a compact component to G if necessary) that G is a $\mathbb{Q}$-group and $\Gamma = G_{\mathbb{Z}}$. Since the space G/Γ is noncompact, it follows that $\operatorname{rank}_{\mathbb{Q}} G \geq 1$, and hence G contains a proper parabolic $\mathbb{Q}$-subgroup $P \subset G$. Since P contains a minimal parabolic subgroup, replacing U by a conjugate subgroup if necessary, we can assume that $U \subset P$. Let $F \subset P$ be the Moore normal subgroup for U, and let $H = (\overline{F\Gamma})_0$. The product $H\Gamma$ is closed, and an H-invariant measure ν_H on $H\Gamma$ is locally finite. The action $(H\Gamma, F)$ is ergodic, and since F is generated by unipotent subgroups, the measure ν_H is finite. It now suffices to note that $H\Gamma$ is a proper subspace of G/Γ, because $H\Gamma$ belongs to the closed subspace $P\Gamma$.

Now let $\operatorname{rank}_{\mathbb{R}} G = 1$ (since G has no compact components, the group G is simple). Then any maximal horospherical flow has a compact orbit according to Theorem 12.9. Since U belongs to a maximal horospherical subgroup, the proof is completed. □

§13. Horospherical flows on homogeneous spaces

1. Let $U = G_a^-$ be the contracting horospherical subgroup for an element $a \in G$ in a semisimple group G. Since $G_a^- = G_{a_s}^-$, where $a_s \in G$ is the semisimple part of a, we can assume from the beginning that the element a is semisimple. From now on, $Q = Q_a$ is the neutral subgroup for the element a, and $U^+ = G_a^+$ is the expanding horospherical subgroup. Recall that Q normalizes both U and U^+. Conjugation by the element a is simultaneously a semisimple and a quasi-unipotent automorphism of Q, and hence determines an isometry on Q.

Assume that the flow $(G/\Gamma, U)$ is ergodic. Then almost all its orbits are dense in G/Γ. What can we say about the remaining U-orbits? It turns out that the structure of an orbit $Ux \subset G/\Gamma$ is closely related to the structure of the positive semiorbit $\{a^n x,\ n \in \mathbb{N}\}$.

THEOREM 13.1 (Dani [**Da86a**]). *Let a be a semisimple element in a semisimple group G, let the horospherical flow $(G/\Gamma, U)$ be ergodic, and let the positive semiorbit $\{a^n x,\ n \in \mathbb{N}\}$ not exit to infinity in G/Γ. Then the contracting leaf Ux is dense in G/Γ.*

For the proof we need the following observation, due to Mostow and Prasad (see [**Da86a**]).

LEMMA 13.2. *Let a be a semisimple element in a Lie group G, and let $H \subset G$ be the subgroup generated by a and U. Then the action $(G/\Gamma, H)$ is minimal whenever the one-generator action $(G/\Gamma, a^{\mathbb{Z}})$ is ergodic.*

Proof. Choose a dense positive semiorbit: $\overline{a^{\mathbb{N}} x} = G/\Gamma$ for some point $x \in G/\Gamma$. Let us prove that the orbit $H\Gamma$ is dense in G/Γ (other orbits are handled in a similar fashion). It suffices to prove that $x \in \overline{H\Gamma}$. Let

$$a^{n_k} x = g_k \Gamma, \qquad \text{where } n_k \to +\infty \text{ and } g_k \to 1,\ k \to \infty.$$

In a neighborhood of the identity element $1 \in G$ there is a decomposition

$$g_k = s_k u_k, \qquad \text{where } s_k \in U^+ Q,\ u_k \in U.$$

The element a^{-1} contracts U^+ and acts isometrically on Q. Since $s_k^{-1} \to 1$, we derive $a^{-n_k} s_k^{-1} a^{n_k} \to 1$ and hence

$$a^{-n_k} u_k \Gamma = a^{-n_k} s_k^{-1} a^{n_k} x \to x.$$

□

Now we turn to the simplest case of our theorem.

LEMMA 13.3. *Under the assumptions of Theorem* 13.1, *suppose in addition that there exists a connected subgroup A with a compact orbit $Ax \subset G/\Gamma$ such that A contains a and normalizes U. Then $\overline{Ux} = G/\Gamma$.*

Proof. Let $\overline{Uy} = G/\Gamma$. By Lemma 13.2, there exist $u_k \in U$ and $n_k \in \mathbb{Z}$ such that $u_k a^{n_k} x \to y$. Since the orbit Ax is compact, we can assume (passing to a subsequence if necessary) that

$$a^{n_k} x = \varepsilon_k b x, \qquad \text{where } \varepsilon_k, b \in A \text{ and } \varepsilon_k \to 1.$$

Clearly, $y \in \overline{Ubx}$, and hence $b\overline{Ux} \supset \overline{Uy} = G/\Gamma$. □

Proof of Theorem 13.1. Without loss of generality, $x = \Gamma \in G/\Gamma$. Let

$$a^{n_k} \gamma_k \to g, \qquad \text{where } n_k \to +\infty,\ \gamma_k \in \Gamma.$$

Take the open set $\Omega = UQU^+ g \subset G$. By the Prasad–Raghunathan Theorem E.5, there exist a Cartan subgroup $A \subset G$ and an element $w \in \Omega$ such that $w^{-1} a w \in A \subset w^{-1} Q w$ and the orbit $A\Gamma$ is compact. By Lemma 13.3, $\overline{Uw\Gamma} = G/\Gamma$. We have

$$U a^{n_k} \gamma_k w^{-1} \to U g w^{-1} = U q h, \qquad \text{where } q \in Q,\ h \in U^+.$$

Thus,

$$a^{n_k}\gamma_k w^{-1} = u_k s_k q h, \quad \text{where } u_k \in U,\ s_k \in QU^+ \text{ and } s_k \to 1.$$

Hence $Ua^{-n_k}(a^{n_k}\gamma_k w^{-1})a^{n_k} \to Uq$. Without loss of generality,

$$w^{-1}a^{n_k}w = \gamma_k' b_k, \quad \text{where } b_k \in A,\ b_k \to b \in A \text{ and } \gamma_k' \in \Gamma \cap A.$$

Hence

$$U\gamma_k\gamma_k' = Uqwb_k^{-1} \to Uqwb^{-1}.$$

But $qwb^{-1} = q'w$, where $q' \in Q$, and then $\overline{U\Gamma} \supset q'\overline{Uw\Gamma} = G/\Gamma$. □

As a corollary we deduce the following result of Bowen [**Bo76**], Dani [**Da76a**], and Veech [**V75,77**]: if Γ is a uniform lattice in a semisimple group G, then a horospherical flow $(G/\Gamma, U)$ is ergodic if and only if it is minimal. For $G = \mathrm{SL}(2,\mathbb{R})$ (that is, for the horocycle flow on a compact surface of curvature -1) this was established by Hedlund [**He36**]. The method of Hedlund was purely geometric; it was generalized by Anosov [**An67**], who proved that the horospherical foliation of a mixing Anosov flow on a compact manifold is minimal.

In particular, if Γ is an irreducible uniform lattice in a semisimple group G without compact components, then any horospherical flow $(G/\Gamma, U)$ is minimal.

Actually, Bowen and later Ellis and Perrizo proved a stronger result.

THEOREM 13.4 [**Bo76**], [**EP**]. *Let a be a semisimple element in a Lie group G, let $U \subset G$ be the corresponding horospherical subgroup, and let Γ be a uniform lattice in G. Then the following conditions are equivalent*:

1) *The one-generator action $(G/\Gamma, a^{\mathbb{Z}})$ is weak mixing.*
2) *The flow $(G/\Gamma, U)$ is ergodic.*
3) *The flow $(G/\Gamma, U)$ is uniquely ergodic.*

Note that Veech [**V77**] proved this result for G a semisimple group and U a maximal horospherical subgroup. A short proof of a stronger version of Theorem 13.4 is given in [**R90b**] (Theorems 4 and 5).

2. Now let us examine the structure of orbits of a horospherical flow on a compact homogeneous space of general origin; that is, we do not require that the flow $(G/\Gamma, U)$ be ergodic and the group G be semisimple. It is natural first of all to find its ergodic decomposition explicitly. We already know that the ergodic decomposition $\widetilde{E}$ is given by $\widetilde{E}_g(U) = \overline{Fg\Gamma}$, where $F \subset G$ is the Moore subgroup for U (in this case it is the smallest normal subgroup of G containing U). Since the one-generator action $(G/\overline{F\Gamma}, a^{\mathbb{Z}})$ preserves a smooth finite measure, we deduce that $U^+ \subset \overline{F\Gamma}$ and hence $A \subset \overline{F\Gamma}$, where A is the Auslander normal subgroup for a generated by U and U^+. Since $F \subset A$, the ergodic decomposition $\widetilde{E}$ is precisely the Pinsker partition $\{gA\Gamma,\ g \in G\}$ for the one-generator action $(G/\Gamma, a^{\mathbb{Z}})$.

Recall that there exists a "more delicate" ergodic partition E into closed subspaces $E_g(U) = \overline{UJg\Gamma}$, $g \in G$, where $J = J(U)$ is the Dani normal subgroup for U, and $J \subset F$. It turns out that for horospherical flows, E and $\widetilde{E}$ are the same:

THEOREM 13.5 [**St91**]. *The ergodic partition E for a horospherical flow is given by $E_g(U) = \overline{UJg\Gamma} = gA\Gamma$, where A is the Auslander normal subgroup generated by U and U^+.*

Proof. As usual, it suffices to handle the point $\Gamma \in G/\Gamma$. Using a standard techniques from Chapter I, it is easy to reduce the proof to the case when the one-generator action $(G/\Gamma, a^{\mathbb{Z}})$ is ergodic and the lattice Γ is Zariski dense in G relative to the Ad-representation. We start with the solvable case. Assume first that the normal subgroup $H = (\overline{A\Gamma})_0 \subset G$ is Abelian. Since $[G, U] \subset [U^+, U]U$ and $[A, A] = 1$, the subgroup U is normal in G. The one-generator action $(G/\overline{U\Gamma}, a^{\mathbb{Z}})$ is measure-preserving; hence $U^+ \subset \overline{U\Gamma}$, and we are done. If the group H is non-Abelian, we take $H_1 = (\overline{U[H, H]\Gamma})_0$. In essence, we have already proved that $H = H_1$, and now Green's criterion 3.6 implies that the flow $(H/\Gamma \cap H, U)$ is ergodic and minimal. It follows that $\overline{U\Gamma} = \overline{A\Gamma}$. In the general case, without loss of generality we can assume that the projection $q(a)$ of $a \in G = L \cdot R$ to the Levi subgroup L is a semisimple element. Since the action of a on $G/\overline{J\Gamma} = G'/\Gamma'$, where $G' = G/(\overline{J\Gamma})_0$, $\Gamma' = \overline{J\Gamma}/(\overline{J\Gamma})_0$, is ergodic, the group G' is solvable, and our theorem follows. Hence $U^+ \cap N \subset \overline{UJ\Gamma}$, where N is the nilradical of G. Since $U^+ = q(U^+)(U^+ \cap N)$ and $q(U^+) \subset J$, the theorem is proved in the general case as well. □

Now that we know the explicit form of the ergodic decomposition E, it is not difficult to modify the proof of Theorem 13.1 and to obtain the following more general result:

THEOREM 13.6 [**St91**]. *Let $q(a)$ be a semisimple element in a Levi subgroup L of a group G. If the positive semiorbit $\{a^n x,\ n \in \mathbb{N}\}$ does not exit to infinity in G/Γ, then $\overline{Ux} = \overline{Ax}$.*

Note that the case of a solvable group G was examined in the course of the proof of Theorem 13.5. It is easily seen that Theorem 13.6 is not a direct corollary of Theorems 13.1 and 13.5, because the element a need not belong to the subgroup $\overline{A\Gamma}$, and hence the flow $(\overline{A\Gamma}, U)$, being ergodic, need not be horospherical.

COROLLARY 13.7 [**St91**]. *If the space G/Γ is compact and the subgroup $U \subset G$ is horospherical, then all the closures $\overline{Ux} \subset G/\Gamma$ are homogeneous and isomorphic to the space $\overline{A\Gamma}/\Gamma$.*

§14. The geodesic and horocycle flows on surfaces of constant negative curvature

First we show that the geodesic and horocycle flows on a surface of constant negative curvature have algebraic origin. Since the natural approach leads to considering *right* actions on *left* homogeneous spaces, in §§14–16 we pass to this system of notation. Then we study ergodic properties of the flows and the structure of their orbits.

1. Let us recall briefly some basic facts from hyperbolic geometry. The Lobachevskii plane $\mathbb{H}^2$ of constant negative curvature -1 in the Poincaré model is the complex half-plane $y = \operatorname{Im}(z) > 0$ with the metric $dl^2 = (dx^2 + dy^2)/y^2$. The line $y = 0$, completed with infinity, will be referred to as the *circle at infinity.* The group $\operatorname{Iso}(\mathbb{H}^2)$ of all isometries of $\mathbb{H}^2$ consists of combinations of the reflection $s(z) = -\bar{z}$ and linear fractional transformations

$$z \longrightarrow \frac{az + b}{cz + d}, \qquad \text{where } ad - bc = 1.$$

The group $\operatorname{Iso}^+(\mathbb{H}^2)$ of all linear fractional transformations consists of all isometries of $\mathbb{H}^2$ preserving orientation, and is the identity component of the group $\operatorname{Iso}(\mathbb{H}^2)$.

Hence if

$$g = \begin{pmatrix} a & b \\ c & d \end{pmatrix} \in \mathrm{SL}(2,\mathbb{R}) \quad \text{and} \quad f_g(z) = \frac{az+b}{cz+d},$$

then the map $g \to f_g$ establishes an epimorphism $f : \mathrm{SL}(2,\mathbb{R}) \to \mathrm{Iso}^+(\mathbb{H}^2)$ with kernel $\mathbb{Z}_2 = \{\pm 1\} \in \mathrm{SL}(2,\mathbb{R})$ (or, equivalently, an isomorphism $\widehat{f} : \mathrm{PSL}(2,\mathbb{R}) = \mathrm{SL}(2,\mathbb{R})/\mathbb{Z}_2 \to \mathrm{Iso}^+(\mathbb{H}^2)$).

In what follows $G = \mathrm{SL}(2,\mathbb{R})$, B is the two-dimensional solvable subgroup of G of all upper-triangular matrices, and $K = \mathrm{SO}(2,\mathbb{R})$ is a maximal compact subgroup of G.

Any point $z = x + iy \in \mathbb{H}^2$ can be written as

$$z = f_{g_z}(z_0), \qquad \text{where } z_0 = i,\ g_z = \begin{pmatrix} \sqrt{y} & x/\sqrt{y} \\ 0 & 1/\sqrt{y} \end{pmatrix} \in B.$$

In particular, the actions of B and G on $\mathbb{H}^2$ are transitive. The stabilizer of the point $z_0 = i$ is just the subgroup $K = \mathrm{SO}(2,\mathbb{R})$. It follows that the map $\phi(gK) = f_g(i)$ establishes the diffeomorphisms

$$\phi : G/K \to \mathbb{H}^2 \quad \text{and} \quad \widehat{\phi} : \mathrm{PSL}(2,\mathbb{R})/\mathrm{PSO}(2,\mathbb{R}) \to \mathbb{H}^2.$$

Up to rescaling, the metric dl coincides with the metric obtained from the G-left-invariant and K-right-invariant Killing metric d_G on G.

We denote by $S\mathbb{H}^2$ the unit tangent bundle over $\mathbb{H}^2$. The elements of $S\mathbb{H}^2$ are the pairs (z, v), where $z \in \mathbb{H}^2$ and $v \in S_z\mathbb{H}^2$ is a unit tangent vector to $\mathbb{H}^2$ at the point z. The differential df_g of the action f_g, $g \in G$, determines a diffeomorphism $S\mathbb{H}^2 \to S\mathbb{H}^2$. Therefore we obtain an induced action df of the group G on $S\mathbb{H}^2$.

The differential df_k for any element $k = \begin{pmatrix} \cos\phi & \sin\phi \\ -\sin\phi & \cos\phi \end{pmatrix} \in K$ determines rotation of the unit tangent circle $S^1 = S_i\mathbb{H}^2$ through the angle 2ϕ. Hence the kernel of the action df on $S\mathbb{H}^2$ is again $\mathbb{Z}_2 \subset G$.

We fix the unit tangent vector $e = (0, 1)$ at the point $z_0 = i$. Then any pair (z, v) can be obtained from the pair (i, e) by a suitable rotation df_k, $k \in K$, of the unit circle $S_{z_0}\mathbb{H}^2$ followed by a transposition df_{g_z} of the point i to the point z. It follows that the map $F(g) = df_g(i, e)$ determines both the covering $F : G \to S\mathbb{H}^2$ with fiber $\mathbb{Z}_2$ and the diffeomorphism $\widehat{F} : \mathrm{PSL}(2,\mathbb{R}) \to S\mathbb{H}^2$. Incidentally, we have established the Iwasava decomposition $G = BK = KB$. We note that, given $g \in \mathrm{Iso}^+(\mathbb{H}^2)$, the map $df_g : S\mathbb{H}^2 \to S\mathbb{H}^2$ corresponds to the left translation

$$L_g : \mathrm{PSL}(2,\mathbb{R}) \to \mathrm{PSL}(2,\mathbb{R}), \qquad L_g(h) = gh, \quad h \in \mathrm{PSL}(2,\mathbb{R})$$

(in other words, $df_g \circ \widehat{F} = \widehat{F} \circ L_g$). The reflection $s : z \to -\overline{z}$ corresponds to an outer automorphism Σ of the group $\mathrm{PSL}(2,\mathbb{R})$:

$$\Sigma(h) = \sigma h \sigma^{-1}, \qquad \text{where } \sigma = \begin{pmatrix} 1 & 0 \\ 0 & -1 \end{pmatrix}.$$

Thus,

$$\Sigma\left(\begin{pmatrix} a & b \\ c & d \end{pmatrix}\right) = \begin{pmatrix} a & -b \\ -c & d \end{pmatrix}.$$

2. One knows that there exists precisely one geodesic line leaving a point $z \in \mathbb{H}^2$ in the direction of a vector $v \in S_z\mathbb{H}^2$. If we apply the operation of parallel translation of tangent vectors along the line, we will obtain a one-parameter flow g_t on $S\mathbb{H}^2$, called the *geodesic flow.*

It is easily seen that the curve $g_t(i, e) = ie^t$ determines the geodesic line in $\mathbb{H}^2$ passing with unit velocity through $z_0 = i$ in the vertical direction $e \in S_{z_0}\mathbb{H}^2$. Since every isometry sends geodesic to geodesic and G acts transitively on $S\mathbb{H}^2$, all geodesics are exhausted by curves of the form $f_g(ie^t)$. Geometrically, these are semicircles (or rays) which meet the circle at infinity $y = 0$ orthogonally at some points x^- and x^+. Here x^- is the limit of the geodesic as $t \to -\infty$ in the Euclidean topology and is called the *negative end* of the geodesic. Similarly, x^+ is called the *positive end* of the corresponding geodesic. Let

$$A_t = \begin{pmatrix} e^{t/2} & 0 \\ 0 & e^{-t/2} \end{pmatrix}, \quad H_t = \begin{pmatrix} 1 & 0 \\ t & 1 \end{pmatrix}, \quad U_t = \begin{pmatrix} 1 & t \\ 0 & 1 \end{pmatrix}.$$

Then $ie^t = f_{A_t}(i)$, and hence any geodesic line in $\mathbb{H}^2$ looks like $\{f_{gA_t}(i)\}$. Thus,

$$g_t(z, v) = df_{gA_t}(i, e), \qquad \text{where } df_g(i, e) = (z, v),$$

and hence *the geodesic flow g_t corresponds to the right action of A_t on G.*

Note that the subgroups $A, U, H \subset G$ are related as follows:

$$A_{-t}U_sA_t = U_{e^{-t}s}, \qquad A_{-t}H_sA_t = H_{e^ts},$$

and hence U and H are the contracting and expanding horospherical subgroups for A. Since for a left-invariant metric d_G in G we have

$$\begin{aligned} d_G(gA_t, gU_sA_t) &= d_G(gA_t, gA_tA_{-t}U_sA_t) \\ &= d_G(1, A_{-t}U_sA_t) = d_G(1, U_{e^{-t}s}) \to 0, \quad t \to +\infty, \end{aligned}$$

the right orbits of U in G form the contracting foliation for the right action of A on G. Hence the curves $df_{gU_s}(i, e)$ in $S\mathbb{H}^2$ form the contracting foliation $\mathcal{U}$ for the geodesic flow on $S\mathbb{H}^2$. They are referred to as *contracting horocycles*, and geometrically they are circles internally tangent to the circle at infinity (or horizontal lines if $v = (0, 1)$). More precisely, the contracting horocycle given by a pair $(z, v) \in S\mathbb{H}^2$ is the circle (possibly of infinite radius) internally tangent to the circle at infinity which passes through $z \in \mathbb{H}^2$ such that v is the inner normal vector.

Given a pair $(z, v) \in S\mathbb{H}^2$, let x^- and x^+ be the ends of the corresponding geodesic on the circle at infinity $\mathbb{R} \cup \infty$. As is easily seen, the contracting horocycle given by the pair (z, v) is tangent to the circle at infinity at the point x^+. Moreover, if a pair $(z', v') \in S\mathbb{H}^2$ is obtained from (z, v) by a translation along the contracting horocycle, then the positive end of the geodesic given by (z', v') is again x^+. Similarly, if (z', v') is obtained from (z, v) by a translation along a geodesic, then the contracting horocycle given by (z', v') is tangent to the circle at infinity at the same point x^+. Thus, the point x^+ corresponds to a family of geodesics which approach each other asymptotically in the positive direction. Clearly, the family is just the orbit $(z, v)B$ of the two-dimensional solvable group $B = A_{\mathbb{R}}U_{\mathbb{R}}$. This correspondence establishes a G-equivariant homeomorphism between the circle at infinity $\mathbb{R} \cup \infty$ and the homogeneous space G/B.

The map $(z, v) \to x^+$ is called the *visual map* and is denoted by

$$\mathrm{Vis} : S\mathbb{H}^2 \to \mathbb{R} \cup \infty.$$

It can be easily written in an algebraic way. More specifically, let $(z, v) = g(i, e)$, where $g = \begin{pmatrix} a & b \\ c & d \end{pmatrix}$. Then $\mathrm{Vis}(g) = \frac{c}{a}$. Indeed, if a matrix $g = \begin{pmatrix} a & b \\ c & d \end{pmatrix}$ decomposes as $g = H_\alpha A_s U_\beta$, then $\alpha = \mathrm{Vis}(g) = \frac{c}{a}$. Only matrices of the form $g = \begin{pmatrix} 0 & b \\ c & d \end{pmatrix} \in G$ do not have such a decomposition; for them we let $\mathrm{Vis}(g) = \infty$. So, algebraically, Vis is the covering $G \to G/B$, given as $\mathrm{Vis}(g) = gB$.

Likewise, the expanding horocycle foliation $\mathcal{H}$ in $S\mathbb{H}^2$ is given by the curves $df_{gH_s}(i, e)$. The *expanding horocycle* given by $(z, v) \in S\mathbb{H}^2$ is the circle internally tangent to the circle at infinity which passes through $z \in \mathbb{H}^2$ and has v as the outer normal vector.

It follows that the geodesic flow g_t on the 3-manifold $S\mathbb{H}^2$ has expanding and contracting foliations, and hence is an Anosov flow. In this case both foliations $\mathcal{U}$ and $\mathcal{H}$ are naturally parametrized by time $s \in \mathbb{R}$, and hence are one-parameter flows on $S\mathbb{H}^2$. They are called the *contracting* and *expanding horocycle* flows. Explicitly,

$$u_s(z, v) = df_{gU_s}(i, e), \quad h_s(z, v) = df_{gH_s}(i, e), \qquad \text{where } df_g(i, e) = (z, v).$$

3. We recall that a *Fuchsian group* is a discrete subgroup $\Gamma \subset \mathrm{Iso}^+(\mathbb{H}^2)$. An arbitrary smooth orientable surface M of constant curvature $k(M) = -1$ is a quotient space of $\mathbb{H}^2$ equipped with a metric of curvature -1 by orbits of some Fuchsian group which has no elements of finite order[5] (that is, elements conjugate to elements in $\mathrm{PSO}(2, \mathbb{R})$). Hence,

$$M = \Gamma\backslash\mathbb{H}^2 = \Gamma\backslash \mathrm{PSL}(2, \mathbb{R})/\mathrm{PSO}(2, \mathbb{R}) \quad \text{and} \quad SM = \Gamma\backslash S\mathbb{H}^2 = \Gamma\backslash \mathrm{PSL}(2, \mathbb{R}).$$

Equivalently, we can write $M = \widetilde{\Gamma}\backslash G/K$ and $SM = \widetilde{\Gamma}\backslash G$, where $\widetilde{\Gamma} \subset G$ is the inverse image of Γ under the epimorphism $\mathrm{SL}(2, \mathbb{R}) \to \mathrm{PSL}(2, \mathbb{R})$ (and $\widetilde{\Gamma}$ has no elements of finite order except for ± 1).

The flows g_t, u_t and h_t on $S\mathbb{H}^2$ cover the corresponding flows on SM (we keep the same notation). Hence the geodesic flow g_t on SM is just the homogeneous flow $(\Gamma\backslash G, A_t)$. The horocycle flows u_t and h_t on SM correspond to the homogeneous flows $(\Gamma\backslash G, U_t)$ and $(\Gamma\backslash G, H_t)$.

A Riemannian metric on M induces a smooth volume measure, denoted by vol. On the other hand, $SM = \Gamma\backslash G$ carries a smooth G-invariant measure ν induced by a bi-invariant Haar measure on G and called the *Liouville measure.* Here $\mathrm{vol}(M) < \infty$ if and only if $\nu(SM) < \infty$, that is, if and only if Γ is a lattice in G. In this case we normalize the Liouville measure ν by the condition $\nu(SM) = 1$.

Assume that $\theta : M_1 \to M_2$ is an isometry of two orientable surfaces of constant negative curvature -1. Then $M_i = \Gamma_i\backslash\mathbb{H}^2$, where $\Gamma_i \subset \mathrm{Iso}^+(\mathbb{H}^2)$, $i = 1, 2$, and θ lifts to an isometry $\widetilde{\theta} : \widetilde{M_1} \to \widetilde{M_2}$. But $\widetilde{M_1} = \widetilde{M_2} = \mathbb{H}^2$, and hence $\widetilde{\theta} \in \mathrm{Iso}(\mathbb{H}^2)$ is an element sending the Γ_1-orbit of any point $z \in H$ to the Γ_2-orbit of the point $\widetilde{\theta}(z)$, $z \in \mathbb{H}^2$. It follows that $\Gamma_2 = \widetilde{\theta}\Gamma_1\widetilde{\theta}^{-1}$. If $\widetilde{\theta} \in \mathrm{Iso}^+(\mathbb{H}^2)$, that is, if the surfaces have equal orientation, then, identifying $\mathrm{Iso}^+(\mathbb{H}^2)$ and $\mathrm{PSL}(2, \mathbb{R})$, one can assume that $\widetilde{\theta} = \alpha \in \mathrm{PSL}(2, \mathbb{R})$. Then the isometry is given by the map

$$\theta : \Gamma_1 z \to \alpha\Gamma_1 z = \Gamma_2\alpha z, \qquad z \in \mathbb{H}^2,$$

[5]If Γ contains an element of finite order, then M is a surface with singularities or, in Thurston's terminology, an *orbifold.* Example: $\Gamma = \mathrm{PSL}(2, \mathbb{Z})$.

and its differential by the map

$$d\theta : \Gamma_1 g \to \alpha\Gamma_1 g = \Gamma_2 \alpha g, \qquad g \in \mathrm{PSL}(2, \mathbb{R}).$$

If the orientations are opposite, then $\widetilde{\theta} = \alpha s$, where $\alpha \in \mathrm{Iso}^+(\mathbb{H}^2)$ and s is the reflection. In this case the isometry of surfaces is the map

$$\theta : \Gamma_1 z \to \alpha\Sigma(\Gamma_1)s(z) = \Gamma_2 \alpha s(z),$$

and its differential is

$$d\theta : \Gamma_1 g \to \alpha\Sigma(\Gamma_1)\Sigma(g) = \Gamma_2 \alpha\Sigma(g),$$

where

$$\Sigma(g) = \sigma g \sigma^{-1}, \qquad \sigma = \begin{pmatrix} 1 & 0 \\ 0 & -1 \end{pmatrix}, \qquad g \in \mathrm{PSL}(2, \mathbb{R}).$$

Therefore two surfaces M_1 and M_2 are isometric if and only if the subgroups $\Gamma_1 = \pi_1(M_1)$ and $\Gamma_2 = \pi_1(M_2)$ are conjugate in $\mathrm{Iso}(\mathbb{H}^2)$.

It is well known that if $\pi_1(M_1)$ and $\pi_1(M_2)$ are isomorphic, then M_1 and M_2 are homeomorphic but not necessarily isometric. The point is that two discrete subgroups (and even lattices) Γ_1 and Γ_2 in $G = \mathrm{SL}(2, \mathbb{R})$ can be isomorphic but not conjugate in G. The family of G-conjugacy classes of all embeddings of Γ into G is equipped with a natural topology and is called the *Teichmuller space* $T(\Gamma)$ of the discrete group Γ (see the survey [**VS**] for more details).

4. From now on $M = \Gamma\backslash G/K$ is a surface of constant negative curvature, where $G = \mathrm{SL}(2, \mathbb{R})$, $K = \mathrm{SO}(2)$, and $SM = \Gamma\backslash G$ is its unit tangent bundle. We assume that $\mathrm{vol}(M) < \infty$, that is, Γ is a lattice in G, and a G-invariant measure ν on SM is normalized by the condition $\nu(SM) = 1$. Let us formulate basic results about the behavior of geodesic and horocycle flows on SM (see also [**R84**], [**R95b**]).

It is well known that the geodesic flow g_t is ergodic (Hopf [**Ho36**]) and mixing (Hedlund [**He39**]). Gelfand and Fomin [**GF**] have proved that it has Lebesgue spectrum of countable multiplicity; see §11 for more details. Moreover, the geodesic flow is a K-flow (Sinai [**Si60**]) and, in particular, it is mixing of all orders. We note that these results are particular cases of results on Anosov flows [**An67**]. The strongest ergodic property of the geodesic flow is that it is Bernoulli (Ornstein and Weiss [**OW**]).

The horocycle flow u_t (everything we say applies also to h_t) is ergodic (Hedlund [**He39**]). It also has a Lebesgue spectrum of countable multiplicity (Parasjuk [**Pa**]), see §11. According to Marcus [**Mar78**], it is mixing of all orders; see §22 for more details. Since the geodesic translate g_t establishes an isomorphism of u_1 and $u_{e^{-t}}$, it follows that the entropy of the horocycle flow is either 0 or ∞. Actually, the horocycle flow has zero entropy (Gurevich [**Gu**]; in the noncompact case see Dani [**Da77**]).

We have already proved in §8 that g_t and u_t are ergodic and weak mixing, using the Mautner phenomenon for unitary representation. Originally, ergodicity of the geodesic flow was proved by Hopf in a more "geometric" way. More specifically, choose $f \in C_c(\Gamma\backslash G)$. By the ergodic theorem, for almost all $x \in \Gamma\backslash G$ one has

$$f^*(x) = \lim_{T\to+\infty} \frac{1}{T}\int_0^T f(xA_s)ds = \lim_{T\to-\infty} \frac{1}{T}\int_0^T f(xA_s)ds,$$

where the limit function $f^* \in L^1(\Gamma\backslash G, \nu)$ is invariant under the subgroup $A_{\mathbb{R}}$ inducing the geodesic flow g_t. On the other hand, since f is uniformly continuous, the first limit is invariant under the subgroup $U_{\mathbb{R}}$ inducing the contracting flow u_t, and the second limit is invariant under the subgroup $H_{\mathbb{R}}$ inducing the expanding flow h_t. A local consideration shows that f^* is constant almost everywhere[6] for every $f \in C_c(\Gamma\backslash G)$. Since the space $C_c(\Gamma\backslash G)$ is dense in $L^1(\Gamma\backslash G, \nu)$, it follows that g_t is ergodic.

In the sequel we will use the mixing property for the geodesic flow. Surely this property follows from the K-property, but for the sake of completeness we will give an elementary proof. For G-invariant Liouville measure this result goes back to Hedlund [**He39**], who deduced the mixing property of the geodesic flow from ergodicity of the horocycle flow. We will prove a more general result (this will be used in §16).

LEMMA 14.1 [**R90b,92**]. *Let Γ be a discrete subgroup of G and let μ be a probability measure on $\Gamma\backslash G$ invariant under $U_{\mathbb{R}}$ and $A_{\mathbb{R}}$. Then the flow $(\Gamma\backslash G, A_{\mathbb{R}}, \mu)$ is mixing whenever the flow $(\Gamma\backslash G, U_{\mathbb{R}}, \mu)$ is ergodic.*

Proof. It suffices to prove that

$$I(\tau) = \int_{\Gamma\backslash G} \phi(x) f(xA_{-\tau}) d\mu \to 0, \qquad \tau \to \infty,$$

for any two bounded uniformly continuous functions ϕ, f on $\Gamma\backslash G$ such that $\int_{\Gamma\backslash G} f d\mu = 0$. We fix $\varepsilon > 0$ and select $0 < \delta < 1$ such that

$$|\phi(x) - \phi(z)| < \varepsilon \quad \text{for any } x, z \in \Gamma\backslash G,\ d(x,z) < \delta.$$

Since the action of $U_{\mathbb{R}}$ on $(\Gamma\backslash G, \mu)$ is ergodic, there exist $t_0 > 1$ and $Y \subset \Gamma\backslash G$ such that

$$\mu(Y) > 1 - \varepsilon \quad \text{and} \quad |S_f(y,t)| < \varepsilon \quad \text{for all } y \in Y,\ t \geq t_0$$

(here $S_f(y,t)$ stands for the time average from the Birkhoff–Khintchine ergodic theorem applied to u_t). We take a sufficiently large $\tau_0 > 0$ such that $e^{-2\tau_0} t_0 = \delta$, and we let $\tau \geq \tau_0$. Then for $Y_\tau = YA_\tau$ we obtain $\mu(Y_\tau) = \mu(Y) > 1 - \varepsilon$. It follows that

$$\begin{aligned}
I(\tau) &= \frac{1}{\delta} \int_0^\delta \Big(\int_{\Gamma\backslash G} \phi(xU_s) f(xU_sA_{-\tau}) d\mu \Big) ds \\
&= \int_{\Gamma\backslash G} \Big(\frac{1}{\delta} \int_0^\delta \phi(xU_s) f(xU_sA_{-\tau}) ds \Big) d\mu \\
&= \int_{\Gamma\backslash G} \frac{\phi(x)}{\delta} \Big(\int_0^\delta f(xA_{-\tau}U_{e^{2\tau}s}) ds \Big) d\mu + \varepsilon_1 \\
&= \int_{\Gamma\backslash G} \frac{\phi(x)}{s_\tau} \Big(\int_0^{s_\tau} f(xA_{-\tau}U_s) ds \Big) d\mu + \varepsilon_1 \\
&= \int_{\Gamma\backslash G} \phi(x) S_f(xA_{-\tau}, s_\tau) d\mu + \varepsilon_1 \\
&= \int_{Y_\tau} \phi(y) S_f(yA_{-\tau}, s_\tau) d\mu + \varepsilon_1 + \varepsilon_2 = \varepsilon_1 + \varepsilon_2 + \varepsilon_3 \to 0,
\end{aligned}$$

[6] The case of variable curvature requires extra arguments; see [**An67**].

where $s_\tau = \delta e^{2\tau} \geq t_0$. Here the estimate on ε_1 follows because ϕ is uniformly continuous, that on ε_2 because ϕ and f are bounded, and that on ε_3 follows from the choice of Y. □

5. Before we formulate results on the structure of $g_{\mathbb{R}}$- and $u_{\mathbb{R}}$-orbits, let us prove two simple assertions. From now on Γ *is again assumed to be a lattice in* G.

LEMMA 14.2. *If a horocycle orbit $xU_{\mathbb{R}} \subset \Gamma\backslash G$ is periodic, then the geodesic orbit $xA_{\mathbb{R}} \subset \Gamma\backslash G$ exits to infinity in the positive direction.*

Proof. Assume that $xA_{t_n} \to y$ as $t_n \to +\infty$. Since the period of the horocycle of the point xA_{t_n} gets shorter as $t_n \to +\infty$, it follows that $y \in \Gamma\backslash G$ would be a fixed point of the horocycle flow. This is impossible, because Γ is discrete. □

This implies, by the way, that if the Fuchsian group Γ contains a parabolic (unipotent) element, then the surface $M = \Gamma\backslash G/K$ and the space $\Gamma\backslash G$ are non-compact. The converse statement is also well known in the theory of Fuchsian groups: if Γ contains no parabolic elements, then the surface $M = \Gamma\backslash G/K$ of finite area is compact.

LEMMA 14.3. *Given a sequence $t_n \to -\infty$ and a point $x \in \Gamma\backslash G$, the set $\bigcup_n xU_{\mathbb{R}}A_{t_n}$ is dense in $\Gamma\backslash G$.*

Proof. Choose $x \in \Gamma\backslash G$. We already know that the flow $(\Gamma\backslash G, A_{\mathbb{R}})$ is topologically mixing. Hence for any sequence $\{t_n \to -\infty\}$ and any neighborhood of the identity element $O(1) \subset G$, the set $\bigcup_n xO(1)A_{t_n}$ is dense in $\Gamma\backslash G$. It follows that for any point $y \in \Gamma\backslash G$ there exists a sequence $\{g_n \in O(1)\}$ such that $xg_nA_{t_n} \to y$. But for a small enough neighborhood $O(1) \subset G$ we have

$$g_n = U_{s_n}A_{\tau_n}H_{r_n}, \qquad \text{where } |s_n|, |\tau_n|, |r_n| < \delta.$$

Hence

$$xg_nA_{t_n} = xU_{s_n}A_{t_n+\tau_n}H_{r'_n}, \qquad \text{where } H_{r'_n} = A_{-t_n}H_{r_n}A_{t_n} \to 1.$$

It follows that $xU_{s_n}A_{t_n+\tau_n} \to y$. Since the number δ can be made arbitrarily small, we deduce that $y \in \overline{\bigcup_n xU_{\mathbb{R}}A_{t_n}}$. □

COROLLARY 14.4. *The action of the triangular group B on a finite volume space $\Gamma\backslash G$ is minimal.*

By duality, we obtain a well-known fact in the theory of Fuchsian groups: the action of a lattice Γ on the circle at infinity G/B is minimal. On the other hand, the Haar measure on G induces a smooth G-quasi-invariant measure μ on the circle at infinity G/B and a finite G-invariant measure ν on $\Gamma\backslash G$. Since the action of B on $(\Gamma\backslash G, \nu)$ is ergodic, the action of Γ on $(G/B, \mu)$ is also ergodic. Hence the action of Γ is ergodic relative to the usual Lebesgue measure on the circle at infinity as well (because the measures are equivalent).

COROLLARY 14.5. *If a geodesic orbit $xA_{\mathbb{R}}$ is periodic, then the horocycle orbit $xU_{\mathbb{R}}$ is dense in $\Gamma\backslash G$.*

Proof. Let $xA_t = x$ for some $t < 0$. Then

$$\Gamma\backslash G = \overline{xU_{\mathbb{R}}A_{nt}} = \overline{xU_{\mathbb{R}}}.$$

□

Clearly, the last three results are particular cases of Lemmas 13.1–13.3.

Let us explain briefly how periodic horocycles in a noncompact manifold $SM = \Gamma\backslash G$ are distributed. Clearly, if an orbit $xU_{\mathbb{R}} \subset \Gamma\backslash G$ is periodic, then so is the orbit of the point xA_s for every $s \in \mathbb{R}$. Hence every periodic horocycle $xU_{\mathbb{R}}$ belongs to the connected cylinder $xB = xA_{\mathbb{R}}U_{\mathbb{R}}$ consisting entirely of periodic horocycles. Clearly, two such cylinders either are disjoint or coincide. By Lemma 14.2, the vertical side $xA_{\mathbb{R}}$ of the cylinder exits to infinity in the positive direction. Hence the semi-cylinder $xA_{\mathbb{R}^+}U_{\mathbb{R}}$ is a smoothly immersed closed submanifold in $\Gamma\backslash G$, which gets thinner approaching infinity. Topologically, the surface M is obtained from a compact surface by gluing those semi-cylinders (after excising the same number of discs). Note that the number of these disjoint cylinders in $\Gamma\backslash G$ is equal to the number of Γ-conjugacy classes of unipotent subgroups in Γ. This number is well known to be finite for any lattice $\Gamma \subset G$ (and is equal to 1 for $\Gamma = \mathrm{SL}(2,\mathbb{Z})$). At the same time the semi-cylinder $xA_{\mathbb{R}^-}U_{\mathbb{R}}$ is dense in $\Gamma\backslash G$ according to Lemma 14.3.

Let us summarize our results on the structure of geodesic orbits on a surface of finite area. Because of ergodicity, almost every orbit of the geodesic flow is dense in SM. Apart from dense and periodic orbits, one can find an orbit whose closure is not a submanifold. This concerns, for example, orbit closures which locally look like the product of a Cantor set and an interval (this is typical, say, for Morse minimal sets [**Mor**]). One observes that an example of locally closed but not closed orbit is given by homoclinic trajectories, which exist for any Anosov flow. In a neighborhood of a homoclinic orbit one can find a nowhere locally-connected compact invariant set containing uncountably many minimal sets, see [**AS**]. Thus, the picture is the same as for Anosov toral automorphisms. If the surface M is noncompact, there also exist orbits exiting to infinity in one or both directions.

For the horocycle flow the situation is much simpler: according to Hedlund [**He36**], an orbit of it is either periodic or dense in SM (if the area of M is finite). In particular, the horocycle flow on a compact surface is minimal; we have already proved this in §13.

Let us consider the structure of finite ergodic measures of the two flows. As for any Anosov flow, there are uncountably many mutually singular g_t-ergodic probability measures on SM (Sinai [**Si72**], Bowen [**Bo75**]). Among them one distinguishes length measures supported on periodic orbits; measures singular to Riemannian volume but at the same time strictly positive; and also measures with nonsmooth support (for instance, measures supported on Morse minimal sets).

If the surface M is compact, the horocycle flow is uniquely ergodic (Furstenberg [**Fu72**]). In the noncompact case, apart from the smooth measure ν on SM, any u_t-ergodic probability measure is the length measure supported on a periodic horocycle (Dani [**Da80b**] for $\Gamma = \mathrm{SL}(2,\mathbb{Z})$; Dani and Smillie [**DS**] for any lattice $\Gamma \subset G$). Later, we will give a proof of this result which is due to Ratner and is of more general nature.

6. Here we consider in more detail orbits of horocycle and geodesic flows for the *modular group* $\Gamma = \mathrm{SL}(2,\mathbb{Z})$. In this case there is only one Γ-conjugacy class of unipotent cyclic subgroups in Γ. Hence all periodic orbits of the horocycle flow

$(\Gamma\backslash G, U_{\mathbb{R}})$ sit on the cylinder $\Gamma B \subset \Gamma\backslash G$. Below we will show that other orbits of the flow are dense in $\Gamma\backslash G$.

We turn to the geodesic flow. Since the horocycles $\Gamma U_{\mathbb{R}}$ and $\Gamma H_{\mathbb{R}}$ are periodic, the geodesic $\Gamma A_{\mathbb{R}}$ exits to infinity in both directions. In addition, given $g \in \mathrm{SL}(2, \mathbb{Q})$, the lattices $\Gamma = \mathrm{SL}(2, \mathbb{Z})$ and $g^{-1}\Gamma g$ are commensurable, and hence $\Gamma g A_{\mathbb{R}}$ also exits to infinity in both directions.

Now we say a few words about positive semiorbits. Every element in the group $G = \mathrm{SL}(2, \mathbb{R})$ decomposes either into the product $g = H_\alpha A_s U_\beta$, or into the product $g = \sigma A_s U_\beta$, where $\sigma = \left(\begin{smallmatrix} 0 & 1 \\ -1 & 0 \end{smallmatrix}\right) \in \Gamma$. In both cases there exist $\alpha, s, \beta \in \mathbb{R}$ and $\gamma \in \Gamma$ such that $g = \gamma H_\alpha A_s U_\beta$. Clearly, while studying divergent and bounded positive semiorbits $\{\Gamma g A_t,\ t > 0\}$, it suffices to consider the case $g = H_\alpha$ (since positive semiorbits of the points ΓH_α and $\Gamma H_\alpha A_s U_\beta$ stay at a bounded distance from each other).

To formulate the result, let us give the following definition. We say that a number $\alpha \in \mathbb{R}$ is *badly approximable* by rational numbers (and write $\alpha \in BA$) if there exists $c > 0$ such that

$$|\alpha - p/q| > \frac{c}{q^2} \qquad \text{for all } p \in \mathbb{Z}, q \in \mathbb{N}$$

(see §29 for more details).

LEMMA 14.6 [**Da85**]. *A positive semiorbit* $\{\Gamma H_\alpha A_t,\ t > 0\}$ *is bounded* $\Longleftrightarrow$ $\alpha \in BA$, *and divergent* $\Longleftrightarrow \alpha \in \mathbb{Q}$.

Proof. We apply the right action of $G = \mathrm{SL}(2, \mathbb{R})$ on $\mathbb{R}^2$. The action of $g \in G$ on $\mathbb{R}^2$ sends the lattice $\mathbb{Z}^2$ to a unimodular lattice $\mathbb{Z}^2 g$. By Mahler's criterion F.1, our semiorbit is bounded $\Longleftrightarrow$ there exists $\delta > 0$ such that

$$\|(p, q) H_\alpha A_t\| > \delta \qquad \text{for all } t > 0 \text{ and } (p, q) \in \mathbb{Z}^2 - \{0\}.$$

Here $\|(x, y)\| = \max(|x|, |y|)$. Suppose that the semiorbit is bounded. Choose $q \in \mathbb{N}$ and select $t > 0$ from the condition $e^{-t/2}|q| = \delta/2$. Then $e^{t/2}|p + \alpha q| > \delta$, whence $|p + \alpha q| > \delta^2/2q$ for all $p \in \mathbb{Z}$. Hence $\alpha \in BA$.

Conversely, let $|p + \alpha q| > \delta^2/q$, $q \in \mathbb{N}$, $p \in \mathbb{Z}$. Then the conditions $e^{t/2}|p + \alpha q| < \delta$ and $e^{-t/2} q < \delta$ can be simultaneously satisfied for $t > 0$ only if $p = q = 0$.

In a similar way, by Mahler's criterion, the semiorbit exits to infinity if and only if for any $\delta > 0$ there exists $T > 0$ such that for all $t > T$ there exist $(p, q) \in \mathbb{Z}^2 - \{0\}$ with $\|(p, q) H_\alpha A_t\| < \delta$. In other words, for any $\delta > 0$ there exists N_0 such that for all $N > N_0$ one can find $p \in \mathbb{Z}$, $q \in \mathbb{N}$ with $q < N$, $|p + \alpha q| < \delta/q$. The theory of continued fractions (see §29) says that this can happen only if $\alpha \in \mathbb{Q}$. □

In terms of the visual map, Lemma 14.6 can be formulated as follows. Let $g = \left(\begin{smallmatrix} a & b \\ c & d \end{smallmatrix}\right) \in G$. Then the positive half of the geodesic $\{\Gamma g A_t, t > 0\}$ exits to infinity if and only if $\mathrm{Vis}(g) = \frac{c}{a} \in \mathbb{Q}$, and it is bounded if and only if $\mathrm{Vis}(g) = \frac{c}{a} \in BA$ (here we agree that $\infty \in \mathbb{Q}$ and $\infty \notin BA$). Incidentally, the decomposition of $\mathbb{R}$ into irrational and rational numbers corresponds to a decomposition (well-known in the theory of Fuchsian groups) of the circle at infinity into *conical* and *parabolic* points (see [**St95c**] for more details).

We see that any point $x \in \Gamma\backslash G$ with its positive semiorbit exiting to infinity is of the form $x = \Gamma H_\alpha A_s U_\beta$, where $\alpha \in \mathbb{Q}$, $s, \beta \in \mathbb{R}$. It is easy to verify that for any rational α one has $x \in \Gamma A_{\mathbb{R}} U_{\mathbb{R}} = \Gamma B$. Thus, the family of divergent $A_{\mathbb{R}^+}$-trajectories corresponds to the cylinder ΓB consisting of periodic U-orbits. By virtue of Theorem 13.1, we conclude that any U-orbit outside ΓB is everywhere dense.

Likewise, the family of divergent $A_{\mathbb{R}^-}$-trajectories forms a cylinder $\Gamma H_{\mathbb{R}} A_{\mathbb{R}}$ consisting of all periodic $H_{\mathbb{R}}$-orbits.

Let us give examples of bounded orbits. The matrix $\gamma_0 = \left(\begin{smallmatrix} 2 & 1 \\ 1 & 1 \end{smallmatrix}\right) \in \Gamma$ reduces to a diagonal form; that is, there exists an element $g_0 \in G$ such that

$$g_0^{-1} \gamma_0 g_0 = A_s, \qquad \text{where } e^{s/2} + e^{-s/2} = 3.$$

Hence the orbit $\Gamma g_0 A_{\mathbb{R}}$ is periodic. For instance, one can choose

$$g_0 = \begin{pmatrix} 1 & -\frac{1}{\sqrt{5}} \\ \frac{\sqrt{5}-1}{2} & \frac{1+\sqrt{5}}{2\sqrt{5}} \end{pmatrix} = H_\alpha U_\beta, \qquad \text{where } \alpha = \frac{\sqrt{5}-1}{2}, \ \beta = -\frac{1}{\sqrt{5}}.$$

This is consistent with Lemma 14.6, since quadratic irrationals are badly approximable (see §29).

Since the lattices Γ and $g^{-1}\Gamma g$ are commensurable for $g \in \mathrm{SL}(2, \mathbb{Q})$, the orbit $\Gamma g g_0 A_{\mathbb{R}}$ is also periodic for every element $g \in \mathrm{SL}(2, \mathbb{Q})$ (in particular, periodic orbits of the geodesic flow are dense in $\Gamma \backslash G$).

Given $s \neq 0$, the positive semiorbit $\Gamma g_0 U_s A_{\mathbb{R}^+}$ is locally closed but not closed, because it winds asymptotically on the periodic geodesic $\Gamma g_0 A_{\mathbb{R}}$. As for the negative semiorbit, its structure is determined by the number s (but, of course, it cannot be periodic for $s \neq 0$). For instance, if we select nonzero numbers t, s, r such that $x = \Gamma g_0 U_s A_t = \Gamma g_0 H_r$, we will obtain a homoclinic orbit $x A_{\mathbb{R}}$ which winds asymptotically on the periodic orbit $\Gamma g_0 A_{\mathbb{R}}$ in both directions.

§15. Rigidity of the horocycle flow

1. Now we turn to the rigidity problem for geodesic and horocycle flows. In what follows, $M_1 = \Gamma_1 \backslash G / K$ and $M_2 = \Gamma_2 \backslash G / K$ are two surfaces of finite area and of constant curvature -1, where $G = \mathrm{SL}(2, \mathbb{R})$, $K = \mathrm{SO}(2)$. The geodesic flows on $SM_1 = \Gamma_1 \backslash G$ and $SM_2 = \Gamma_2 \backslash G$ will be denoted by $g_t^{(1)}$ and $g_t^{(2)}$, and the horocycle flows by $u_t^{(1)}$ and $u_t^{(2)}$. In addition, we normalize G-invariant Liouville measures on SM_1 and SM_2 by the condition $\nu_1(SM_1) = \nu_2(SM_2) = 1$.

Since the entropy of the translate g_1 is equal to 1 and the geodesic flow is Bernoulli, Ornstein's theorem [**O**] shows that the geodesic flows on surfaces M_1 and M_2 are measure-theoretically isomorphic. It follows that *the geodesic flow is not measure rigid*: measure-theoretic isomorphism of $g_t^{(1)}$ and $g_t^{(2)}$ implies neither isometry nor diffeomorphism of M_1 and M_2. We note also that geodesic flows on compact diffeomorphic surfaces M_1 and M_2 are topological equivalent (Anosov [**An67**]). Indeed, M_2 can be viewed as M_1 with a perturbed metric of constant negative curvature. If the change of metric is small enough, the Anosov flow $g_t^{(2)}$ is a small C^1-perturbation of the flow $g_t^{(1)}$. Hence, according to Anosov's structural stability theorem, the flows are topologically equivalent. It now suffices to observe that the space of curvature -1 metrics on the compact surface M_1 is the Teichmuller space $T(\Gamma_1)$ of the group $\pi_1(M_1) = \Gamma_1$, and hence is connected. However, M_1 and M_2 can fail to be isometric even if they have the same curvature (this is the case if

Γ_1 and Γ_2 are isomorphic but not conjugate in G). Hence *the geodesic flow is not topologically rigid either.*

As for the horocycle flow u_t, any two flows $u_t^{(1)}$ and $u_t^{(2)}$ are Kakutani equivalent; that is, there exists a measurable map $\phi : SM_1 \to SM_2$ that sends $u_t^{(1)}$-orbits to $u_t^{(2)}$-orbits preserving their orientations (moreover, it can be made C^∞-smooth along $u_t^{(1)}$-orbits). This follows from Ratner's result [**R78**] that the horocycle flow u_t is Kakutani equivalent to a rectilinear irrational flow on the 2-torus (note that any two irrational flows on the 2-torus are Kakutani equivalent according to Katok [**K75**]). It is interesting to notice that the diagonal flow $u_t \times u_t$ does not possess this property [**R79**].

Now we turn to the question of when two horocycle flows can be measure-theoretically isomorphic. First, it is clear that any isometry $\theta : M_1 \to M_2$ of two equally oriented surfaces induces the isomorphism $d\theta$ of flows $g_t^{(1)}$, $u_t^{(1)}$, $h_t^{(1)}$ and $g_t^{(2)}$, $u_t^{(2)}$, $h_t^{(2)}$ respectively. Indeed, the map

$$d\theta(\Gamma_1 g) = \Gamma_2 \alpha g, \quad \text{where } \widetilde{\theta} = f_\alpha \in \mathrm{Iso}^+(\mathbb{H}^2),\ \alpha \in G \text{ and } \alpha\Gamma_1\alpha^{-1} = \Gamma_2,$$

has the property

$$d\theta(\Gamma_1 gh) = d\theta(\Gamma_1 g)h \quad \text{for all } g, h \in G.$$

Second, one can take the differential $d\theta$ of an isometry followed by translation along a constant horocycle $u_s^{(2)}$. Surprisingly, according to Ratner's theorem on measure rigidity of the horocycle flow, proved in the early eighties, any measure-theoretic isomorphism of $u_t^{(1)}$ and $u_t^{(2)}$ comes from this construction!

THEOREM 15.1 [**R82a**]. *Let $G = \mathrm{SL}(2,\mathbb{R})$ and let Γ_1, Γ_2 be two lattices in G. Assume that $\phi : (\Gamma_1\backslash G, U_t) \to (\Gamma_2\backslash G, U_t)$ is a measure-theoretic conjugacy of horocycle flows. Then there exist an element $\alpha \in G$ and a number $s \in \mathbb{R}$ such that*

$$\alpha\Gamma_1\alpha^{-1} \subset \Gamma_2 \quad \text{and} \quad \phi(\Gamma_1 g) = \Gamma_2 \alpha g U_s \text{ almost everywhere.}$$

COROLLARY 15.2 [**R82a**]. *Let $M_1 = \Gamma_1\backslash G/K$ and $M_2 = \Gamma_2\backslash G/K$ be two oriented surfaces of finite area and of curvature -1. Assume that $\phi : (SM_1, \nu_1) \to (SM_2, \nu_2)$ is a measure-theoretic isomorphism of horocycle flows. Then $\phi = u_s^{(2)} \circ d\theta$, where $u_s^{(2)}$ is a constant translation along a horocycle and $d\theta$ is the differential of some orientation-preserving isometry $\theta : M_1 \to M_2$.*

COROLLARY 15.3 [**R83**]. *Assume that $M = \Gamma\backslash G/K$ is an oriented surface of constant negative curvature and of finite area, and that a map $\phi : (SM, \nu) \to (SM, \nu)$ establishes a measure-theoretic isomorphism of u_t and u_{-t} for all t. Then $\beta\Gamma\beta^{-1} = \Gamma$ for some $\beta \in \mathrm{Iso}(\mathbb{H}^2)$, $\det(\beta) = -1$.*

For the proof we note that the map $\Sigma(\Gamma g) = \Sigma(\Gamma)\Sigma(g)$ (where Σ is taken from §14.3) induces an isomorphism of u_t and u_{-t}. Hence $\Sigma \circ \phi$ induces an isomorphism of u_t with itself. It follows that $(\sigma\alpha)\Gamma(\sigma\alpha)^{-1} = \Gamma$ for some $\alpha \in G$.

Thus, the horocycle flow is isomorphic to its inverse only if there exists an isometry of M changing the orientation. On the contrary, the geodesic flow is always isomorphic to its inverse (it suffices to take a diffeomorphism of SM that is fixed on M and is an inversion on each tangent space).

Later, we will derive the rigidity theorem from the classification of ergodic joinings of the horocycle flow.

Remark. The rigidity phenomenon of the horocycle flow has been generalized in different directions (see [**St97**] for more details). Feldman and Ornstein [**FO**] used Ratner's methods to prove an analogue for the *uniformly* parametrized horocycle foliation of the geodesic flow on a compact surface of *variable* negative curvature. The result is that any measure-theoretic isomorphism of horocycle flows on two surfaces induces a C^1-smooth isomorphism of geodesic flows. If one of the surfaces has constant curvature, then according to Katok [**K82**] the surfaces are isometric. It was proved independently by Otal [**Ot**] and Croke [**Cr**] that in the general case a C^1-smooth isomorphism of geodesic flows is the differential of an isometry as well. Hence a complete analogue of Corollary 15.2 holds in this case.

For $n > 2$ the horospherical foliation on the unit tangent bundle SM of a compact manifold M of constant negative curvature has no natural parametrization. Nevertheless, Flaminio [**F**] managed to prove that an invertible Liouville measure-preserving map $\phi : SM_1 \to SM_2$ that is an isometry when restricted to leaves of the horospherical foliations is the differential of a global isometry $\theta : M_1 \to M_2$. An analogue of this result for a certain class of manifolds of constant curvature and of infinite volume was proved in [**FS**].

Witte [**Wi85**] proved rigidity of unipotent ergodic flows on homogeneous spaces of finite volume. We discuss this result in more detail in §18. Note that all the generalizations of the rigidity theorem use the method of [**R82a**].

2. Rigidity of the horocycle flow is also expressed by the fact that quotient flows and joinings of it are of algebraic origin. This was also discovered by Ratner.

Let us consider the horocycle flow $u_t = (\Gamma\backslash G, U_t)$ on a space with finite G-invariant measure ν. If a discrete subgroup $\Gamma' \subset G$ contains Γ, the horocycle flow $u'_t = (\Gamma'\backslash G, U_t)$ is a quotient flow of u_t. As Ratner proved, every quotient flow for u_t arises from this algebraic construction:

THEOREM 15.4 [**R82b**]. *Let S_t be a quotient flow of the horocycle flow $u_t = (\Gamma\backslash G, U_t)$, Γ being a lattice in G. Then there exists a lattice $\Gamma' \subset G$ such that $\Gamma \subset \Gamma'$ and the flow S_t is isomorphic to the flow $u'_t = (\Gamma'\backslash G, U_t)$.*

Since there are finitely many lattices $\Gamma' \subset G$ containing the lattice $\Gamma \subset G$, it follows that there are finitely many nonisomorphic quotient flows of the horocycle flow u_t. In particular, if the lattice Γ is maximal, then u_t has no nontrivial quotient flows at all. This reveals another striking difference from the geodesic flow. Indeed, a flow S_t is a quotient of a Bernoulli flow T_t if and only if S_t is also Bernoulli and the entropy $e(S_t)$ is at most $e(T_t)$. It follows that there are uncountably many nonisomorphic quotient flows of the geodesic flow.

The theorem on quotients, as well as the rigidity theorem, is easily deduced from the classification of ergodic joinings of the horocycle flow. We recall that a *joining* of two horocycle flows $u_t^{(1)} = (\Gamma_1\backslash G, U_t, \nu_1)$ and $u_t^{(2)} = (\Gamma_2\backslash G, U_t, \nu_2)$ is a $U_t \times U_t$-invariant probability measure on $(\Gamma_1\backslash G) \times (\Gamma_2\backslash G)$ which projects to G-invariant measures ν_1 and ν_2. First we give an algebraic construction of ergodic joinings for $u_t^{(1)}$ and $u_t^{(2)}$.

Assume that $c \in G$ is such that $\Gamma = \Gamma_1 \cap c^{-1}\Gamma_2 c$ is a lattice in G. Let $\Omega_c = \{(\Gamma_1 g, \Gamma_2 cg),\ g \in G\}$. Then the set Ω_c is invariant under the diagonal subgroup $\Delta_G = \{(g, g),\ g \in G\}$, and Δ_G acts transitively on Ω_c with stationary subgroup Δ_Γ. A Δ_G-invariant probability measure ν_0 on $\Omega_c = \Delta_\Gamma \backslash \Delta_G$ is a $U_t \times U_t$-ergodic joining of the flows $u_t^{(1)}$ and $u_t^{(2)}$. A more general construction arises as a translate of the set Ω_c and the measure ν_0 by any element from $G \times G$ commuting with the subgroup $\Delta_U = \{U_t \times U_t,\ t \in \mathbb{R}\}$. It is clear that any such translate is of the form

$$\Omega_{c',s} = \{(\Gamma_1 g, \Gamma_2 c' g U_s),\ g \in G\} \quad \text{for some } s \in \mathbb{R} \text{ and } c' \in c\mathbb{Z}_2.$$

The following result of Ratner shows that all nontrivial ergodic joinings of $u_t^{(1)}$ and $u_t^{(2)}$ occur in a similar way.

THEOREM 15.5 [**R83**]. *Let a measure m on $(\Gamma_1\backslash G) \times (\Gamma_2 \backslash G)$ be an ergodic joining for $u_t^{(1)} = (\Gamma_1\backslash G, U_t, \nu_1)$ and $u_t^{(2)} = (\Gamma_2 \backslash G, U_t, \nu_2)$. Then either $m = \nu_1 \times \nu_2$, or there exist an element $c \in G$ and a number $s \in \mathbb{R}$ such that*

1) *$\Gamma = \Gamma_1 \cap c^{-1}\Gamma_2 c$ is a lattice in G,*
2) *$m(\Omega_{c,s}) = 1$, and*
3) *the flow $((\Gamma_1\backslash G) \times (\Gamma_2\backslash G), U_t \times U_t, m)$ is measure-theoretically isomorphic to the horocycle flow $(\Gamma\backslash G, U_t, \nu_0)$.*

COROLLARY 15.6 [**R83**]. *There are finitely many nonisomorphic ergodic joinings of two horocycle flows. If the lattice Γ_1 is uniform or arithmetic, and the lattice Γ_2 is not, then the flows $u_t^{(1)}$ and $u_t^{(2)}$ are disjoint.*

Note that there are uncountably many nonisomorphic ergodic joinings of any two Bernoulli (in particular, geodesic) flows.

We indicate another interesting corollary of Theorem 15.5. If the lattice Γ_1 is uniform and arithmetic, and the lattice Γ_2 is uniform and nonarithmetic, then the one-parameter unipotent flow $((\Gamma_1\backslash G) \times (\Gamma_2\backslash G), U_t \times U_t)$ is uniquely ergodic. This follows from the classification of joinings and the unique ergodicity of the horocycle flows on $\Gamma_1\backslash G$ and $\Gamma_2\backslash G$. It is of interest in this example that the one-parameter subgroup $\Delta_U = \{U_t \times U_t,\ t \in \mathbb{R}\}$ *is not horospherical* (a horospherical one is the two-dimensional subgroup $U_\mathbb{R} \times U_\mathbb{R}$). Until Ratner proved Theorem 15.5, the only examples of uniquely ergodic flows on compact homogeneous spaces of semisimple Lie groups were given by horospherical flows (see §13).

Glasner and Weiss [**GW**] have constructed an example of two maximal uniform arithmetic lattices Γ_1 and Γ_2 such that Γ_1 and Γ_2 are not conjugate and $\Gamma_1 \cap \Gamma_2$ is of finite index in Γ_1 and Γ_2. Therefore minimal horocycle flows $u_t^{(1)}$ and $u_t^{(2)}$ are not disjoint. This answers Furstenberg's question negatively: *minimal flows with no common quotients are not always disjoint.*

Now we turn to counting the number of nonisomorphic ergodic self-joinings of the horocycle flow $u_t = (\Gamma\backslash G, U_t, \nu)$. Given a lattice $\Gamma \subset G$, let $\overline{\Gamma}$ be the subgroup of all elements $c \in G$ such that the intersection $\Gamma \cap c^{-1}\Gamma c$ is a lattice in G. According to Margulis (see [**Ma91c**]), the subgroup $\overline{\Gamma}$ is dense in G if and only if the lattice Γ is arithmetic. Otherwise the subgroup $\overline{\Gamma}$ is discrete and hence contains Γ as a subgroup of finite index.

COROLLARY 15.7 [**R83**]. *If a lattice $\Gamma \subset G$ is arithmetic, then there are countably many nonisomorphic ergodic self-joinings for the horocycle flow $u_t = (\Gamma\backslash G, U_t, \nu)$. If a lattice $\Gamma \subset G$ is nonarithmetic, then there are finitely many nonisomorphic ergodic self-joinings for u_t, and if in addition Γ is maximal, then every nontrivial self-joining for u_t is diagonal.*

Following Ratner, let us show how the rigidity theorem for a horocycle flow can be deduced from the classification of ergodic joinings. Let

$$\phi : (\Gamma_1\backslash G, U_t, \nu_1) \to (\Gamma_2\backslash G, U_t, \nu_2)$$

be a measure-preserving conjugacy of horocycle flows. We consider the graph

$$\Omega = \{(x, \phi(x)),\ x \in \Gamma_1\backslash G\} \subset (\Gamma_1\backslash G) \times (\Gamma_2\backslash G)$$

along with a probability measure m given as

$$m(A \times B) = \int_A m_{\phi(x)}(B) d\nu_1(x),$$

where the measure m_y, $y \in \Gamma_2\backslash G$, is concentrated at the point y. Then the set Ω and the measure m are $U_t \times U_t$-invariant, and m is a joining of our two flows. Since the projection $p_1 : \Omega \to \Gamma_1\backslash G$ is one-to-one, the measure m is ergodic. It follows from the classification of ergodic joinings that there exist an element $c \in G$ and a number $s \in \mathbb{R}$ such that the intersection $\Gamma = \Gamma_1 \cap c^{-1}\Gamma_2 c$ is a subgroup of finite index in Γ_1 and $c^{-1}\Gamma_2 c$. In addition, $\phi(\Gamma_1 g) = \Gamma_2 c g U_s$. Finally since p_1 is one-to-one, we conclude that $\Gamma = \Gamma_1$, and therefore $\Gamma_1 \subset c^{-1}\Gamma_2 c$.

Remark. Actually, all the results of the chapter were proved by Ratner in a stronger form: not only for the horocycle flow $(\Gamma\backslash G, U_\mathbb{R})$ but also for the horocycle one-generator action $(\Gamma\backslash G, U_\mathbb{Z})$ (see also [**R84**]). For instance, in the rigidity Theorem 15.1 it suffices to require that the map $\phi : \Gamma_1\backslash G \to \Gamma_2\backslash G$ be such that $\phi \circ U_1 = U_1 \circ \phi$.

§16. Classification of ergodic measures of the horocycle flow

1. Here we prove the Dani measure conjecture for the horocycle flow $(\Gamma\backslash G, U_\mathbb{R})$, where Γ is assumed to be an *arbitrary discrete subgroup* of $G = \mathrm{SL}(2, \mathbb{R})$. More specifically, we will prove that an ergodic probability measure for the flow either is supported on a periodic orbit, or coincides with G-invariant Liouville measure (and then Γ is a lattice in G). The case where Γ is a lattice in G was examined by Dani and Smillie [**DS**]. We present Ratner's proof [**R92**], which does not involve the fact that $U_\mathbb{R}$ is a horospherical subgroup; see also [**Gh**].

We consider the Lie group $G = \mathrm{SL}(2, \mathbb{R})$ equipped with a left-invariant metric d_G. Let us introduce the following notation:

$$U_s = \begin{pmatrix} 1 & s \\ 0 & 1 \end{pmatrix}, \qquad A_\tau = \begin{pmatrix} e^\tau & 0 \\ 0 & e^{-\tau} \end{pmatrix}, \qquad H_b = \begin{pmatrix} 1 & 0 \\ b & 1 \end{pmatrix}.$$

We recall that the diagonal subgroup $A_\mathbb{R}$ normalizes the unipotent subgroups $U_\mathbb{R}$ and $H_\mathbb{R}$. Therefore $B = A_\mathbb{R} U_\mathbb{R}$ and $W = A_\mathbb{R} H_\mathbb{R}$ are two-dimensional solvable subgroups of G.

With respect to the left-invariant metric d_G, the leaf $xU_\mathbb{R}$ of a point $x \in G$ is contracting for the right action of the diagonal subgroup $A_\mathbb{R}$ on G. Similarly, the partition of G into orbits $xH_\mathbb{R}, x \in G$, forms the expanding foliation. Thus, the right action of $A_\mathbb{R}$ on G is an Anosov flow.

However, here we are concerned with the right action of the unipotent subgroup $U_\mathbb{R}$ on G. We agree for simplicity that the $U_\mathbb{R}$-direction is vertical. Then the transversal leaf xW of $x \in G$ can be naturally called *horizontal*. Let us introduce the notation

$$A(\delta) = \{A_\tau, |\tau| \le \delta\}, \quad H(\delta) = \{H_b, |b| \le \delta\}, \quad U(\delta) = \{U_s, |s| \le \delta\}$$

and let

$$W(\delta) = A(\delta)H(\delta), \quad B(\delta) = A(\delta)U(\delta).$$

We fix an arbitrary point $x \in G$ and consider the horizontal leaf xW. In time s the point x moves to the point $xU_s \in G$. It is intuitively clear that if the time s is not large and the size δ of the rectangle $xW(\delta)$ is small enough, then the trajectory $yU_{\mathbb{R}}$ of an arbitrary point $y \in xW(\delta)$ meets the horizontal leaf xU_sW at some moment $\alpha(y,s)$. We let

$$y = xA_\tau H_b \quad \text{and} \quad yU_{\alpha(y,s)} = xU_s A_{\tau(y,s)} H_{b(y,s)}.$$

Thus, $\alpha(y,s)$ is the time required for the point $y \in xW$ to meet the horizontal leaf xU_sW of the point xU_s, and $\tau(y,s)$ and $b(y,s)$ are the A- and H-coordinates of the point $yU_{\alpha(y,s)}$ in the horizontal leaf xU_sW (where $\tau(y,0)=\tau$, $b(y,0)=b$).

It is straightforward to verify that

$$\alpha(y,s) = \frac{s}{e^{2\tau} - sb}, \qquad \tau(y,s) = \ln(e^\tau - sbe^{-\tau}), \qquad b(y,s) = b(1 - bse^{-2\tau}).$$

Hence the functions $\alpha(y,s)$, $\tau(y,s)$, $b(y,s)$ are defined for $sb < e^{2\tau}$. Note that if $b = 0$, that is, $y \in xA_{\mathbb{R}}$, then $\tau(y,s) = \tau, b(y,s) = 0, \alpha(y,s) = s/e^{2\tau}$, and all these three functions are defined everywhere. But if $b \neq 0$, then there exists a finite critical value $s_{\text{cr}} = s_{\text{cr}}(y)$ such that the trajectory $yU_{\mathbb{R}}$ does not intersect the horizontal leaf $xU_{s_{\text{cr}}}W$ of the point $xU_{s_{\text{cr}}}$. Instead, it asymptotically approaches the leaf in both directions (but from different sides).

The following lies at the heart of this phenomenon: topologically, the group G is not homeomorphic to the vector 3-space and does not decompose into the product $U_{\mathbb{R}}A_{\mathbb{R}}H_{\mathbb{R}}$. Therefore, the element $y^{-1}xUs_{\text{cr}}$ can fail to belong to the product $U_{\mathbb{R}}W$, and thus the intersection $xU_{s_{\text{cr}}}W \cap yU_{\mathbb{R}}$ may be empty.

2. Divergence of points xU_s and yU_s in the metric d_G can be estimated coordinatewise. Assume that originally (for $s = 0$) a point y is at a distance τ from a point x in the A-direction and at a distance b in the H-direction. Then within time s they will diverge at the distance $\tau(y,s)$ in the A-direction, at the distance $b(y,s)$ in the H-direction and at the distance $s - \alpha(y,s)$ in the U-direction. The dynamics of the $U_{\mathbb{R}}$-action on G is such that divergence of close points occurs in the U-direction much faster than in the transversal A- and H-directions. For instance, if $y = xA_\tau$, then $\tau(y,s) \equiv \tau$, $b(y,s) \equiv 0$; that is, there is no divergence in the A- and H-directions at all. But $\alpha(y,s) = s/e^{2\tau}$, and hence divergence of points within time s in the U-direction is proportional to s.

If $y = xA_\tau H_b$, where $b \neq 0$, it makes sense to estimate the behavior of the functions $\alpha(y,s)$, $\tau(y,s)$ and $b(y,s)$ as $s \to s_{\text{cr}}$. We let $r(s) = 1 - s/s_{\text{cr}}$. Then

$$\alpha(y,s) = 1/br - 1/b, \quad \tau(y,s) = \tau + \ln(r), \quad b(y,s) = br.$$

Thus, $b(y,s) \to 0$, $\tau(y,s)$ has a singularity of order $\ln(r)$, and $\alpha(y,s)$ has a singularity of order $1/r$ as $s \to s_{\text{cr}}$.

Now we fix a parameter $0 < \theta < 1$ and consider $y \in xW(\delta)$, where $\delta < \theta/10$. Let $s_\theta = s_\theta(\tau,b) > 0$ be the first moment when the absolute value of the divergence

$\tau(y,s)$ of points x and $y = xA_\tau H_b, b \neq 0$, reaches θ; that is, $|\tau(y,s_\theta)| = \theta$ (sooner or later this happens for any point y with $b \neq 0$). Explicitly,

$$s_\theta = \frac{e^\tau(e^\tau - e^{-\theta\,\mathrm{sgn}(b)})}{b}, \quad \alpha(y,s_\theta) = \frac{s_\theta}{e^{-\theta\,\mathrm{sgn}(b)+\tau}},$$
$$\tau(y,s_\theta) = -\theta\,\mathrm{sgn}(b), \quad b(y,s_\theta) = be^{-\theta\,\mathrm{sgn}(b)-\tau},$$

where $\mathrm{sgn}(b)$ is the signature of the parameter $b \neq 0$. If we make the size δ of the original rectangle $xW(\delta)$ small, we obtain

$$s_\theta \to +\infty, \quad \alpha(y,s_\theta)/s_\theta \to e^{\theta\,\mathrm{sgn}(b)}, \quad b(y,s_\theta) \to 0.$$

Thus, the closer the points y and x are, the longer time is required for them to move apart in the A-direction at a distance θ. Moreover, divergence of these points in the U-direction at this moment tends to ∞, and in the H-direction to 0 (as $y \to x$). We need an exact formulation of this observation, the so-called *Ratner property* [**R90b,95b**], which modifies the H-property from [**R82a**] and plays an essential role in the proof of the Dani measure conjecture.

LEMMA 16.1 (R-property of the horocycle flow). *There exist constants $0 < \eta < 1$ and $C > 1$ such that if*

$$0 < \theta < 1, \qquad 0 < \delta < \theta/10, \qquad y \in xW(\delta), \qquad |\tau(y,s_\theta)| = \theta,$$

where $|\tau(y,s)| < \theta$, $0 < s < s_\theta$, then

$$\theta/2 \le |\tau(y,s)| \le \theta, \quad |b(y,s)| \le C\theta/s_\theta \qquad \text{for all } s \in [(1-\eta)s_\theta, s_\theta].$$

Proof. Let

$$xW^+(\delta) = \{y = xA_\tau H_b, |\tau| < \delta, \ 0 < b < \delta\}.$$

The function s_θ increases monotonically in θ for all $y \in xW^+(\delta)$. Let us prove that there exists $0 < \eta < 1$ such that $s_{\theta/2} \ge (1-\eta)s_\theta$ for all $0 < \theta < 1$, $y \in xW^+(\delta)$. It suffices to show that $\eta = \inf\{1 - s_{\theta/2}/s_\theta\} > 0$, where the infimum is taken over all $0 < \theta < 1$, $y \in xW^+(\theta/10)$. But for all $|\tau| \le \theta/10$ we have

$$1 - \frac{s_{\theta/2}}{s_\theta} = \frac{e^{-\theta/2} - e^{-\theta}}{e^\tau - e^{-\theta}} \ge \frac{e^{-\theta/2} - e^{-\theta}}{e^{\theta/10} - e^{-\theta}} = \frac{e^{\theta/2} - 1}{e^{1.1\theta} - 1}.$$

Since $(e^{\theta/2} - 1)/(e^{1.1\theta} - 1) \to 10/22$ as $\theta \to 0$, it follows that $\eta > 0$.

Let us prove now that $C = \sup\{|b(y,s)|s_\theta/\theta\} < \infty$, where the supremum is taken over all $0 < \theta < 1$, $y \in xW^+(\theta/10)$, $(1-\eta)s_\theta < s < s_\theta$. But for all $|\tau| \le \theta/10$ we have

$$\frac{b(y,s)s_\theta}{\theta} = \frac{(1 - bse^{-2\tau})(e^{2\tau} - e^{\tau-\theta})}{\theta} \le \frac{e^{2\tau} - e^{\tau-\theta}}{\theta} \le \frac{e^{\theta/5} - e^{-1.1\theta}}{\theta}.$$

Since $(e^{\theta/5} - e^{-1.1\theta})/\theta \to 13/10 < \infty$ as $\theta \to 0$, it follows that $C < \infty$, and we are done. The lower half $xW^-(\theta/10)$ is considered in a similar way. □

This lemma again reflects the property of the horocycle flow that the divergence of two close points x and y occurs faster in the flow direction than in transversal directions. Indeed, if we fix $0 < \theta < 1$ and make $\delta \to 0$, we obtain

$$s_\theta \to \infty, \quad s_\theta - \alpha(y,s_\theta) \to \infty, \quad |\tau(y,s_\theta)| = \theta, \qquad b(y,s_\theta) \to 0;$$

that is, when the divergence in the A-direction is fixed, the divergence in the U-direction tends to infinity, and that in the H-direction to 0.

We need also the following result: given a point $y \in xW$ close enough to x, the function $\phi_y(s) = \alpha(y, s)$ on the segment $[0, s_\theta]$ almost does not distort the Lebesgue measure λ.

LEMMA 16.2. *For any $\varepsilon > 0$ there exists $\theta = \theta(\varepsilon) > 0$ such that, given any $y \in xW(\theta)$ and any Borel set $C \subset [0, s_\theta]$, one has*

$$\left| \frac{\lambda(C)}{\lambda(\phi_y(C))} - 1 \right| < \varepsilon.$$

Proof. It suffices to prove that for arbitrarily small $\varepsilon > 0$ there exists $\theta > 0$ such that for all $y \in xW(\theta)$ and $s \in [0, s_\theta]$ one has $|\phi_y'(s) - 1| < \varepsilon$. But

$$\phi_y(s) = s/(e^{2\tau} - sb), \quad \phi_y'(s) = e^{2\tau}/(e^{2\tau} - sb)^2.$$

Since the function $\phi_y'(s)$ is monotone and

$$\phi_y'(0) = e^{-2\tau}, \quad \phi_y'(s_\theta) = e^{2\tau + 2\theta \operatorname{sgn}(b)},$$

it now suffices to select θ from the condition $e^{4\theta} - 1 < \varepsilon$. □

3. Now we formulate and prove the classification of finite ergodic measures for the horocycle flow.

THEOREM 16.3. *Let Γ be a discrete subgroup of $G = \mathrm{SL}(2, \mathbb{R})$ and let μ be an ergodic $U_\mathbb{R}$-invariant probability measure on $X = \Gamma\backslash G$. Then either*
1) Γ is a lattice in G and the measure μ is G-invariant, or
2) μ is supported on a periodic orbit $U_\mathbb{R}$.

Let us proceed to the proof. From now on d_X denotes the metric on X induced by a left-invariant metric d_G on G, and $O_\varepsilon(x) = \{z \in X \,:\, d_X(z, x) < \varepsilon\}$ stands for the ε-neighborhood of a point $x \in X$. Let

$$\Lambda = \Lambda(\mu) = \{g \in G \,:\, \mu(E) = \mu(Eg) \text{ for every Borel } E \subset X\}.$$

Then Λ is the stabilizer of the measure μ, and Λ is a closed subgroup of G containing U. There are two possibilities:
1) $A_\mathbb{R} \not\subset \Lambda$,
2) $A_\mathbb{R} \subset \Lambda$.

Consider the first case. Since $A_\mathbb{R} \not\subset \Lambda$, it follows that $B = A_\mathbb{R} U_\mathbb{R} \not\subset \Lambda$.

LEMMA 16.4. *If $A_\mathbb{R} \not\subset \Lambda$, then there exists a subset $Y \subset X$ such that $\mu(Y) = 1$ and $Y \cap Yq = \varnothing$ for any $q \in B - \Lambda$.*

Proof. First we find the set in question for a single element $q \in B - \Lambda$. We put $\mu_q(E) = \mu(Eq)$ for $q \in B - \Lambda$. Then the measure μ_q is a $U_\mathbb{R}$-invariant ergodic probability measure which is different from μ. Hence the measures μ_q and μ are mutually singular; that is, there exists a set $E \subset X$ such that $\mu(E) = 1$ and $\mu_q(E) = \mu(Eq) = 0$. Then for $Y_q = E - Eq$ one has $\mu(Y_q) = 1$ and $Y_q \cap Y_q q = \varnothing$.

Consider a compact set $K \subset Y_q$ such that $\mu(K) \geq 0.99$. Since $K \cap Kq = \varnothing$, there exists $\varepsilon > 0$ such that $d_X(K, Kq) \geq \varepsilon$. Since μ is ergodic, the set $Y_{q,\varepsilon} \subset X$

consisting of points $x \in X$ such that

$$S_{\chi_K}(x,t) = \frac{1}{t}\int_0^t \chi_K(xU_s)ds \to \mu(K)$$

is of full μ-measure (from here on, χ_K stands for the characteristic function of a subset $K \subset X$).

Let us prove that

$$Y_{q,\varepsilon} \cap Y_{q,\varepsilon} g = \varnothing \quad \text{for any } g \in qB(\varepsilon).$$

Suppose on the contrary that there exist an element $g \in qB(\varepsilon)$ and a point $x, y \in Y_{q,\varepsilon}$ such that $x = yg$. Let $g = A_\tau U_r$. Then $yU_s g = xU_{e^{-2\tau}s}$ for any $s \in \mathbb{R}$. Since $x, y \in Y_{q,\varepsilon}$, it follows that for large enough $t > 1$ we have

$$S_{\chi_K}(y,t) \geq 0.9 \quad \text{and} \quad S_{\chi_K}(x, e^{-2\tau}t) \geq 0.9.$$

Hence one can always find $s \in [0,t]$ such that

$$z = yU_s \in K \quad \text{and} \quad zg = xU_{e^{-2\tau}s} \in K.$$

But $zg = zqp$, where $p \in B(\varepsilon)$, and therefore $d_X(zg, zq) < \varepsilon$, where $zg \in K$ and $zq \in Kq$, a contradiction.

Finally, we represent the set $B - \Lambda$ as a countable union $B - \Lambda = \bigcup_i B(\varepsilon_i, q_i)$ and define $Y = \bigcap_i Y_{q_i,\varepsilon_i}$. Then $\mu(Y) = 1$ and $Y \cap Y(B - \Lambda) = \varnothing$. □

Now we can handle the first possibility.

LEMMA 16.5. *If $A_{\mathbb{R}} \not\subset \Lambda$, then there exists a point $x \in X$ such that $\mu(xU_{\mathbb{R}}) = 1$, and hence the measure μ is supported on a periodic orbit.*

Proof. Let $\theta > 0$ satisfy $A_\tau \notin \Lambda$ for all $0 < |\tau| < \theta$ and $\theta \leq \theta(0.1)$ (see Lemma 16.2). We take constants $C > 1, \eta < 1$ from the R-property, and a set $Y \subset X$ from Lemma 16.4. Then there exist a compact set $K \subset Y$ and a number $\delta > 0$ such that

$$\mu(K) > 1 - \eta/1000 \quad \text{and} \quad d_X(K, KA_\tau) \geq \delta \quad \text{for all } \theta/2 \leq |\tau| \leq \theta.$$

Since μ is ergodic, there exist a set $F \subset X$ of positive μ-measure and a number $t_0 \geq 1$ such that

$$S_{\chi_K}(x,t) \geq 1 - \eta/100 \quad \text{for all } x \in F,\ t \geq t_0.$$

Finally, we select small enough $0 < \xi < \theta$ such that for all $x \in G$ and $y \in W(x,\xi)$ we have $s_\theta \geq 10t_0$ and $C\theta/s_\theta < \delta/10$.

Let us prove that the set F is locally contained in the two-dimensional vertical B-leaf: $F \cap O_\varepsilon(x) \subset xB(\xi)$ for each $x \in F$. Without loss of generality we assume the opposite: there exist

$$x, y \in F, \quad y = xA_\tau H_b, \quad |\tau| < \xi, \quad 0 < |b| < \xi.$$

We have $s_\theta > t_0$ and $\alpha(y, s_\theta) > t_0$. So $S_{\chi_K}(x, s_\theta) \geq 1 - \eta/100$, and similarly $S_{\chi_K}(y, \alpha(y, s_\theta)) \geq 1 - \eta/100$. In view of Lemma 16.2 and the choice of θ, the function $\alpha(y,s)$ almost does not distort the Lebesgue measure on $\mathbb{R}$. Hence one can find a number $s \in [(1-\eta)s_\theta, s_\theta]$ such that $xU_s \in K$ and $yU_{\alpha(y,s)} \in K$. But

$$yU_{\alpha(y,s)} = xU_s A_{\tau(y,s)} H_{b(y,s)},$$

and by the R-property

$$\theta/2 \le |\tau(y,s)| \le \theta \quad \text{and} \quad |b(y,s)| \le C\theta/s_\theta \le \delta/10.$$

Since $xU_s \in K$, it follows that $xU_sA_{\tau(y,s)} \in KA_{\tau(y,s)}$. Consequently,

$$d_X(K, KA_{\tau(y,s)}) \le \delta/10, \quad \text{where } \theta/2 \le |\tau(y,s)| \le \theta.$$

This contradicts the choice of δ and shows that $F \cap O_\varepsilon(x) \subset xB(\xi)$.

On the other hand, since

$$B(\xi) \cap (B - \Lambda) \subset U \quad \text{and} \quad Y \cap Y(B-\Lambda) = \varnothing,$$

we have $Y \cap xB(\xi) \subset xU_{\mathbb{R}}$ for each $x \in Y$. Because $\mu(F) > 0$ and $\mu(Y) = 1$, there exists a point $x \in F \cap Y$ such that $\mu(F \cap O_\varepsilon(x)) > 0$ for any $\varepsilon > 0$. In particular,

$$0 < \mu(F \cap O_\xi(x)) < \mu(xB(\xi)) = \mu(xB(\xi) \cap Y) < \mu(xU_{\mathbb{R}}).$$

Now in view of ergodicity we conclude that $\mu(xU_{\mathbb{R}}) = 1$. □

Now we turn to the second possibility. We have already proved in §14 that the flow $(\Gamma\backslash G, A_{\mathbb{R}})$ is mixing with respect to μ whenever μ is $A_{\mathbb{R}}$-invariant. The last step in the proof of the theorem is the following lemma.

LEMMA 16.6. *Let $A_{\mathbb{R}} \subset \Lambda$. If ν is a G-invariant Borel measure on $X = \Gamma\backslash G$, then $\nu(X) < \infty$ and $\mu = \nu/\nu(X)$.*

Proof. Choose an arbitrary continuous function f on X with compact support. Since μ is $A_{\mathbb{R}}$-ergodic, the set $C_f \subset X$ of points $y \in X$ such that

$$S_{f,n}(y) = \frac{1}{n}\sum_{i=0}^{n-1} f(yA_{-i}) \to f_\mu = \int_X f\,d\mu, \quad n \to \infty,$$

is of full μ-measure. Since $H_{\mathbb{R}}$-orbits are the contracting horocycles for the negative direction of geodesics, it follows that $d_X(zA_{-n}, yA_{-n}) \to 0$, $n \to +\infty$, for any element $z \in yH_{\mathbb{R}}$. The function f is uniformly continuous, and hence from $S_{f,n}(y) \to f_\mu$ it follows that $S_{f,n}(z) \to f_\mu$, $z \in yH$. Consequently, $C_fH = C_f$. On the other hand, the measure μ is B-invariant and $\mu(C_f) = 1$. It follows that C_f is of full ν-measure. More specifically, let us decompose the measure μ in the neighborhood

$$O(x) = xB(\delta/2)H(\delta) \cap xH(\delta/2)B(\delta)$$

into conditional measures μ_y on leaves $yB(\delta/2)$, $y \in xH(\delta/2)$. Since μ is B-invariant, almost every μ_y is smooth, i.e., equivalent to the Lebesgue measure on $yB(\delta/2)$. All intersections $C_f \cap yB(\delta/2)$ are obtained from $C_f \cap xB(\delta/2)$ by analytic maps which send sets of zero Lebesgue measure to sets of zero Lebesgue measure. Hence the volume of $C_f \cap O(x)$ is equal to the volume of $O(x)$ (see [**R92**] for more details).

Assume now that the function f is in addition nonnegative and $f_\mu > 0$. Then, by the Fatou lemma,

$$f_\mu \nu(X) = f_\mu \nu(C_f) = \int_{C_f} f_\mu\,d\nu \le \lim_{n\to\infty} \int_{C_f} S_{f,n}\,d\nu = \int_{C_f} f\,d\nu = \int_X f\,d\nu < \infty.$$

We have proved that $\nu(X) < \infty$. Since the sequence $S_{f,n}$ is uniformly bounded, we can apply the Lebesgue dominated convergence theorem:

$$f_\nu = \int_X f d\nu = \int_{C_f} f d\nu = \int_{C_f} S_{f,n} d\nu \to \int_{C_f} f_\mu d\nu = f_\mu \nu(X).$$

Hence $f_\mu = f_\nu / \nu(X)$ for any nonnegative continuous function f with compact support. Thus, $\mu = \nu/\nu(X)$, as required. $\square$

§17. Classification of ergodic measures of unipotent flows

1. Here we discuss the fundamental Ratner measure theorem which states that all ergodic probability measures for a unipotent flow on a homogeneous space have an algebraic origin. We return to the *left* actions on *right* homogeneous spaces.

THEOREM 17.1 [**R91a**]. *Assume that G is a Lie group, U is a unipotent subgroup of G, Γ is a discrete subgroup of G, and μ is an ergodic U-invariant probability measure for the action $(G/\Gamma, U)$. Then there exist a point $x \in G/\Gamma$ and a closed subgroup $H \subset G$ such that μ is an H-invariant measure supported on the closed orbit $Hx \subset G/\Gamma$.*

The case of a one-parameter subgroup $U \subset G$ occupies the main part of the proof. We only outline this rather complicated proof (recall that the case $G = \mathrm{SL}(2,\mathbb{R})$ was examined in §16).

Let $a \in G$ and $x \in G$, and let a subgroup $\Gamma \subset G$ be discrete. One says that the semi-orbit $\{a^n x\Gamma, n \in \mathbb{Z}^+\}$ is *essentially divergent* if there exist elements $\gamma_n \in \Gamma - \{1\}$ such that

$$a^n x \gamma_n (a^n x)^{-1} \to 1, \qquad n \to +\infty.$$

It is easy to see that any essentially divergent semiorbit $a^n x\Gamma$ is divergent (that is, exits to infinity in G/Γ). Indeed, suppose that $a^{n_k} x\Gamma \to z\Gamma$, $n_k \to +\infty$. Then there exists a sequence $\theta_k \in \Gamma$ such that $a^{n_k} x\theta_k \to z$. But

$$(a^{n_k} x)\gamma_{n_k}(a^{n_k} x)^{-1} = (a^{n_k} x\theta_k)\theta_k^{-1}\gamma_{n_k}\theta_k(a_{n_k} x\theta_k)^{-1} \to 1,$$

and hence $\theta_k^{-1}\gamma_{n_k}\theta_k \to 1$. Since Γ discrete and $\gamma_{n_k} \neq 1$, we arrive at a contradiction.

Obviously, a divergent sequence can fail to be essentially divergent (example: $\Gamma = \mathbb{Z} \subset \mathbb{R}^2$). Actually, a discrete subgroup $\Gamma \subset G$ such that the two properties are equivalent is a lattice (see [**Rag**]).

Given an element $a \in G$, let us consider the set $D(a) \subset G/\Gamma$ of all essentially divergent positive semi-orbits. If μ is a finite a-invariant measure on G/Γ, then by the Poincaré recurrence theorem, $\mu(D(a)) = 0$.

We recall that an element $\mathfrak{a} \in \mathfrak{g}$ is *semisimple* if the operator $\mathrm{ad}_\mathfrak{a}$ is diagonalizable over $\mathbb{C}$. If the operator is diagonalizable over $\mathbb{R}$, the element $\mathfrak{a} \in \mathfrak{g}$ is called $\mathbb{R}$-*diagonalizable.*

A nilpotent element $\mathfrak{u} \in \mathfrak{g}$ is said to be a *horocycle* if there exist an $\mathbb{R}$-diagonalizable element $\mathfrak{a} \in \mathfrak{g}$ and a nilpotent element $\mathfrak{u}^* \in \mathfrak{g}$ such that

$$[\mathfrak{u}, \mathfrak{u}^*] = \mathfrak{a}, \qquad [\mathfrak{a}, \mathfrak{u}] = -2\mathfrak{u}, \qquad [\mathfrak{a}, \mathfrak{u}^*] = 2\mathfrak{u}^*.$$

In this case the element $\mathfrak{a}$ is said to be *diagonal* for u. Likewise, the element $a = \exp(\mathfrak{a}) \in G$ is said to be *diagonal* for the *horocycle* element $u = \exp(\mathfrak{u}) \in G$.

By definition, the elements $\mathfrak{a}, \mathfrak{u}, \mathfrak{u}^*$ generate a subalgebra $\mathrm{sl}_2(\mathfrak{u}, \mathfrak{a}) \simeq \mathrm{sl}(2, \mathbb{R})$. In turn, the subgroups $\exp(\mathbb{R}\mathfrak{a})$, $\exp(\mathbb{R}\mathfrak{u})$ and $\exp(\mathbb{R}\mathfrak{u}^*)$ generate a subgroup $\mathrm{SL}_2(u, a)$ of G locally isomorphic to $\mathrm{SL}(2, \mathbb{R})$. Note that by the Jacobson–Morozov lemma, any nilpotent element in a semisimple Lie algebra is a horocycle. It is clear also that a nilpotent element in an arbitrary Lie algebra $\mathfrak{g}$ need not be a horocycle (example: $\mathfrak{g}$ is solvable).

Let

$$G^-(a) = \{g \in G : \ a^n g a^{-n} \to 1, \ n \to +\infty\}$$

be the contracting horospherical subgroup for a diagonal element $a = \exp(\mathfrak{a}) \in G$. Then $u = \exp(\mathfrak{u}) \in G^-(a)$. Note that a diagonal element for a horocycle element is not unique in general. However, one can prove (see [**R91a**]) that if $a, a' \in G$ are diagonal for $u \in G$, then $a' = cac^{-1}$ for some $c \in G^-(a) \cap Z_G(u)$ (here $Z_G(u)$ stands for the centralizer of u in G).

Example 17.2. Let $G = \mathrm{SL}(3, \mathbb{R})$ and

$$a = \begin{pmatrix} 1/2 & 0 & 0 \\ 0 & 1 & 0 \\ 0 & 0 & 2 \end{pmatrix}, \qquad u = \begin{pmatrix} 1 & 1 & 0 \\ 0 & 1 & 0 \\ 0 & 0 & 1 \end{pmatrix},$$

$$v = \begin{pmatrix} 1 & 0 & 0 \\ 0 & 1 & 1 \\ 0 & 0 & 1 \end{pmatrix}, \qquad c = \begin{pmatrix} 1 & 0 & 1 \\ 0 & 1 & 0 \\ 0 & 0 & 1 \end{pmatrix}.$$

Then the element a is diagonal for horocycle elements $u, v \in G$, and the unipotent subgroup of strictly upper-triangular matrices is the contracting horospherical subgroup $G^-(a)$. The element $a' = cac^{-1}$ is also diagonal for u and v.

Note that if an element $a \in G$ is diagonal for $u \in G$, then the set $D(a)$ of essentially divergent semiorbits in G/Γ is invariant under the u-translation. Therefore, given an ergodic u-invariant probability measure μ, either $\mu(D(a)) = 0$ or $\mu(D(a)) = 1$. In addition, $D(a) = D(cac^{-1})$ for every $c \in G^-(a)$. It follows that the set $D(a)$ is independent of the choice of a diagonal element for $u \in G$.

Let μ be an ergodic probability measure on G/Γ for a horocycle element $u \in G$, and let $\Lambda = \Lambda(\mu) \subset G$ be its stabilizer. Using the R-property (see [**R90b,95b**] for a discussion in the general case), Ratner proved that $\mu(D(a)) = 0$ or 1 according as there does or does not exist an element $c \in G^-(a)$ such that $cac^{-1} \in \Lambda$. If such an element $c \in G$ exists, one can prove that $c\,\mathrm{SL}_2(u, a)c^{-1} \subset \Lambda$ (and hence the subgroup Λ is not amenable) and the measure μ is algebraic: $\mu(\Lambda_0 x) = 1$ for some point $x \in G/\Gamma$. Thus, the following theorem holds.

THEOREM 17.3 [**R91a**]. *Let an element $a \in G$ be diagonal for a horocycle element $u \in G$, and let μ be a u-ergodic probability measure on G/Γ. Then either*

1) $\mu(D(a)) = 1$, *or*

2) $c\,\mathrm{SL}_2(u, a)c^{-1} \subset \Lambda(\mu)$ *for some* $c \in G^-(a)$, *and the measure μ is algebraic.*

Note that for $G = \mathrm{SL}(2, \mathbb{R})$ (see §16) the first possibility occurs when the geodesic flow does not preserve the measure μ (and the measure is supported on a periodic horocycle orbit). In the second case μ is the G-invariant measure on G/Γ induced by Haar measure on G.

The main measure theorem for a one-parameter unipotent subgroup U is proved using induction on $\dim G$ in the following way. One considers the stabilizer Λ for a U-ergodic probability measure μ. Then there are two possibilities. If the identity component Λ_0 is amenable, then the maximal connected subgroup $N \subset \Lambda_0$ unipotent in G is normal in Λ_0 and contains U. In this case, using a generalized R-property for the unipotent subgroup $N \subset G$, one manages to prove that μ is algebraic. If Λ_0 is not amenable, it contains a horocycle element h, a diagonal element a and a subgroup $\mathrm{SL}_2(h, a)$. Since $a \in \Lambda$, it follows that $\mu(D(a)) = 0$. Hence, by Theorem 17.3, μ-almost every ergodic component for the horocycle element $h \in G$ is algebraic. Considering the quotient space modulo this partition allows one to lower the dimension of the Lie group involved and to apply the induction argument.

To prove Theorem 17.1 in the case $\dim U > 1$, we recall the following result.

THEOREM 17.4 (Pugh–Shub, [**PS**]). *Let a connected Abelian Lie group U act ergodically on a probability space (X, μ). Then almost every element of U (relative to Haar measure on U) acts ergodically on (X, μ).*

Ratner has generalized this result to the class of connected nilpotent Lie groups (see also [**Da89**]), thus proving Theorem 17.1 for connected unipotent subgroups $U \subset G$. A more general conjecture of Margulis [**Ma91b**] says that Theorem 17.1 can be generalized to the class of arbitrary subgroups $U \subset G$ generated by unipotent elements in G. Ratner proved this conjecture in a weaker form:

THEOREM 17.5 (Ratner's measure theorem [**R91a**]). *Let G be a Lie group with a discrete subgroup $\Gamma \subset G$. Let a subgroup $U \subset G$ be such that $U = \bigcup_{i=1}^{\infty} u_i U_0$, where all elements u_i are unipotent in G, the group U/U_0 is finitely generated, and the identity component U_0 is generated by one-parameter unipotent subgroups. Then any U-ergodic probability measure on G/Γ is algebraic.*

It is easy to derive this result for connected subgroups $U \subset G$ from the one-parameter case. Indeed, such a group U is of the form $U = PM$, where the subgroup $M \subset G$ is unipotent, and $P \subset G$ is semisimple and has no compact components. Let $V \subset P$ be a maximal connected unipotent subgroup. Then VM is a maximal connected unipotent subgroup of U, and the Moore normal subgroup for VM in U coincides with U. But according to the Mautner phenomenon 2.1, any U-ergodic finite measure is VM-ergodic, and we can find an ergodic one-parameter subgroup of VM.

The case when U is generated by unipotent elements (and is not as in Theorem 17.5) is technically more involved. It was recently examined by Shah. Thus, in full scope the generalized measure conjecture is settled in the following theorem.

THEOREM 17.6 [**Sh98**]. *Let G be a Lie group with a closed subgroup $\Gamma \subset G$, and let a subgroup $U \subset G$ be generated by unipotent elements. Then any U-ergodic probability measure on G/Γ is algebraic.*

2. Theorem 17.6 ensures the algebraicity of ergodic probability measures for actions $(G/\Gamma, U)$, where U is generated by unipotent elements and $\Gamma \subset G$ is an arbitrary closed subgroup. In explicit form the condition for Γ to be discrete was removed by Witte [**Wi94**] by using the Density Theorem E.13. But essentially this was done already in the author's paper [**St88**], where the theory of one-parameter homogeneous flows was reduced to the case of a discrete isotropy subgroup.

We describe the idea of this reduction for our case (combining [**Wi94**] and [**St88**]) using the semisimple splitting construction. Let $\mathcal{A}(G) = (L \times \mathcal{A}(T_R)) \cdot R$

be a ψ-algebraic hull of a connected simply connected Lie group $G = L \cdot R$ from §5. We recall that every Ad-unipotent element $u \in G$ is ψ-unipotent in $\mathcal{A}(G)$ (that is, the element $\psi(u)$ is unipotent in the algebraic group $\psi(\mathcal{A}(G))$).

Let a closed subgroup U be generated by unipotent elements, and let μ be a U-invariant ergodic finite measure on $\mathcal{A}(G)/\Gamma$. We set

$$X = \{g \in \mathcal{A}(G) : \; g^{-1}Ug \subset \mathcal{A}(\Gamma)\}.$$

Then $X\Gamma = X$. Let us prove that the measure μ is supported on the closed subset $X\Gamma \subset G/\Gamma$. Indeed, if $g \notin X$, then there exists a unipotent element $u \in U$ such that $g^{-1}ug \notin \mathcal{A}(\Gamma)$. The ψ-algebraic hull of the subgroup $\{g^{-1}u^k g, \; k \in \mathbb{Z}\}$ is a one-parameter unipotent subgroup. Since the product of two ψ-algebraic subgroups is locally closed and $g \notin X$, it follows that the orbit $\{u^k g\mathcal{A}(\Gamma), \; k \in \mathbb{Z}\} \subset \mathcal{A}(G)/\mathcal{A}(\Gamma)$ is locally closed and nonperiodic. Then the orbit $\{u^k g\Gamma, \; k \in \mathbb{Z}\} \subset \mathcal{A}(G)/\Gamma$ is locally closed and nonperiodic as well. But the family of such orbits is of zero μ-measure, because μ-almost all orbits of the subgroup $\{u^k, \; k \in \mathbb{Z}\}$ are recurrent in G/Γ. Since U contains a dense countable subgroup generated by unipotent elements, we conclude that $\operatorname{supp}(\mu) \subset X\Gamma$.

Therefore the measure μ is supported on a union of U-invariant closed subsets of the form $X(g) = g\mathcal{A}(\Gamma)/\Gamma$, $g \in X$. Since μ is ergodic, there exists an element $g \in X$ such that μ is supported on $X(g) \subset \mathcal{A}(G)/\Gamma$. The flow $(X(g), U)$ is isomorphic to the flow $(\mathcal{A}(\Gamma)/\Gamma, g^{-1}Ug)$. But the identity component Γ_0 is normal in $\mathcal{A}(\Gamma)$, and one can use the classification of ergodic measures on the homogeneous space $(\mathcal{A}(\Gamma)/\Gamma_0)/(\Gamma/\Gamma_0)$ with discrete isotropy subgroup.

3. In conclusion we outline the proof of the measure theorem for algebraic groups, following Margulis and Tomanov [**MT94**] (note that the algebraicity condition was removed in [**MT96**]).

Let $G \subset \mathrm{GL}(n, \mathbb{R})$ be an algebraic group over $\mathbb{R}$ and $\Gamma \in G$ a discrete subgroup If an element $g \in G$ normalizes a connected subgroup $A \subset G$, let $\alpha(g, A)$ stand for the *deformation ratio* of Haar measure m on A by the element g, that is,

$$m(gXg^{-1}) = \alpha(g, A)m(X) \quad \text{for any measurable } X \subset A.$$

We say that an element $s \in G$ is of *class* $\mathcal{A}$ if its algebraic hull in G is one-dimensional and diagonalizable over $\mathbb{R}$. Given an element $s \in G$ of class $\mathcal{A}$, we define

$$W^+ = G^+(s), \qquad W^- = G^-(s), \qquad Z = Z_G(s).$$

The element s^{-1} expands W^-, and hence $\alpha(s^{-1}, V) > 1$ for any s-invariant subgroup $V \subset W^-$. The following entropy estimate is important in its own right, and considerably simplifies the proof of the measure theorem.

THEOREM 17.7 [**MT94**]. *Assume that μ is an s-invariant ergodic probability measure on G/Γ, and $V \subset W^-$ is normalized by an element s of class $\mathcal{A}$. Let $h = h(s, \mu)$ be the entropy of μ with respect to s, and $\alpha = \alpha(s^{-1}, V)$. Then:*

1) *If μ is V-invariant, then $h \geq \log_2 \alpha$.*

2) *If there exists a set $M \subset G/\Gamma$ such that $\mu(M) = 1$ and $M \cap W^- x \subset Vx$ for all $x \in M$, then $h \leq \log_2 \alpha$, and equality implies that the measure μ is V-invariant.*

Note that the proof of the theorem uses results of [**LY**].

Now let $U \subset W^+$ be a unipotent algebraic subgroup. An important role in [**MT94**] is played by the so-called *quasi-regular map*

$$\phi : U \to N_G(U)$$

(cf. [**R90b,c**]). Let us give an idea of the construction. Assume for simplicity that s normalizes U. Let $V \subset W^+$ be an s-invariant complement to U in W^+. Then the product $L = W^- ZV$ is a cross-section for G/U (that is, locally $L \cap U = 1$ and $G = LU$). We fix a sequence $g_n \in G - N_G(U)$ converging to 1, and define a map $\widetilde{\phi}_n : U \to L$ by the rule $ug_n \in \widetilde{\phi}_n(u)U$. By the Chevalley Theorem D.1, there exist an action of G on a linear space E and a vector $q \in E$ such that $U = \{h \in G : \ hq = q\}$. We fix relatively compact neighborhoods $A = O(1) \subset U$ and $D = O(q) \subset E$. Now we set $A_n = s^n A s^{-n}$ and define integers r_n as follows:

$$A_{r_n} g_n q \not\subset D \qquad \text{and} \qquad A_k g_n q \subset D \quad \text{for all } k < r_n.$$

Let $\phi_n : U \to L$ be defined in the following way:

$$\phi_n(u) = \widetilde{\phi}_n(s^{r_n} u s^{-r_n}).$$

It is easy to see that the maps $\phi'_n : U \to E$ given by $\phi'_n(u) = \phi_n(u)q$ are polynomials of uniformly bounded degrees. Since $\phi'_n(A_{-1}) \subset D$ and $\phi'_n(A) \not\subset D$, we can assume that ϕ'_n converges to a nonconstant polynomial map $\phi' : U \to E$. At the same time, ϕ_n converges to a map $\phi : U \to L$. One can verify that $\mathrm{Im}(\phi) \subset N_G(U)$ and the group $F \subset G$ generated by U and $\mathrm{Im}(\phi)$ is large enough: the factor-group F/U is noncompact.

Example 17.8. Let us treat $G = \mathrm{SL}(2, \mathbb{C})$ as an $\mathbb{R}$-algebraic group. We introduce the following notation:

$$s = \begin{pmatrix} \sqrt{2} & 0 \\ 0 & 1/\sqrt{2} \end{pmatrix}, \quad U_t = \begin{pmatrix} 1 & t \\ 0 & 1 \end{pmatrix}, \quad V_t = \begin{pmatrix} 1 & it \\ 0 & 1 \end{pmatrix},$$
$$U_t^- = \begin{pmatrix} 1 & 0 \\ t & 1 \end{pmatrix}, \quad V_t^- = \begin{pmatrix} 1 & 0 \\ it & 1 \end{pmatrix}.$$

It is clear that $W^+ = U_{\mathbb{R}} \times V_{\mathbb{R}}$, $W^- = U_{\mathbb{R}}^- \times V_{\mathbb{R}}^-$, $L = W^- Z V_{\mathbb{R}}$, where Z stands for the subgroup of diagonal matrices. Let e', e'' be the standard $\mathbb{C}$-basis in $\mathbb{C}^2$. Then $e_1 = e'$, $e_2 = ie'$, $e_3 = e''$, $e_4 = ie''$ form an $\mathbb{R}$-basis. We denote by ρ the standard action of G on $\mathbb{C}^2$, and define $\rho' = \rho \oplus (\rho \wedge \rho)$. Then, given $q = e_1 \oplus (e_1 \wedge e_3)$, one obtains $U = \{h \in G : \ \rho'_h q = q\}$. It is easily seen that the following data satisfies our construction:

$$g_n = \begin{pmatrix} 1 & 0 \\ 1/n & 1 \end{pmatrix}, \qquad r_n = \log_2 n, \qquad \phi(U_t) = \begin{pmatrix} 1+t & 0 \\ 0 & 1/(1+t) \end{pmatrix}.$$

In addition, there is another combination:

$$g_n = \begin{pmatrix} 1 & 0 \\ i/n & 1 \end{pmatrix}, \qquad r_n = \frac{1}{2} \log_2 n, \qquad \phi(U_t) = \begin{pmatrix} 1 & -it^2 \\ 0 & 1 \end{pmatrix}.$$

More generally, it is easy to verify that $\mathrm{Im}(\phi) \subset W^+$ whenever $g_n \in W^+ Z V^- - W^+ Z$.

Henceforth we say that a sequence $\{g_n\}$ *satisfies the condition* $(\star)$ if the sequence $\{s^{-r_n} g_n s^{r_n}\}$ is relatively compact (for $G = \mathrm{SL}(2,\mathbb{C})$ this is always the case).

Let μ be a U-invariant ergodic probability measure on G/Γ. Another concept introduced in [**MT94**] is the notion of a *set of uniform convergence* for the sequence $\{A_n\}$ (cf. [**R90b,c**]). This is a set $M \subset G/\Gamma$ such that for any $\varepsilon > 0$ and $f \in C_c(G/\Gamma)$ there exists $N = N(\varepsilon, f)$ for which

$$|\frac{1}{m(A_n)}\int_{A_n} f(ux)dm_u - \int_{G/\Gamma} f d\mu| < \varepsilon \quad \text{for all } x \in M,\ n > N,$$

where m is Haar measure on U (note that the sets A_n grow uniformly enough in all directions, and we can apply an analogue of the Birkhoff–Khintchine ergodic theorem to the averaging sequence $\{A_n\} \subset U$). It is proved in [**MT94**] that, given $\delta > 0$, there exists a set M_δ of uniform convergence for $A_n = s^n A s^{-n}$ with $\mu(M_\delta) > 1 - \delta$, regardless of the choice of a relatively compact open subset $A \subset U$. Use of quasi-regular maps $\phi : U \to N_G(U)$ in the proof of the measure theorem is based on the following result.

LEMMA 17.9 [**MT94**]. *Let M be a set of uniform convergence for a sequence $A_n \subset U$ as above. Assume that $x_n \in M$ converges to $x \in M$, $g_n \in G - N_G(U)$ converges to $1 \in G$ and satisfies the condition $(\star)$, and $g_n x_n \in M$ for all n. Then the measure μ is* $\mathrm{Im}(\phi)$*-invariant.*

We say that a measure μ is *Zariski dense* in G/Γ if there is no proper algebraic subgroup $P \subset G$ such that $U \subset P$ and $\mu(Px) = 1$ for some $x \in G/\Gamma$. Given $M \subset G/\Gamma$, we define

$$\Psi(M) = \{g \in G : \ gM \cap M \neq \varnothing\}.$$

Let A, B, C, D be algebraic subgroups of G, and $D \subset AB$. We say that the product AB is *transversal to* C if locally $G = ABC$ and $AB \cap C = 1$. The basic tool for the use of Lemma 17.9 is the following result.

LEMMA 17.10 [**MT94**]. *Let μ be Zariski dense in G/Γ and invariant under C. Assume that the product AB is transversal to C. Then there is $\delta > 0$ such that for any measurable set $M \subset G/\Gamma$ with $\mu(M) > 1 - \delta$ there exists a sequence $g_n \in \Psi(M) \cap (AB - D)$ converging to $1 \in G$.*

Now assume that we want to prove the algebraicity of an ergodic U-invariant measure μ on G/Γ for a unipotent algebraic subgroup $U \subset G$. Clearly, without loss of generality *the measure μ may be assumed to be Zariski dense.* Another assumption is that *U is not normal in G* (otherwise the proof is almost trivial). Finally, without loss of generality we assume that *the subgroup U is maximal among algebraic unipotent subgroups preserving μ.* We have to prove that the measure μ is G-invariant.

From now on we again assume for simplicity that $U \subset W^+$ for some element $s \in G$ of class $\mathcal{A}$. Let $X \subset G$ be the Zariski closure of the product $W^- Z U$. We introduce the subgroup

$$U^- = \{g \in W^- : \ Xg = X\},$$

which plays the role of "the counterpart" to U. A rather delicate analysis of the local structure of μ, based on Ratner's ideas, provides us with the following information.

LEMMA 17.11 [**MT94**]. *Assume that s preserves the measure μ. Then there exists a conull set $M \subset G/\Gamma$ such that $M \cap W^- x \subset U^- x$ for all $x \in M$.*

We finish the proof by considering the case $G = \mathrm{SL}(2, \mathbb{C})$. It suffices to prove that the measure μ is W^+-invariant (which contradicts the maximality of U). Let D and S be the one-dimensional subgroups of diagonal matrices with real entries and complex entries of absolute value 1, respectively. Then $Z = D \times S$ and $N_G(U) = DW^+$. Let M_δ be a set of uniform convergence for the sequence $A_n \subset U$ with $\mu(M_\delta) > 1 - \delta$. It follows from 17.10 that, given a small enough $\delta > 0$, there exists a sequence $g_n \in \Psi(M_\delta) \cap (W^- ZV - DV)$ converging to 1. Lemma 17.9 provides us with a quasi-regular map $\phi : U \to N_G(U)$ such that $\mathrm{Im}(\phi)$ preserves μ. Let $s' \in D$ be an element of class $\mathcal{A}$ generated by U and $\mathrm{Im}(\phi)$, and expanding U. It follows from the Mautner phenomenon that the measure μ is s'-ergodic. Applying Lemma 17.11 and Theorem 17.7, we see that

$$\log_2 \alpha(s', U) \le h(s', \mu) \le \log_2 \alpha({s'}^{-1}, U^-) = \log_2 \alpha(s', U).$$

Now Theorem 17.7 implies that μ is U^--invariant. Finally, we take the product W^+ZV^- transversal to U^- and, using Lemma 17.10, construct a sequence $g_n \in \Psi(M) \cap (W^+ZV^- - W^+Z)$ converging to 1 for some set M of uniform convergence with respect to $\{A_n\} \subset U$. This gives us a quasi-regular map $\phi : U \to W^+$, and Lemma 17.9 implies that the measure μ is W^+-invariant. □

§18. Rigidity, quotients, and joinings of unipotent flows

1. In this section we show how Ratner's theorems on rigidity, quotients and joinings of horocycle flows are generalized to the class of ergodic unipotent flows on homogeneous spaces of finite volume. First we give some preliminary definitions.

Let G/Γ and H/Λ be homogeneous spaces with volume measures ν_1 and ν_2. Then a measure-preserving map $\phi : (G/\Gamma, \nu_1) \to (H/\Lambda, \nu_2)$ is said to be *affine for* $g \in G$ if there exists an element $\widetilde{g} \in H$ such that

$$\phi(gg'\Gamma) = \widetilde{g}\phi(g'\Gamma) \quad \text{for } \nu_1\text{-almost all } g' \in G.$$

A measure-preserving map $\phi : (G/\Gamma, \nu_1) \to (H/\Lambda, \nu_2)$ is called *G-affine* if it is affine for every $g \in G$.

An explicit form of G-affine maps is given by the following statement.

PROPOSITION 18.1 (cf. [**Wi85**]). *Let Γ and Λ be lattices in Lie groups G and H, and let $\phi : G/\Gamma \to H/\Lambda$ be a volume-preserving map. If the lattice $\Lambda \subset H$ is faithful, then the following conditions are equivalent*:

1) *ϕ is G-affine.*

2) *The graph* $\mathrm{Graph}(\phi) = \{(x, \phi(x)),\ x \in G/\Gamma\}$ *is a homogeneous subspace* (mod 0) *in $G/\Gamma \times H/\Lambda$.*

3) *There exist an element $c \in H$ and an epimorphism $\sigma : G \to H$ such that $\sigma(\Gamma) \subset \Lambda$ and $\phi(g\Gamma) = c\sigma(g)\Lambda$ almost everywhere in G/Γ.*

Proof. $1 \Rightarrow 2$. According to Zimmer [**Z**], the map ϕ can be modified on a zero measure set in G/Γ so that the equality $\phi'(gg'\Gamma) = \widetilde{g}\phi(g'\Gamma)$ holds for all $g, g' \in G$. Since $\phi(g'\Gamma) = \phi(1g'\Gamma) = \widetilde{1}\phi(g'\Gamma)$ and the map ϕ is surjective, the element $\widetilde{1}$ belongs to the intersection of Λ with the center $Z(H)$. Since the lattice Λ is faithful, we derive that $\widetilde{1} = 1$. Likewise one proves that $\widetilde{gg'} = \widetilde{g}\widetilde{g}'$. Hence the set $F = \{(g, \widetilde{g}),\ g \in G\}$ is a subgroup of $G \times H$ which acts transitively on $\mathrm{Graph}(\phi)$.

$2 \Rightarrow 3$. Let $p : G \times H \to G$ and $p^* : G/\Gamma \times H/\Lambda \to G/\Gamma$ be natural projections. Assume that $\mathrm{Graph}(\phi) = cF(\Gamma \times \Lambda)$ for an element $c = (c_1, c_2) \in G \times H$ and a subgroup $F \subset G \times H$. By definition of the graph, the restriction $p^* : \mathrm{Graph}(\phi) \to G/\Gamma$ is bijective and the restriction $p : F \to G$ is an epimorphism with kernel $F \cap H \subset \Lambda$. But the lattice Λ is faithful, so that $p : F \to G$ is an isomorphism. Let $\lambda : G \to F$ be the inverse isomorphism, and let σ be the composite of λ and the projection $q : G \times H \to H$. Assume that

$$(g\Gamma, \phi(g\Gamma)) = (c_1, c_2) f(\Gamma \times \Lambda) \quad \text{for some } g \in G,\ f \in F.$$

Then $f = (c_1^{-1} g, \sigma(c_1^{-1} g))$, and hence

$$\phi(g\Gamma) = c_2 \sigma(c_1)^{-1} \sigma(g) \Lambda \quad \text{for all } g \in G.$$

$3 \Rightarrow 1$. Define $\widetilde{g} = c\sigma(g)c^{-1}$. Then

$$\phi(gg'\Gamma) = c\sigma(g'g)\Lambda = c\sigma(g)c^{-1}c\sigma(g')\Lambda = \widetilde{g}\phi(g'\Gamma),$$

as required. □

Therefore a G-affine map is a homogeneous cover followed by a constant translation. Such a map can also be viewed as a map of the form $\phi(g\Gamma) = \alpha(g)c\Lambda$, where $\alpha : G \to H$ is an epimorphism such that $\alpha(\Gamma) \subset c\Lambda c^{-1}$. Indeed, it suffices to define $\alpha(g) = c\sigma(g)c^{-1}$ for all $g \in G$.

Now we can give an algebraic formulation of Theorem 15.1 on rigidity of the horocycle flow.

THEOREM 18.2. *Let $G = \mathrm{SL}(2, \mathbb{R})$ and let Γ_1, Γ_2 be two lattices in G. Assume that a measure-preserving map $\phi : G/\Gamma_1 \to G/\Gamma_2$ is U_t-affine for some $t > 0$. Then it is G-affine.*

Witte proved that the result holds true in the class of ergodic unipotent actions.

THEOREM 18.3 (on rigidity, [**Wi85**]). *Let Γ, Λ be lattices in connected Lie group G, H, and let $\phi : (G/\Gamma, u_1^{\mathbb{Z}}) \to (H/\Lambda, u_2^{\mathbb{Z}})$ be a measure-theoretic conjugacy of two ergodic unipotent one-generator actions. Then ϕ is G-affine. In particular, if the map ϕ is (almost everywhere) finite-to-one, then the groups G and H are locally isomorphic. If the map ϕ is invertible (that is, it is a measure-theoretic isomorphism) and both the lattices Γ and Λ are faithful, then $\phi(g\Gamma) = \alpha(g)c\Lambda$ for an element $c \in G$ and an isomorphism $\alpha : G \to H$ such that $\alpha(\Gamma) = c\Lambda c^{-1}$.*

The direct proof of the theorem is far from trivial. Following [**R82a**], Witte managed to find a proper formulation and to prove the R-property (see §15) for unipotent translations on arbitrary homogeneous spaces. However, now (after the proof of the Dani measure conjecture) it can be easily deduced from the classification of ergodic joinings of unipotent translations (or from the measure theorem itself). The deduction is completely similar to the proof of the rigidity theorem for horocycle flows in §15.

2. Now we turn to quotients of ergodic unipotent flows. The algebraicity of quotients (as well as the rigidity theorem) for ergodic nilflows was already known to Parry [**P71,73**]. This result and Ratner's theorem on quotients of the horocycle flow were generalized by Witte.

Let Γ be a lattice in a Lie group G, let $U \subset G$ be a unipotent subgroup, and let the flow $(G/\Gamma, U)$ be ergodic. Suppose that $\Lambda \subset G$ is a closed subgroup of G containing Γ. Then the flow $(G/\Lambda, U)$ is a quotient of the original flow $(G/\Gamma, U)$. If $K \subset G$ is a compact subgroup of G that commutes with U, the action of U on the double-coset space $K\backslash G/\Lambda$ is also a quotient flow for $(G/\Gamma, U)$. Finally, instead of K one can take any closed subgroup of the group $\mathrm{Aff}(G/\Lambda)$ of all G-affine maps of the space G/Λ, which commutes with U and for which the orbit space $K\backslash G/\Lambda$ is countably separated (or, equivalently, almost all K-orbits on G/Λ are locally closed). The following theorem shows that these are all measurable quotient flows for an ergodic unipotent flow.

THEOREM 18.4 (on quotients, [**Wi94**]). *Let a unipotent flow $(G/\Gamma, U)$ on a finite volume space be ergodic. Then every measurable quotient flow of it is isomorphic to a flow $(K\backslash G/\Lambda, U)$, where $\Lambda \subset G$ contains Γ, and $K \subset \mathrm{Aff}(G/\Lambda)$ commutes with U and has the space $K\backslash G/\Lambda$ measurable.*

We briefly present the proof of the theorem. Let $\psi : (G/\Gamma, \mu) \to (X, \nu)$ be a measurable U-conjugacy. We introduce

$$\Lambda = \{\lambda \in G \mid \psi(g\lambda\Gamma) = \psi(g\Gamma), g \in G\}.$$

It is easy to prove that Λ is a closed subgroup of G and hence the space G/Λ has G-invariant probability measure $\overline{\mu}$. Moreover, the map ψ determines a measurable U-conjugacy $\overline{\psi} : (G/\Lambda, \overline{\mu}) \to (X, \nu)$ by the formula $\overline{\psi}(g\Lambda) = \psi(g\Gamma)$, $g \in G$. Let K stand for the subgroup of all elements $\phi \in \mathrm{Aff}(G/\Lambda)$ that commute with U and keep the leaves of the map $\overline{\psi}$ invariant, i.e., $\overline{\psi}(\phi(s)) = \overline{\psi}(s)$ for almost all $s \in G/\Lambda$. Let us prove that the space X is isomorphic to $K\backslash G/\Lambda$.

To do this, we consider the set

$$M = \{(s\Lambda, t\Lambda) \in (G/\Lambda) \times (G/\Lambda) \mid \overline{\psi}(s\Lambda) = \overline{\psi}(t\Lambda)\}.$$

It suffices to prove that K acts transitively on the leaves of $\overline{\psi}$; that is, for any pair $(s\Lambda, t\Lambda) \in M$ there exists $\phi \in K$ such that $\phi(s\Lambda) = t\Lambda$. We decompose the measure $\overline{\mu}$ into probability measures $\overline{\mu}_x$ supported on the leaves $\psi^{-1}(x)$, $x \in X$. This means that, given a Borel set $A \subset G/\Lambda$, one has $\overline{\mu}(A) = \int_X \overline{\mu}_x(A) d\nu(x)$. Now we introduce a probability measure ρ on $(G/\Lambda) \times (G/\Lambda)$ by the formula

$$\rho(A \times B) = \int_X \overline{\mu}_x(A)\overline{\mu}_x(B) d\nu(x).$$

It is straightforward to verify that the measure ρ is supported on the set M, and both M and ρ are invariant under the diagonal subgroup $\Delta_U = \{(u, u),\ u \in U\}$. Moreover, the measure ρ is a self-joining of $\overline{\mu}$. It follows from the classification of joinings or from Ratner's measure theorem that the ergodic component of the measure ρ containing the point $(s\Lambda, t\Lambda) \in M$ is supported on a closed orbit $H(s\Lambda, t\Lambda)$. Moreover, it is H-invariant for some subgroup $H \subset G \times G$ (where H projects epimorphically onto each of the two copies of G, and contains Δ_U).

We must prove that the homogeneous space $H(s\Lambda, t\Lambda)$ is the graph of an affine map $\phi : G/\Lambda \to G/\Lambda$. For a pair $(g, g') \in H$, define $\phi(gs\Lambda) = g't\Lambda$. The map ϕ is well defined if

$$H \cap (s\Lambda s^{-1} \times G) \subset s\Lambda s^{-1} \times t\Lambda t^{-1}.$$

Let us prove this relation, say, for $s = t = 1 \in G$. Choose $\lambda \in \Lambda$ and $(\lambda, h) \in H$. Since $H(\Lambda, \Lambda) \subset M$, then for all $(g, g') \in H$ we have

$$\overline{\psi}(g'h\Lambda) = \overline{\psi}(g\lambda\Lambda) = \overline{\psi}(g\Gamma) = \overline{\psi}(g'\Gamma),$$

and hence $h \in \Lambda$. Now it follows from the definition that the map ϕ is affine and commutes with U. □

Remark. For the horocycle flow (when $G = \mathrm{SL}(2,\mathbb{R})$) the only subgroup $K \subset \mathrm{Aff}(G/\Gamma)$ that has a "good" orbit space $K\backslash G/\Lambda$ and commutes with the flow is the subgroup of translations along the center $\mathbb{Z}_2 \subset \mathrm{SL}(2,\mathbb{R})$. In this case we observe that $K\backslash G/\Lambda = G/\Lambda'$, where $\Lambda' = \mathbb{Z}_2\Lambda$, thus avoiding consideration of double cosets and obtaining Theorem 15.4.

3. Now we formulate and give a simplified deduction of the classification of ergodic joinings for arbitrary unipotent flows with the help of Ratner's measure theorem.

THEOREM 18.5 (classification of joinings [**R90a,c**]). *Let G_i, $i = 1, 2$, be connected Lie groups with lattices Γ_i and unipotent subgroups $u_t^{(i)}$. Assume that a measure μ on $(G_1/\Gamma_1) \times (G_2/\Gamma_2)$ is an ergodic joining for the flows $(G_1/\Gamma_1, u_t^{(1)}, \nu_1)$ and $(G_2/\Gamma_2, u_t^{(2)}, \nu_2)$. Then there exist a closed subgroup $\Lambda \subset G_1 \times G_2$ and an element $c \in G_2$ such that*

1. *the subgroups $\Lambda_i = \Lambda \cap G_i$ are normal in G_i,*
2. *there exists a continuous epimorphism $\alpha : G_1 \to \Lambda_2\backslash G_2$ with kernel Λ_1,*
3. *the intersection $\alpha(\Gamma_1) \cap \Lambda_2 c\Gamma_2 c^{-1}$ is a lattice in $\Lambda_2\backslash G_2$, and*
4. *μ is a Λ-invariant measure supported on a closed orbit $\Lambda(\Gamma_1, c\Gamma_2)$.*

If $\overline{\alpha}(g) \subset G_2, g \in G_1$, stands for the inverse image of an element $\alpha(g) \in \Lambda_2\backslash G_2$ in G_2, then $\operatorname{supp}\mu = \Lambda(\Gamma_1, c\Gamma_2) = \{(g\Gamma_1, \overline{\alpha}(g)c\Gamma_2), g \in G_1\}$.

Proof. Let

$$p_i : G_1 \times G_2 \to G_i \quad \text{and} \quad p_i^* : (G_1/\Gamma_1) \times (G_2/\Gamma_2) \to G_i/\Gamma_i$$

be natural projections. It follows from Ratner's measure theorem that there exist a point $x \in (G_1/\Gamma_1) \times (G_2/\Gamma_2)$ and a closed subgroup $\Lambda \subset G_1 \times G_2$ such that the measure μ is invariant under Λ and is supported on the closed orbit Λx. Since μ is a joining, then $p_i^*(\mu) = \nu_i$, and hence $p_i(\Lambda) = G_i$, $i = 1, 2$. The subgroups $\Lambda_i = \Lambda \cap G_i$ are normal in Λ, and $p_i(\Lambda) = G_i$; hence each Λ_i is normal in G_i, $i = 1, 2$.

We denote by $\overline{p}_1 : \Lambda \to G_1$ the group epimorphism obtained by the restriction of p_1 onto Λ. For each pair $(g, \overline{g}) \in \Lambda$ we let

$$\alpha(g) = \Lambda_2\overline{g} \in \Lambda_2 p_2 \overline{p}_1^{-1}(g) \in \Lambda_2\backslash G_2.$$

Then $\alpha : G_1 \to \Lambda_2\backslash G_2$ is a well defined group epimorphism with kernel Λ_1. Note that $\overline{\alpha}(g) = p_2\overline{p}_1^{-1}(g)$. It is easy to see that $\Lambda(H_1, H_2) \cap G_2 = \overline{\alpha}(H_1)H_2$ for any subgroups $H_1 \subset G_1$, $H_2 \subset G_2$.

Since $p_1^*(\Lambda x) = G_1/\Gamma_1$, we can assume without loss of generality that $x = (\Gamma_1, c\Gamma_2)$ for some $c \in G_2$, and hence

$$\Lambda(\Gamma_1, c\Gamma_2) = \{(g\Gamma_1, \overline{\alpha}(g)c\Gamma_2), g \in G_1\}.$$

Since the orbit $\Lambda(\Gamma_1, c\Gamma_2)$ is closed, the product $\Lambda(\Gamma_1 \times c\Gamma_2 c^{-1}) \subset G_1 \times G_2$ is also closed. It follows that the intersection

$$\Lambda(\Gamma_1 \times c\Gamma_2 c^{-1}) \cap G_2 = \overline{\alpha}(\Gamma_1) c\Gamma_2 c^{-1} \subset G_2$$

and the product $\Lambda_2 \alpha(\Gamma_1) c\Gamma_2 c^{-1} \subset \Lambda_2 \backslash G_2$ are closed as well. Since Γ_1 and Γ_2 are countable, the subgroups $\alpha(\Gamma_1)$ and $\Lambda_2 c\Gamma_2 c^{-1}$ are discrete in $\Lambda_2 \backslash G_2$ and hence are lattices therein. Finally, fibers of the fibration of homogeneous spaces of finite volume

$$p_1^* : \Lambda(\Gamma_1, c\Gamma_2) \to G_1/\Gamma_1$$

are isomorphic to the homogeneous space $\overline{\alpha}(\Gamma_1)/\overline{\alpha}(\Gamma_1) \cap \Lambda_2 c\Gamma_2 c^{-1}$. Hence the space is also of finite volume. It follows that $\alpha(\Gamma_1)/\alpha(\Gamma_1) \cap \Lambda_2 c\Gamma_2 c^{-1}$ is finite, and we are done. □

As a corollary of the classification of joinings, we can conclude that the existence of a nontrivial ergodic joining of two unipotent flows on homogeneous spaces of simple Lie groups implies that the groups involved are locally isomorphic. Note also that this general classification is consistent with the classification 15.5 of nontrivial ergodic joinings of horocycle flows.

§19. Uniform distribution of unipotent trajectories

We proceed to the proof of the Raghunathan topological conjecture.

1. Let $u_{\mathbb{R}} = \{u_t,\ t \in \mathbb{R}\}$ be a one-parameter subgroup of G, let Γ be a discrete subgroup of G, and let $x \in G/\Gamma$. We say that the trajectory $u_{\mathbb{R}}x$ is *uniformly distributed* (in its closure) if there exists a (closed) subgroup $H \subset G$ such that the orbit $Hx \subset G/\Gamma$ is closed and carries an H-invariant probability measure ν_H, and for any function $f \in C_c(G/\Gamma)$ (that is, continuous function with compact support) we have

$$\frac{1}{T}\int_0^T f(u_t x)dt \to \int_{G/\Gamma} f d\nu_H,\ T \to \infty.$$

The following result holds.

THEOREM 19.1 (Ratner's uniform distribution theorem [**R91b**]). *If Γ is a lattice in a connected Lie group G, then every unipotent orbit in G/Γ is uniformly distributed.*

If $u_{\mathbb{R}}x$ is uniformly distributed in the homogeneous space Hx, then for sufficiently large $T > 0$, the fraction of time spent by the segment $u_{[0,T]}x$ inside a "good enough"[7] open set $W \subset Hx$, is approximately $\nu_H(W)$. In particular, the trajectory $u_{\mathbb{R}}x$ is dense in Hx. Hence the Raghunathan conjecture follows from the uniform distribution. This important property of unipotent flows was first proved by Ratner in [**R91b**]. Simultaneously Shah [**Sh91**] settled a particular case (the group G is semisimple, $\operatorname{rank}_{\mathbb{R}} G = 1$ and the subgroup $u_{\mathbb{R}}$ is regular). Both the proofs are based on Ratner's measure theorem. First we present the common part of these proofs.

Let Γ be an arbitrary discrete subgroup of G, and let $X = G/\Gamma$. We introduce the *singular set* $Y \subset X$ of those points $x \in X$ for which there exists a proper subgroup $H \subset G$ such that $u_{\mathbb{R}} \subset H$ and the closed orbit Hx carries an H-invariant

[7] This means that $\nu_H(\partial W) = 0$ (then the characteristic function for W is well approximated by continuous functions)

probability measure. Using induction on $\dim G$, it suffices to consider the case $Y \neq X$ and to prove the uniform distribution for points $x \in X - Y$. Let $\overline{X} = X \cup \infty$ be the one-point compactification if the space X is noncompact, and $\overline{X} = X$ otherwise. Given $T > 0$, we define a probability measure μ_T on $\overline{X}$ by the rule

$$\int_X f d\mu_T = \frac{1}{T}\int_0^T f(u_t x)dt, \qquad f \in C_c(X).$$

We recall that the set of probability measures on a compact set is compact with respect to the weak* topology. Hence to prove the uniform distribution of the orbit $u_{\mathbb{R}}x$ of a point $x \in X - Y$, it suffices to show that μ is the G-invariant probability measure on X whenever $\mu_{T_i} \to \mu$ as $T_i \to \infty$.

It is easy to see that the limit measure μ is $u_{\mathbb{R}}$-invariant. We decompose μ into $u_{\mathbb{R}}$-ergodic components. By Ratner's measure theorem, all the components are algebraic. Suppose we know that $\mu(\infty) < 1$ and $\mu(Y) = 0$. Then $\mu(X) > 0$, and there exists an ergodic component of $\mu|_X$ which is not supported on Y. Since the ergodic measure is algebraic, it must be G-invariant, and hence Γ is a lattice in G. By Dani's Theorem 12.2, a unipotent trajectory on a homogeneous space of finite volume spends almost all its life in a compact subset. This means that $\mu(X) = 1$, and hence the measure μ is G-invariant.

As for the condition $\mu(\infty) < 1$, it can fail (for instance, in the case where the orbit $u_{\mathbb{R}}x$ exits to infinity in G/Γ). However, if Γ is a lattice in G, then by Dani's theorem $\mu(\infty) = 0$, and there is nothing to worry about. The condition $\mu(Y) = 0$ holds whenever $x \notin Y$, but the proof is far from trivial. There are two approaches to proving the equality. Ratner's approach, as well as the proof of the measure theorem, exploits the R-property of unipotent trajectories. Our subsequent account follows the paper of Dani and Margulis [**DM93**], which studies how unipotent trajectories behave near the singular set $Y \subset X$. A similar idea had already been used in [**DM90a**], where the Raghunathan conjecture was settled for regular unipotent subgroups of $G = \mathrm{SL}(3, \mathbb{R})$. Then this idea was refined by Shah [**Sh91**]. Our choice is motivated by the fact that this approach allows one to prove a stronger result. Henceforth l denotes the Lebesgue measure on $\mathbb{R}$.

THEOREM 19.2 [**DM93**]. *Let Γ be a discrete subgroup of a Lie group G, let a subgroup $u_{\mathbb{R}}$ be unipotent in G, let $Y \subset G/\Gamma$ be the singular set for $u_{\mathbb{R}}$, and let $x \notin Y$. Then for any $\varepsilon > 0$ there exists a neighborhood Ω of the singular set $Y \subset G/\Gamma$ such that*

$$l\{t \in [0, T] : \ u_t x \in \Omega\} \leq \varepsilon T \quad \textit{for all } T \geq 0.$$

Thus, we have to prove that our unipotent trajectory spends only a small part of its life inside a (small enough) neighborhood of the singular set $Y \subset X$. The theorem implies the equality $\mu(Y) = 0$ for the limit measure μ obtained from the trajectory $u_{\mathbb{R}}x$, $x \in X - Y$. Here it is important that the theorem is valid for any discrete subgroup $\Gamma \subset G$ (not necessarily lattice). Hence the closure of a unipotent trajectory is obviously algebraic provided that $\mu(X) > 0$. We have derived the following important corollary (it is not formulated explicitly in [**DM93**]).

THEOREM 19.3. *Let Γ be a discrete subgroup of G, let a subgroup $u_{\mathbb{R}} \subset G$ be unipotent, and let $x \in G/\Gamma$. Assume that there exist $\varepsilon > 0$ and a compact set*

$K \subset G/\Gamma$ such that

$$l\{t \in [0,T] : u_t x \in K\} > \varepsilon T \quad \textit{for all } T > 0.$$

Then the closure $\overline{u_{\mathbb{R}} x}$ is algebraic, and the trajectory is uniformly distributed.

This result proves, in particular, Ratner's conjecture that the compact closure of a unipotent trajectory on G/Γ is algebraic whenever the subgroup $\Gamma \subset G$ is discrete. But for an arbitrary subgroup $\Gamma \subset G$, the compact closure of a unipotent trajectory need not be smooth. Indeed, let $G = \mathrm{SL}(2,\mathbb{R}) \times \mathbb{R}/\mathbb{Z}$, and let $g_{\mathbb{R}}$ be the diagonal subgroup of the cylinder $U \times \mathbb{R}/\mathbb{Z}$, where $U \subset \mathrm{SL}(2,\mathbb{R})$ is the unipotent subgroup of strictly upper-triangular matrices. Now if $B \subset \mathrm{SL}(2,\mathbb{R})$ is the two-dimensional subgroup of upper-triangular matrices, then the space G/B is diffeomorphic to the 2-torus. The torus has a single periodic $g_{\mathbb{R}}$-orbit, and the remaining orbits are locally closed and winding on the periodic orbit in both directions.

2. We proceed to the proof of Theorem 19.2. First we consider in more detail the structure of the singular set $Y \subset X$ associated to the unipotent subgroup $u_{\mathbb{R}} \subset G$. Let $\mathcal{H}$ denote the family of all proper connected subgroups $H \subset G$ such that $\Gamma \cap H$ is a Zariski dense lattice in H relative to the Ad-representation of the group G. In addition, for any two closed subgroups $H, W \subset G$ we introduce the analytic set

$$X(H,W) = \{g \in G : g^{-1} W g \subset H\}.$$

LEMMA 19.4. $Y = \bigcup_{H \in \mathcal{H}} X(H, u_{\mathbb{R}})\Gamma \subset G/\Gamma$.

Proof. Let $g\Gamma \in X(H, u_{\mathbb{R}})\Gamma$, where $H \in \mathcal{H}$. Then the subspace $gH\Gamma \subset G/\Gamma$ is closed, has finite volume and contains the trajectory $u_{\mathbb{R}} g\Gamma$. It follows that $g\Gamma \in Y$.

Conversely, let $g\Gamma \in Y$. Then there exists a proper connected subgroup $H \subset G$ such that $u_{\mathbb{R}} \subset H$ and the closed subspace $Hg\Gamma$ is of finite volume. Consequently, $g^{-1} u_{\mathbb{R}} g \subset g^{-1} H g$, and $\Gamma \cap g^{-1} H g$ is a lattice in $g^{-1} H g$. Let H' be the subgroup of the connected Lie group $g^{-1} H g$ generated by all one-parameter subgroups Ad-unipotent in G. Then $g^{-1} u_{\mathbb{R}} g \subset H'$, H' is normal in $g^{-1} H g$, and, by Dani's version E.13 of the Borel density theorem, the algebraic closure of the subgroup $\mathrm{Ad}(\Gamma \cap g^{-1} H g)$ contains $\mathrm{Ad}(H')$. Hence if $F = (\overline{H'\Gamma})_0$, then $g^{-1} u_{\mathbb{R}} g \subset F$ and $\Gamma \cap F$ is a lattice in F which is Zariski dense in the Ad-representation of G. It follows that $g \in X(F, u_{\mathbb{R}})$, where $F \in \mathcal{H}$. $\square$

THEOREM 19.5. *The family $\mathcal{H}$ is countable.*

Note that this important result appeared for the first time in Ratner's paper [**R91a**], where it was used in the proof of Theorem 19.1, and also (in a particular case) in Shah's paper [**Sh91**].

We give the main idea of the proof from [**DM93**]. Assume for simplicity that the Ad-representation of the group G is faithful and the image $\mathrm{Ad}(G)$ is closed in the Euclidean topology of the algebraic group $\mathrm{Aut}(\mathfrak{g})$ (in particular, the image $\mathrm{Ad}(H)$ is closed for every closed subgroup $H \subset G$). The intersection $\Gamma \cap H$ for $H \in \mathcal{H}$, being a lattice in a connected Lie group H, is a finitely generated group. Since Γ is countable, it contains countably many finitely generated subgroups. Hence it suffices to show that, given a finitely generated subgroup $\Delta \subset \Gamma$, there exist at most countably many connected subgroups $H \subset G$ such that Δ is a lattice in H and the Zariski closure of the group $\mathrm{Ad}(H)$ coincides with the Zariski closure of the group $\mathrm{Ad}(\Delta)$ (which will be denoted by $\mathcal{A}$ as well). By a well-known property of algebraic groups, the last condition implies that $\overline{[\mathcal{A},\mathcal{A}]} \subset \mathrm{Ad}(H)$. Moreover, since Δ is a lattice in H, the quotient space $\mathrm{Ad}(H)/\overline{\mathrm{Ad}(\Delta)[\mathcal{A},\mathcal{A}]}$ is compact. Now the theorem follows because the Abelian group $\mathcal{A}/\overline{[\mathcal{A},\mathcal{A}]}$ has countably many connected subgroups containing $\overline{\mathrm{Ad}(\Delta)[\mathcal{A},\mathcal{A}]}$ as a uniform subgroup. □

Having at our disposal Lemma 19.4 and Theorem 19.5, it now suffices to consider the behavior of unipotent trajectories near the singular subset $X(H,u_{\mathbb{R}})\Gamma \subset G/\Gamma$, $H \in \mathcal{H}$. With each closed subset $X(H,u_{\mathbb{R}})$ one associates an affine manifold $A = A_H$ in a vector space $V = V_H$, where a piece of an Ad-unipotent trajectory close to $X(H,u_{\mathbb{R}})\Gamma \subset G/\Gamma$ corresponds to a countable family of pieces of trajectories of a unipotent subgroup of $\mathrm{GL}(V)$ close to $A \subset V$. Recall that any affine manifold in a vector space V over $\mathbb{R}$ is given as the set of zeros of a polynomial. In turn, the coordinates of a unipotent trajectory are also polynomials in t. Using standard properties of polynomials (see Lemma 12.3), one can prove the following result on the behavior of unipotent trajectories near an affine manifold in a vector space.

LEMMA 19.6 [**DM93**]. *Let A be an affine manifold in a real vector space V, let $C \subset A$ be a compact subset, and let $\varepsilon > 0$. Then there exists a compact set $D \subset A$ (containing C) such that for any neighborhood $\Phi = \Phi(D) \subset V$ there exists a small neighborhood $\Psi = \Psi(C) \subset \Phi$ such that for any unipotent subgroup $\{u_t\} \subset \mathrm{GL}(V)$, any point $v \in V - \Phi$ and any $T > 0$, the following holds*:

$$l\{t \in [0,T] : \ u_t v \in \Psi\} < \varepsilon l\{t \in [0,T] : \ u_t v \in \Phi\}.$$

If compact sets $C \subset D \subset V$ satisfy the property as in the lemma, one says that the *relative size* of the compact subset C in D does not exceed ε. For instance, Lemma 12.3 allows one to estimate the relative size of the ball $D_\varepsilon \subset \mathbb{R}^n$ in the ball D_1.

3. First we demonstrate how to reduce the proof of Theorem 19.2 to Lemma 19.6 on the elementary example $G = \mathrm{SL}(2,\mathbb{R})$, $U = \{u_t = \left(\begin{smallmatrix} 1 & t \\ 0 & 1 \end{smallmatrix}\right), t \in \mathbb{R}\}$. Assume that Γ is not a uniform lattice in G (otherwise the singular set is empty and there is nothing to prove). It is clear that then $\mathcal{H}$ consists of all unipotent one-parameter subgroups having nontrivial intersection with Γ. Since all unipotent one-parameter subgroups of $\mathrm{SL}(2,\mathbb{R})$ are conjugate to U, it follows that $\mathcal{H} = \{H_i = g_i^{-1}Ug_i,\ i = 1,2,\dots\}$ for some countable family of elements $\{g_i\}$ in G. Therefore $X(H_i,U) = N(U)g_i$, where $N(U)$ is the normalizer of U in G (obviously, $N(U)$ is the product of the two-dimensional connected solvable subgroup B of upper-triangular matrices and the center $\mathbb{Z}_2 \subset G$). It follows that the singular set Y of all periodic horocycles in G/Γ is the union $Y = \bigcup_i N(U)g_i\Gamma$, $i = 1,2,\dots$. Any set $N(U)g_i\Gamma$ has one (if $\mathbb{Z}_2 \subset \Gamma$) or two (if $\mathbb{Z}_2 \not\subset \Gamma$) connected components, and each of them is an immersed cylinder with no self-intersections. If Γ is a nonuniform lattice, then the cylinder $N(U)g_i\Gamma$ is everywhere dense in G/Γ.

Let us treat the Lie algebra $\mathrm{sl}(2,\mathbb{R})$ as a vector space V with the basis

$$e_1 = \begin{pmatrix} 0 & 1 \\ 0 & 0 \end{pmatrix}, \qquad e_2 = \begin{pmatrix} 1 & 0 \\ 0 & -1 \end{pmatrix}, \qquad e_3 = \begin{pmatrix} 0 & 0 \\ 1 & 0 \end{pmatrix}.$$

Assume for simplicity that the subgroup U intersects Γ in a lattice (that is, $U \in \mathcal{H}$), and consider the set $X(U,U) = N(U) = \mathbb{Z}_2 \times B$. Let $\rho = \mathrm{Ad}$ be the adjoint representation of G in V, $p = e_1 \in V$ and $\eta(g) = \rho(g)p$. Then the orbit $\eta(G) = \rho(G)p$ is the half of the cone $x_1x_3 + x_2^2 = 0$ given by the inequality $x_1 > 0$. The set $\eta(X(U,U))$ is the half-line $x_2 = x_3 = 0,\ x_1 > 0$. Clearly, the Zariski closure of the half-line is the line $A = \{x_2 = x_3 = 0\}$ and $X(U,U) = \eta^{-1}(A) = \{g \in G : \eta(g) \in A\}$. Since $\Gamma_U = \Gamma \cap (\mathbb{Z}_2 \times U)$ is a uniform lattice in the stabilizer $\mathbb{Z}_2 \times U$ of the point $p \in V$, it follows that the image $\eta(\Gamma)$ is discrete in $\eta(G)$. For the same reason, given any compact set $D \subset A$, the set $\eta^{-1}(D)\Gamma \subset G/\Gamma$ is compact and hence there exists a (sufficiently small) neighborhood $\Phi = \Phi(D) \subset V$ such that the set $\eta^{-1}(\Phi)\Gamma \subset G/\Gamma$ has no self-intersections. Hence the map $g\Gamma \to \eta(g),\ g \in \eta^{-1}(\Phi)$, is well defined.

Now we choose $x = g\Gamma \notin Y$. For any $\varepsilon > 0$ we want to construct a neighborhood Ω of the "cylinder" $X(U,U)\Gamma \subset G/\Gamma$ such that

$$l\{t \in [0,T] : u_t x \in \Omega\} \le \varepsilon T \qquad \text{for all } T > 0.$$

Since the closed set $X(U,U)$ is σ-compact, it suffices to construct, for any $\varepsilon > 0$ and any compact set $C \subset X(U,U)$, the neighborhood Ω of the set $C\Gamma \subset G/\Gamma$ in question, and then to use the countability trick $\varepsilon = \sum_{i=1}^{\infty} \varepsilon/2^i$. Let C be a compact subset in $X(U,U)$, let $D \subset A$ be a compact set containing $\eta(C)$ as a subset with relative size $\le \varepsilon$, and let $\Phi = \Phi(D) \subset V$ be a neighborhood for which the map $\eta^{-1}(\Phi)\Gamma \to \Phi$ is well defined. Consider a small neighborhood $\Psi = \Psi(\eta(C)) \subset \Phi$ as in Lemma 19.6. Let us prove that $\Omega = \eta^{-1}(\Psi)\Gamma$ is what we need.

We fix $T > 0$ and for each $q \in \eta(\Gamma) = \rho(\Gamma)p$ define

$$I(q) = \{t \in (0,T) : \rho(u_t g)q \in \Phi\}, \quad J(q) = \{t \in (0,T) : \rho(u_t g)q \in \Psi\}.$$

Then $J(q) \subset I(q)$, and the sets $I(q_1)$, $I(q_2)$ are disjoint for $q_1 \ne q_2$ by the choice of Φ. Since

$$u_t g\Gamma \in \Omega \iff \exists \gamma \in \Gamma : \eta(u_t g \gamma) \in \Psi,$$

it suffices to prove the inequality $l(\bigcup_{q \in \eta(\Gamma)} J(q)) \le \varepsilon T$. Let (a,b) be a connected component of $I(q)$. Then, putting $v = \rho(u_a g)q$ and $T = b - a$ in Lemma 19.6, we obtain $l(J(q) \cap (a,b)) \le \varepsilon(b-a)$, and hence $l(J(q)) \le \varepsilon l(I(q))$ for all $q \in \eta(\Gamma)$. Now our inequality follows because the sets $I(q)$ are disjoint. Finally, considering in a similar way all other "cylinders" $X(U,U)g_i\Gamma$ and using the countability trick, we are able to construct the neighborhood in question of the singular set $Y \subset G/\Gamma$.

4. We proceed to the general case. Let G be a connected Lie group. With every subgroup $H \in \mathcal{H}$ of dimension h we associate the vector space $V = V_H = \bigwedge^h \mathfrak{g}$ (the h-th exterior power of the Lie algebra $\mathfrak{g}$) and the representation $\rho = \rho_H = \bigwedge^h \operatorname{Ad} : G \to \mathrm{GL}(V_H)$ (the h-th exterior power of the adjoint representation). Let $\xi_1, \dots, \xi_h$ be a basis of the subalgebra $\mathfrak{h} \subset \mathfrak{g}$ of the subgroup $H \subset G$. Then

$$\bigwedge^h \mathfrak{h} = \mathbb{R}p, \qquad \text{where } p = p_H = \xi_1 \wedge \cdots \wedge \xi_h \in V.$$

Also we denote by $\eta = \eta_H : G \to V$ the map $\eta(g) = \rho(g)p,\ g \in G$. Clearly, $\eta(H) \subset \mathbb{R}p$. Finally, for $H \in \mathcal{H}$ let $\Gamma_H = N_\Gamma(H)$ be the normalizer of H in Γ.

LEMMA 19.7. *If $H \in \mathcal{H}$, then $\Gamma_H = \{\gamma \in \Gamma : \rho(\gamma)p = \pm p\}$.*

Proof. If $\rho(\gamma)p = \pm p$ for $\gamma \in \Gamma$, then $\operatorname{Ad}(\gamma)$ normalizes $\mathfrak{h}$ and hence $\gamma \in \Gamma_H$. Conversely, if $\gamma \in \Gamma_H$, then γ normalizes the lattice $\Gamma \cap H \subset H$, and hence the conjugation $h \to \gamma h \gamma^{-1}$ preserves biinvariant Haar measure on H and (up to the sign) the volume form on $\mathfrak{h}$. $\square$

LEMMA 19.8. *Let $H \in \mathcal{H}$ and let $A = A_H \subset V$ be the Zariski closure of the set $\eta(X(H, u_{\mathbb{R}}))$. Then $X(H, u_{\mathbb{R}}) = \eta^{-1}(A) = \{g \in G : \ \eta(g) \in A\}$.*

Proof. Clearly, $X(H, u_{\mathbb{R}}) \subset \eta^{-1}(A)$. Conversely, let $\mathfrak{u} \subset \mathfrak{g}$ be the (one-dimensional) Lie subalgebra for $u_{\mathbb{R}} \subset G$. Then

$$g \in X(H, u_{\mathbb{R}}) \Longleftrightarrow \mathfrak{u} \subset \mathrm{Ad}_g(\mathfrak{h}) \Longleftrightarrow \eta(g) \wedge \mathfrak{u} \equiv 0 \Longleftrightarrow \eta(g) \in A_0,$$

where $A_0 = \{v \in V : \ v \wedge \mathfrak{u} \equiv 0\}$ iz Zariski closed and hence contains A. $\square$

Now we consider the structure of singular subsets $X(H, u_{\mathbb{R}})\Gamma \subset G/\Gamma$ for $H \in \mathcal{H}$. If no such set has self-intersections, the proof for $G = \mathrm{SL}(2, \mathbb{R})$ goes through without any great modifications in the general case. But if it has self-intersections, the proof gets more complicated because it is not possible to construct a well-defined map of a neighborhood $\eta^{-1}(D) \subset G/\Gamma$ to a neighborhood of the compact set $D \subset A$. This difficulty was first overcome by Shah in [**Sh91**].

Let M be a subset of G. We say that $g \in M$ is a point of (H, Γ)-*self-intersection* if there exists an element $\gamma \in \Gamma - \Gamma_H$ such that $g\gamma \in M$.

Equivalently, $g \in M$ is not a point of (H, Γ)-self-intersection if from $g\Gamma = g'\Gamma$, $g' \in M$, it follows that $g\Gamma_H = g'\Gamma_H$.

We are interested in self-intersections of the set $X(H, u_{\mathbb{R}})$. Assume that $u_{\mathbb{R}} \subset H$. Then $N(H) \subset X(H, u_{\mathbb{R}})$, and if $N(H) = X(H, u_{\mathbb{R}})$, then $X(H, u_{\mathbb{R}})$ has no (H, Γ)-self-intersections. In general, however, the inclusion $N(H) \subset X(H, u_{\mathbb{R}})$ is strict, and self-intersections may exist.

Example 19.9. Let $G = \mathrm{SL}(2, \mathbb{C})$, $\Gamma = \mathrm{SL}(2, \mathbb{Z} + i\mathbb{Z}), u_t = \left(\begin{smallmatrix} 1 & t \\ 0 & 1 \end{smallmatrix}\right)$ and $H = \mathrm{SL}(2, \mathbb{R})$. Then $X(H, u_{\mathbb{R}}) = u'_{\mathbb{R}}H$, where $u'_t = \left(\begin{smallmatrix} 1 & it \\ 0 & 1 \end{smallmatrix}\right)$ and $\Gamma_H = \mathrm{SL}(2, \mathbb{Z})$. It is easy to check that the set of (H, Γ)-self-intersections of $X(H, u_{\mathbb{R}})$ is $u_{\mathbb{R}} \times u'_{\mathbb{R}}$ (in other words, the set $X(H, u_{\mathbb{R}})\Gamma \subset G/\Gamma$ intersects itself in the torus $(u_{\mathbb{R}} \times u'_{\mathbb{R}})\Gamma$).

Fortunately, one can deal with self-intersections by using induction on the dimension and the following result.

LEMMA 19.10. *Let $H \in \mathcal{H}$ and $h = \dim H$. Then the set of (H, Γ)-self-intersections of $X(H, u_{\mathbb{R}})$ belongs to the union of all $X(H', u_{\mathbb{R}})$, where $H' \in \mathcal{H}$ and $\dim H' \leq h - 1$.*

Proof. Let $g \in X(H, u_{\mathbb{R}})$ be a point of (H, Γ)-self-intersection. Then there exists an element $\gamma \in \Gamma - \Gamma_H$ such that $g^{-1}u_{\mathbb{R}}g \subset H \cap \gamma H \gamma^{-1}$. Since $\gamma \notin \Gamma_H$, we conclude that $\dim(H \cap \gamma H \gamma^{-1}) \leq h - 1$. The subspace $(H \cap \gamma H \gamma^{-1})\Gamma$ is open and closed in the intersection of two closed subspaces $H\Gamma$ and $\gamma H \Gamma$ (because $\Gamma \subset G$ is discrete). Hence it is closed. Note that the intersection of two homogeneous subspaces of finite volume can fail to have finite volume (for instance, $H\Gamma$ and $k^{-1}Hk\Gamma$, where $k = \left(\begin{smallmatrix} i & 0 \\ 0 & -i \end{smallmatrix}\right)$ from Example 19.9, intersect each other in a one-dimensional noncompact manifold). We avoid this obstacle as follows. Let H_γ be the smallest (connected) subgroup of $H \cap \gamma H \gamma^{-1}$ containing all one-parameter subgroups unipotent in G. Then H_γ is normal in $H \cap \gamma H \gamma^{-1}$, and hence H_γ acts ergodically on $H'/(\Gamma \cap H')$, where $H' = (\overline{H_\gamma \Gamma})_0$, relative to a locally finite H'-invariant measure. Since $H'\Gamma$ is a closed subspace of the finite volume space $H\Gamma$ and H_γ is generated by unipotent one-parameter subgroups, then by Theorem 12.14

the H'-invariant measure on $H'\Gamma$ is finite. Hence $g \in X(H', u_{\mathbb{R}})$, where $H' \in \mathcal{H}$ and $\dim H' \leq h - 1$. □

Now we formulate without proof one of the main results from [**DM93**].

THEOREM 19.11. *Let Γ be a discrete subgroup of G, and let $H \in \mathcal{H}$. Then the orbit $\eta(\Gamma)$ is discrete in V.*

To illustrate the theorem we consider the case when G is a closed subgroup of $\mathrm{SL}(n, \mathbb{R})$, and $\Gamma \subset G \cap \mathrm{SL}(n, \mathbb{Z})$. Then a connected subgroup H intersecting Γ in a lattice corresponds to a rational subalgebra $\mathfrak{h} \subset M(n, \mathbb{R})$. Hence the vector $p \in V$ is collinear to a vector with rational coordinates, whereas $\rho(\Gamma)$ consists of integer matrices. It follows that the orbit $\eta(\Gamma) = \rho(\Gamma)p$ is discrete.

COROLLARY 19.12. *Let $H \in \mathcal{H}$, let A be the Zariski closure of $\eta(X(H, u_{\mathbb{R}}))$ in V, and let D be a compact set in A. Denote by Y_D the set of (H, Γ)-self-intersections of $\eta^{-1}(D)$, and let K be a compact set in $G - Y_D\Gamma$. Then there exists a neighborhood Φ of D in V such that $\eta^{-1}(\Phi) \cap K\Gamma$ has no self-intersections.*

Proof. Assume for a contradiction that there exist sequences $\{g_i\}$, $\{g_i'\}$ of elements in G such that

$$g_i\Gamma = g_i'\Gamma \in K\Gamma, \ g_i\Gamma_H \neq g_i'\Gamma_H \quad \text{and} \quad \eta(g_i) \to d \in D, \ \eta(g_i') \to d' \in D.$$

Since K is a compact set, one can assume that

$$g_i = h_i\gamma_i, \ g_i' = h_i\gamma_i', \quad \text{where } \gamma_i, \gamma_i' \in \Gamma \text{ and } h_i \to h \in K.$$

Then $\eta(\gamma_i) = \rho(h_i^{-1})\eta(g_i) \to \rho(h^{-1})d$, and similarly $\eta(\gamma_i') \to \rho(h^{-1})d'$. Since the set $\eta(\Gamma)$ is discrete, it follows that $\eta(\gamma_i) = \rho(h^{-1})d$, $\eta(\gamma_i') = \rho(h^{-1})d'$; that is, $h\gamma_i, h\gamma_i' \in \eta^{-1}(D) \subset X(H, u_{\mathbb{R}})$ for large enough i. Since $h\Gamma \in K\Gamma$ and $K\Gamma \cap Y_D = \varnothing$, no element $h\gamma_i$ is a point of (H, Γ)-self-intersection. Hence from $h\gamma_i\Gamma = h\gamma_i'\Gamma$ it follows that $\gamma_i\Gamma_H = \gamma_i'\Gamma_H$, and then $g_i\Gamma_H = h_i\gamma_i\Gamma_H = h_i\gamma_i'\Gamma_H = g_i'\Gamma_H$. Contradiction. □

Proof of Theorem 19.2. We recall that the singular set $Y \subset X$ is a countable union of sets $X(H, u_{\mathbb{R}})\Gamma$, $H \in \mathcal{H}$. In addition, the spaces G/Γ and all $X(H, u_{\mathbb{R}})$ are σ-compact, and the family $\mathcal{H}$ is countable. Hence it suffices to construct, for any $x \in X - Y$, $H \in \mathcal{H}$, $\varepsilon > 0$, and any compact sets $C \subset X(H, u_{\mathbb{R}})$, $K \subset G/\Gamma$, a neighborhood Ω of the compact set $C\Gamma \subset G/\Gamma$ such that

$$(*) \qquad l\{t \in [0, T] : \ u_t x \in \Omega \cap K\} \leq \varepsilon T \quad \text{for all } T \geq 0.$$

We shall do this by induction on $\dim H$. For $\dim H = 1$ the statement is trivial. Assume it has been proved for $\dim H \leq n - 1$. We fix

$$x \in X - Y, \quad H \in \mathcal{H}, \quad \varepsilon > 0, \quad C \subset X(H, u_{\mathbb{R}}), \quad K \subset G/\Gamma.$$

Let $Y_H \subset G$ be the set of (H, Γ)-self-intersections of the set $X(H, u_{\mathbb{R}})$. Then, by Lemma 19.10 and inductive hypothesis, there exists a neighborhood Ω' of the set $Y_H\Gamma \subset G/\Gamma$ such that

$$(**) \qquad l\{t \in [0, T] : \ u_t x \in \Omega' \cap K\} \leq \varepsilon T/2 \quad \text{for all } T \geq 0.$$

From now on we let

$$V = V_H, \quad \rho = \rho_H, \quad p = p_H, \quad \eta = \eta_H,$$

and let $A = A_H$ be the Zariski closure of the set $\eta(X(H, u_{\mathbb{R}}))$. By Lemma 19.6, there exists a compact set $D \subset A$ containing $\eta(C)$ as a subcompact of relative size at most $\varepsilon/4$. We fix $g\Gamma = x \in X - Y$. Since the image $\eta(\Gamma) \subset V$ is discrete, the set $\eta(g\Gamma) = \rho(g)\eta(\Gamma)$ is discrete as well. Note that

$$X(H, u_{\mathbb{R}}) = \eta^{-1}(A) \quad \text{and} \quad g\gamma \notin X(H, u_{\mathbb{R}}) \quad \text{for all } \gamma \in \Gamma,$$

and hence $\eta(g\Gamma) \cap A = \varnothing$. It follows that there exists a neighborhood $\Phi' = \Phi'(D) \subset V$ which is disjoint from $\eta(g\Gamma)$.

Let K' be a compact set in G such that $K'\Gamma = K_1 = K - \Omega' \subset G/\Gamma$. By Corollary 19.12, there exists a neighborhood $\Phi'' = \Phi''(D) \subset V$ such that the set $\eta^{-1}(\Phi'') \cap K'\Gamma$ has no (H, Γ)-self-intersections. We let $\Phi = \Phi' \cap \Phi''$.

Now we introduce a neighborhood $\Psi = \Psi(\eta(C)) \subset \Phi \subset V$ for which

$$(***) \qquad l\{t \in [0, T] : \ \rho(u_t)v \in \Psi\} < \varepsilon l\{t \in [0, T] : \ \rho(u_t)v \in \Phi\}/4$$

for all $v \in V - \Phi$, $T > 0$. Let us prove that the neighborhood $\Omega = \eta^{-1}(\Psi)\Gamma$ of the compact set $C\Gamma \subset G/\Gamma$ is what we need.

We fix $T > 0$ and for every $q \in \eta(\Gamma) = \rho(\Gamma)p$ we define

$$\begin{aligned} I(q) &= \{t \in (0, T) : \ \rho(u_t g)q \in \Phi\}, \\ J(q) &= \{t \in (0, T) : \ \rho(u_t g)q \in \Psi \text{ and } u_t g\Gamma \in K_1\}, \\ I'(q) &= \{t \in (0, T) : \ \exists a \geq 0 : \ [t, t + a] \subset I(q) \text{ and } u_{t+a} g\Gamma \in K_1\}. \end{aligned}$$

Then $J(q) \subset I'(q) \subset I(q)$ for all $q \in \eta(\Gamma)$. Let us show that, given any $q_1, q_2 \in \eta(\Gamma)$, either $q_1 = \pm q_2$ or the sets $I'(q_1), I'(q_2)$ are disjoint. Let

$$q_1 = \eta(\gamma_1), \qquad q_2 = \eta(\gamma_2), \qquad t \in I'(q_1) \cap I'(q_2).$$

Then there exists a number $a \geq 0$ such that

$$[t, t + a] \subset I(q_1) \cap I(q_2) \quad \text{and} \quad u_{t+a} g\Gamma \in K_1 = K'\Gamma \subset G/\Gamma.$$

Since $\eta^{-1}(\Phi) \cap K'\Gamma$ has no (H, Γ)-self-intersections, it follows that $\gamma_1\Gamma_H = \gamma_2\Gamma_H$ and, by Lemma 19.7, $q_1 = \rho(\gamma_1)p = \pm\rho(\gamma_2)p = \pm q_2$.

Now let $q \in \eta(\Gamma)$ and let (a, b) be a connected component of $I(q)$. Then $\rho(u_a g)q \notin \Phi$ and if $J(q) \cap (a, b) \neq \varnothing$, then $I'(q)$ contains an interval (a, c), $a < c \leq b$, such that $J(q) \cap (a, b) \subset (a, c]$. Applying (***) to $v = \rho(u_a g)q \in V - \Phi$ and $T = c - a > 0$, we obtain $l(J(q) \cap (a, b)) \leq \varepsilon l(I'(q) \cap (a, b))/4$. Therefore $l(J(q)) \leq \varepsilon l(I'(q))/4$. Hence

$$l\{t \in [0, T] : \ u_t g\Gamma \in \Omega \cap K_1\} = l\Big(\bigcup_{q \in \eta(\Gamma)} J(q)\Big) \leq \frac{\varepsilon}{4} \sum_{q \in \eta(\Gamma)} l(I'(q)) \leq \frac{\varepsilon T}{2},$$

where the last inequality follows because the sets $I'(q_1)$ and $I'(q_2)$ are disjoint for $q_1 \neq \pm q_2$. Comparing this with (**), we obtain (*). Theorem 19.2 has been proved. Incidentally, we have proved Theorems 19.1 and 19.3 (and, in particular, the Raghunathan topological conjecture for one-parameter unipotent subgroups).□

5. Now we discuss the Raghunathan topological conjecture for arbitrary subgroups $U \subset G$ generated by unipotent elements (that is, in the general form suggested by Margulis). Ratner has proved the following result.

THEOREM 19.13 (Ratner's topological theorem [**R91b**]). *Let Γ be a lattice in a connected Lie group G, and let a subgroup $U \subset G$ be as in Theorem 17.5. Then all U-orbits on G/Γ are homogeneous subspaces of finite volume.*

We only show how to prove the theorem in the case when U is a connected unipotent subgroup (the general case is technically more involved). Choose $x = g\Gamma \in G/\Gamma$. We know that for any subgroup $u_{\mathbb{R}} \subset U$ there exists a connected subgroup $H \subset G$ such that $\overline{u_{\mathbb{R}}x} = Hx$ and Hx carries finite H-invariant measure. Since the unipotent flow $(g^{-1}Hg/\Gamma \cap g^{-1}Hg, g^{-1}u_{\mathbb{R}}g)$ has a dense orbit, it must be ergodic, and by Theorem E.13 the lattice $\Gamma \cap g^{-1}Hg$ is Zariski dense in $g^{-1}Hg$ in the Ad-representation of the group G. Thus, by Theorem 19.5, the family of connected subgroups $H \subset G$ such that Hx is of finite volume and $\overline{u_{\mathbb{R}}x} = Hx$ for some subgroup $u_{\mathbb{R}} \subset U$ is at most countable. It follows that the connected subgroup U belongs to a countable union of closed subgroups $H_i \subset G$. From dimension arguments there exists $H_i \subset G$ such that $U \subset H_i$, and then obviously $\overline{Ux} = H_i x$.

The case when the subgroup U is more general than in Theorem 19.13 was "almost completely" examined by Shah.

THEOREM 19.14 [**Sh98**]. *Let a space G/Γ be of finite volume and let a subgroup $U \subset G$ be generated by unipotent elements. Then for any point $x \in G/\Gamma$ there exists a subgroup $F \subset G$ such that $\overline{Ux} = Fx$ and F_0x has finite F_0-invariant measure.*

The fact that Fx has finite F-invariant measure was proved by Eskin and Margulis (personal communication). Hence one can regard the generalized Raghunathan conjecture as completely proved. In particular, results of Shah, Eskin and Margulis solve the following problem mentioned in [**R94**]. Let G be a simple Lie group and let $\Lambda \subset G$ be a nonuniform lattice generated by unipotent elements. Then every Λ-orbit on a space G/Γ of finite volume has an algebraic closure. In particular, if the product of two lattices in G is closed, they are commensurable (the same can be proved for any two lattices in a connected Lie group). Note that for $G = \mathrm{PSL}(2, \mathbb{C})$ this was proved earlier by Shah.

6. To conclude, we show how the Raghunathan conjecture reduces to the case when $\Gamma \subset G$ is a lattice (Ratner's results were obtained precisely in this form). Let $G = L \cdot R$ be a Levi decomposition of a connected, simply connected Lie group G. Consider the ψ-algebraic hull $\mathcal{A}(G) = (L \times \mathcal{A}(T_R)) \cdot R$. According to Theorem 6.14, $\mathcal{A}(\Gamma)$ contains all unipotent elements, and hence $U \subset \mathcal{A}(\Gamma)$. Now while studying the orbit $U\Gamma$ we can concentrate our attention on the finite volume subspace H/Γ, where $H = G \cap \mathcal{A}(\Gamma)$. Here the identity component Γ_0 is normal in $\mathcal{A}(\Gamma)$ and hence in H. It follows that we can use Theorem 19.13 for the lattice $(\Gamma \cap H)/\Gamma_0$ in the Lie group H/Γ_0 (H may not be connected in this case, but this involves no difficulties).

§20. Various problems of convergence in the space of measures

Here we present without proof some generalizations of Theorem 19.1 on the uniform distribution of unipotent orbits. All of them are based on Ratner's measure Theorem 17.1 and do not use Theorem 19.1.

1. A point $x \in G/\Gamma$, a subgroup $u_{\mathbb{R}} \subset G$ and a number $T > 0$ determine a probability measure $\mu_{T,x,u}$ on G/Γ by the formula

$$\mu_{T,x,u}(f) = \frac{1}{T}\int_0^T f(u_t x)dt, \qquad f \in C_c(G/\Gamma)$$

(that is, $\mu_{T,x,u}(f)$ is the time average of the function f along the piece of trajectory $u_{[0,T]}x$). We recall that the singular set $Y = Y(u_{\mathbb{R}}) \subset G/\Gamma$ for a unipotent flow $(G/\Gamma, u_{\mathbb{R}})$ (that is, the union of all proper closed invariant homogeneous subspaces) is a countable union of locally analytic sets: $Y = \bigcup_{H\in\mathcal{H}} X(H, u_{\mathbb{R}})\Gamma$. Theorem 19.1 states that in the weak* topology $\mu_{T,x,u}$ converges to G-invariant measure ν as $T \to \infty$ for any point x outside the singular set Y.

Note that for the horocycle flow $(\mathrm{SL}(2,\mathbb{R})/\Gamma, u_{\mathbb{R}})$ with a uniform lattice Γ, each trajectory is uniformly distributed in $\mathrm{SL}(2,\mathbb{R})/\Gamma$ (and this is equivalent to unique ergodicity). Rather delicate estimates for the asymptotics of the convergence $\mu_{T,x,u} \to \nu$ in this case have been found by Burger [**Bu**] with the use of the theory of representations.

It turns out that the convergence takes place for any sequence $x_n \in G/\Gamma$ converging to a nonsingular point $x \in G/\Gamma$. The first part of the following theorem was also independently proved by Ratner [**R94**].

THEOREM 20.1 [**DM93**]. a) *Let $T_n \to +\infty$ and let $x_n \to x$, where $x \in G/\Gamma$ is nonsingular for a unipotent subgroup $u_{\mathbb{R}} \subset G$. Then $\mu_{T_n,x_n,u} \to \nu$ as $n \to \infty$.*

b) *Moreover, given a function $f \in C_c(G/\Gamma)$, a compact set $K \subset G/\Gamma$, and a number $\varepsilon > 0$, there exist finitely many subgroups $H_i \in \mathcal{H}$, $i = 1,\dots,k$, and compact sets $C_i \subset X(H_i, u_{\mathbb{R}})$ such that $|\mu_{T,x,u}(f) - \nu(f)| < \varepsilon$ uniformly on any compact subset $F \subset K - \bigcup_i C_i\Gamma$, and all $T \geq T_0 = T_0(f)$.*

As an example, let us consider $G = \mathrm{SL}(2,\mathbb{R})$. Choose a sequence of periodic horocycles $u_{\mathbb{R}}x_n \subset G/\Gamma$ with periods $T_n \to \infty$, together with the corresponding ergodic measures μ_{T_n}. It is not difficult to prove (using topological mixing of the geodesic flow), that, given an arbitrarily small neighborhood $O_\varepsilon(x)$ of a point $x \in G/\Gamma$, there exists N such that $u_{\mathbb{R}}x_n \cap O_\varepsilon(x) \neq \varnothing$ for all $n \geq N$. Hence there exist points $x'_n \in u_{\mathbb{R}}x_n$ converging to x. Now, applying the previous result, we obtain Sarnak's uniform distribution theorem [**Sa81**]: the periodic horocycle $u_{\mathbb{R}}x_n = u_{\mathbb{R}}x'_n$ gets more and more uniformly distributed in G/Γ as $T_n \to \infty$ (an elementary proof of the fact can be also found in Ratner's paper [**R92**]).

Actually, Sarnak has also found the asymptotics of the convergence $\mu_T \to \nu$ up to $o(T^{-1/2})$ for all nonuniform lattices $\Gamma \subset G$. For $\Gamma = \mathrm{SL}(2,\mathbb{Z})$ it has the especially simple form (found for the first time by Zagier [**Za**]):

$$\mu_T(f) = \nu(f) + o(T^{-1/2}),\ T \to +\infty, \qquad \text{for any } f \in C_c^\infty(G/\Gamma).$$

Zagier also found a remarkable connection between the rate of the convergence $\mu_T(f) \to \nu(f)$ as $T \to +\infty$ and the famous Riemann hypothesis in number theory: the latter holds if and only if

$$\mu_T(f) = \nu(f) + o(T^{-3/4+\varepsilon}) \quad \text{for all } \varepsilon > 0, \qquad f \in C_c^\infty(G/\Gamma).$$

It was proved in [**Ve**] that $\alpha = -1/2$ is the best possible constant if one finds the asymptotics of the form $o(T^\alpha)$ for characteristic functions of the simplest open subsets of G/Γ (like cubic neighborhoods). Certainly, this does not disprove the Riemann hypothesis, because the functions involved are discontinuous.

2. The result on the uniform distribution of closed orbits can be also proved for actions of a semisimple subgroup $H \subset G$.

THEOREM 20.2 [**DM93**]. *Let $H \subset G$ be a maximal proper connected subgroup, H being semisimple and with no compact components. Assume that every orbit Hx_n is closed and hence carries a unique H-invariant probability measure μ_n. Let $x_n \to x$, $n \to \infty$, where the orbit Hx is not closed. Then $\mu_n \to \nu$ as $n \to \infty$.*

Proof. It is clear that H has a unipotent subgroup $u_{\mathbb{R}}$ that is not contained in a proper subgroup normal in H. Then, by Moore's ergodicity criterion (see §8), every flow $(Hx_n, u_{\mathbb{R}})$ is ergodic. Let $x = g\Gamma$ for some $g \in G$. Let us prove that $hg\Gamma \in G/\Gamma$ is not a singular point (in the sense of §19) for the flow $(G/\Gamma, u_{\mathbb{R}})$ for almost every $h \in H$. If this is not the case, then using the structure of the singular set one finds a proper subgroup $L \subset G$ such that $\Gamma \cap L$ is a lattice in L, and $hg \in X(L, u_{\mathbb{R}}) = \{f \in G : f^{-1}u_{\mathbb{R}}f \subset L\}$ for a set $h \in H$ of positive measure. Since the set $X(L, u_{\mathbb{R}})$ is analytic, the intersection $Hg \cap X(L, u_{\mathbb{R}})$ can have positive measure only if $Hg \subset X(L, u_{\mathbb{R}})$, and hence $h^{-1}u_{\mathbb{R}}h \subset gLg^{-1}$ for all $h \in H$. In view of the choice of $u_{\mathbb{R}}$ and the maximality of H, it follows that $H = gLg^{-1}$. But then the orbit $Hg\Gamma = gL\Gamma$ is closed, which contradicts the hypothesis on $x = g\Gamma$.

Therefore the point hx is not singular for $u_{\mathbb{R}}$ for almost all $h \in H$. For every n, the orbit $u_{\mathbb{R}}hx_n$ is uniformly distributed for almost all $h \in H$ by the Birkhoff–Khintchine ergodic theorem. It follows that there exists an element $h \in H$ such that every orbit $u_{\mathbb{R}}hx_n$ is uniformly distributed in Hx_n relative to μ_n and the point $hx \in G/\Gamma$ is not singular for $u_{\mathbb{R}}$. Now applying Theorem 20.1 to the sequence $hx_n \to hx$, we obtain the desired result. □

Note that the theorem generalizes that of Burger and Sarnak [**BS**] on the uniform distribution of the so-called Hecke points coming from number theory. To obtain the result of [**BS**] it suffices to take a noncompact simple group S with a lattice Λ and consider the case where H is the diagonal subgroup of the Lie group $G = S \times S$ with $\Gamma = \Lambda \times \Lambda$ as a lattice.

3. As was shown by Dani and Margulis, the uniform distribution Theorem 19.1 can be extended even further if one varies not only the point $x \in G/\Gamma$ but also the subgroup $u_{\mathbb{R}}$:

THEOREM 20.3 [**DM93**]. *Let a sequence of unipotent subgroups $u_{\mathbb{R}}^{(n)}$ converge to a unipotent subgroup $u_{\mathbb{R}}$ (this means that $u_t^{(n)} \to u_t$ as $n \to \infty$ for all $t \in \mathbb{R}$). Let $x_n \to x$, where the limit point $x \in G/\Gamma$ is nonsingular for $u_{\mathbb{R}} \subset G$. Then $\mu_{T_n, x_n, u^{(n)}} \to \nu$ as $T_n \to +\infty$.*

4. Now let Γ be an arbitrary discrete subgroup of G, $X = G/\Gamma$, and let $\mathcal{P}(X)$ be the affine space of Borel probability measures on X with the topology of weak* convergence. Following Mozes and Shah [**MS**], we denote by $Q(X)$ the subset of all measures $\mu \in \mathcal{P}(X)$, for which there exists a one-parameter unipotent subgroup $u_{\mathbb{R}} \subset G$ such that μ is an invariant ergodic measure for the flow $(X, u_{\mathbb{R}})$.

THEOREM 20.4 [**MS**]. *The set $Q(X)$ is closed in $\mathcal{P}(X)$.*

This means that, given $\mu_n \in Q(X)$ such that $\mu_n \to \mu \in \mathcal{P}(X)$, the measure μ is also ergodic relative to a unipotent subgroup; that is, $\mu \in Q(X)$.

Assume that Γ is a lattice in G, $X = G/\Gamma$, and let X^c be the one-point compactification of X (if X is noncompact). Then the space $\mathcal{P}(X^c)$ is compact (relative to the weak* topology). According to Theorem 12.2, a unipotent trajectory

$u_{\mathbb{R}}x \subset X$ spends almost all its life in a compact set $K \subset X$. This result can be generalized if one varies the point x in a compact set $C \subset X$, and chooses the same compact set K for all unipotent subgroups $u_{\mathbb{R}} \subset X$.

THEOREM 20.5 [**DM93**]. *Let C be a compact set in a space G/Γ of finite volume. Then for any $\varepsilon > 0$ there exists a compact set $K \subset G/\Gamma$ (containing C) such that*

$$l\{t \in [0,T] : u_t x \in K\} \geq (1-\varepsilon)T$$

for all $T > 0$, all $x \in C$, and all unipotent subgroups $u_{\mathbb{R}} \subset G$.

This result can be applied to the proof of Theorem 20.3, and enables one to get the following statement.

COROLLARY 20.6 [**MS**]. *Let $X = G/\Gamma$ be a space of finite volume and let ν_∞ be the probability measure on X^c concentrated at infinity (if X is noncompact). Then the set $Q(X) \cup \nu_\infty \subset \mathcal{P}(X^c)$ is compact.*

Indeed, let $\mu_n \in Q(X)$ and let $\mu_n \to \mu \in \mathcal{P}(X^c)$. If $\mu \neq \nu_\infty$, then there exists a compact set $C \subset X$ with $\mu(C) > 0$. Without loss of generality, $\mu_n(C) > 0$ for all n. Given any $\varepsilon > 0$, we take a compact set $K = K(\varepsilon, C)$ from Theorem 20.5. Suppose that the measure μ_n is ergodic relative to a unipotent subgroup $u_{\mathbb{R}}^{(n)} \subset G$. It follows from the Birkhoff–Khintchine ergodic theorem and Theorem 20.5 that $\mu_n(K) > 1 - \varepsilon$ for all n. Hence $\mu(K) \geq 1 - \varepsilon$. Therefore $\mu(X) = 1$, that is, $\mu \in \mathcal{P}(X)$, and it suffices to apply the property of $Q(X)$ being closed in $\mathcal{P}(X)$.

An example of a sequence $\{\mu_n \in Q(X)\}$ converging to ν_∞, is given by normalized measures on a sequence of periodic horocycles of decreasing length in $\mathrm{SL}(2,\mathbb{R})/\mathrm{SL}(2,\mathbb{Z})$.

Choose a point $x \in X$, and let $Q(x)$ be the set of measures in $Q(X)$ with their supports containing x. It follows immediately from Corollary 20.6 that the set $Q(x)$ is compact.

5. In a series of problems related to number theory (see §31), one is concerned with convergence of ergodic measures for reductive algebraic subgroups (note that such a subgroup need not be generated by unipotent elements). The general setting here is as follows: let $H \subset G \subset \mathrm{SL}(n,\mathbb{R})$ be two reductive algebraic subgroups over $\mathbb{Q}$ with no nontrivial $\mathbb{Q}$-characters. Then, by Theorem E.6 of Borel and Harish-Chandra, $G_{\mathbb{Z}} = G \cap \mathrm{SL}(n,\mathbb{Z})$ and $H_{\mathbb{Z}} = H \cap \mathrm{SL}(n,\mathbb{Z})$ are lattices in G and H. Given a probability measure $\mu \in \mathcal{P}(G/G_{\mathbb{Z}})$, let $g\mu$, $g \in G$, stand for the translate $g\mu(E) = \mu(g^{-1}E)$. We obtain a continuous action of the group G on the space $\mathcal{P}(G/G_{\mathbb{Z}})$ equipped with the weak* topology. By μ_H and μ_G we denote the H- and G-invariant probability measures on the spaces $HG_{\mathbb{Z}} \simeq H/H_{\mathbb{Z}}$ and $G/G_{\mathbb{Z}}$. Let $\{g_i\}$ be a sequence of elements in G. The question arises: which measures in $\mathcal{P}(G/G_{\mathbb{Z}})$ can be limit measures for the sequence $\{g_i\mu_H\}$? In particular, when does $g_i\mu_H \to \mu_G$? In which case is the family $\{g\mu_H,\ g \in G\}$ relatively compact?

The answers to these questions are found in papers of Eskin, Mozes, and Shah [**EMS96,97**]. We first consider the last question. Let $\pi : G \to G/G_{\mathbb{Z}}$ be the natural projection. Since H is a reductive $\mathbb{Q}$-group, so is the centralizer $Z_G(H)$. Hence the subspace $\pi(Z_G(H)) \subset G/G_{\mathbb{Z}}$ is closed. If it is not compact, there exists a sequence $z_i \in Z_G(H)$ such that $\pi(z_i) \to \infty$. Equivalently, for any two compact sets $K \subset G/G_{\mathbb{Z}}$, $C \subset \pi(H)$, the intersection $z_iC \cap K$ is empty for all but finitely many i. Hence $z_i\mu_H(K) \to 0$, $i \to \infty$, for any compact set $K \subset G/G_{\mathbb{Z}}$; that is, $z_i\mu_H \to 0$. We have shown that the family $\{g\mu_H,\ g \in G\}$ is not relatively compact if $\pi(Z_G(H))$ is noncompact. On the other hand, the following holds.

THEOREM 20.7 [**EMS97**]. *Let $H \subset G \subset \mathrm{SL}(n, \mathbb{R})$ be two reductive algebraic $\mathbb{Q}$-groups, and let $H_{\mathbb{Z}}$ and $G_{\mathbb{Z}}$ be lattices in H and G. If $\pi(Z_G(H))$ is compact, then for any $\varepsilon > 0$ there exists a compact set $K \subset G/G_{\mathbb{Z}}$ such that $g\mu_H(K) > 1 - \varepsilon$ for all $g \in G$.*

Thus, we have established a criterion for the family of probability measures $\{g\mu_H,\ g \in G\}$ to be relatively compact. In the proof one needs to generalize Theorem 20.5 of Dani and Margulis on the homecoming of unipotent trajectories to a compact set. Note that the compactness condition for $\pi(Z_G(H))$ is equivalent to either of the following two conditions: 1) H is not contained in a proper parabolic $\mathbb{Q}$-subgroup of G; 2) any $\mathbb{Q}$-subgroup of G that contains H, is reductive.

The answer to the second question (but not its proof) turns out to be very simple.

THEOREM 20.8 [**EMS96**]. *Let $H \subset G \subset \mathrm{SL}(n, \mathbb{R})$be two algebraic $\mathbb{Q}$-groups, and let $H_{\mathbb{Z}}$ and $G_{\mathbb{Z}}$ be lattices in H and G. Then any limit probability measure for the family $\{g\mu_H,\ g \in G\}$ is algebraic.*

Moreover, if $g_i\mu_H \to \mu \in \mathcal{P}(G/G_{\mathbb{Z}})$, then there exist an algebraic $\mathbb{Q}$-group $L \subset G$ and sequences $\gamma_i \in G_{\mathbb{Z}}$ and $c_i \to c \in G$ such that μ is a cLc^{-1}-invariant measure on $c\pi(L)$, $\gamma_i H \gamma_i^{-1} \subset L$ and $g_i H = c_i \gamma_i H$ for all but finitely many i.

The proof of the theorem is based on the following observation [**EMS96**]: either there exists a sequence $c_i \to c$ such that $g_i\mu_H = c_i\mu_H$ (and then $\mu = c\mu_H$), or μ is invariant under a unipotent subgroup of G (and then one can use a well-developed technique of unipotent flows). Note that in the result one does not require H and G to be reductive.

Actually, the second part of the theorem solves the first problem as well. Sometimes there is no need to involve intermediate subgroups $L \subset G$.

COROLLARY 20.9 [**EMS96**]. *Under the assumptions of Theorem 20.8, let H be a maximal connected algebraic $\mathbb{Q}$-subgroup of G. Then $g_i\mu_H \to \mu_G$ whenever $g_iH \to \infty$ in G/H.*

6. Now we discuss the paper of Shah [**Sh94**], which contains the quite surprising result that the algebraicity of orbit closures and the uniform distribution hold if in place of unipotent subgroups one considers polynomial curves. The idea is that in the proof of Theorem 19.1 one only uses the polynomial divergence of unipotent trajectories. We say that $\theta : \mathbb{R} \to \mathrm{SL}(n, \mathbb{R})$ is a *polynomial* if every coordinate function $\theta_{i,j}$ is polynomial.

THEOREM 20.10 [**Sh94**]. *Let $G \subset \mathrm{SL}(n, \mathbb{R})$ be a closed subgroup with a lattice $\Gamma \subset G$, and let $\theta : \mathbb{R} \to G \subset \mathrm{SL}(n, \mathbb{R})$ be a polynomial such that $\theta(0) = 1 \in G$. Then there exists a closed subgroup $F \subset G$ such that the orbit $F\Gamma$ carries a probability F-invariant measure ν_F, and $\mu_{T,\Gamma,\theta} \to \nu_F$ as $T \to \infty$.*

The proof is based on the following idea. Let μ be a limit measure for a sequence $\{\mu_{T_n,\Gamma,\theta}\}$. Then this measure turns out to be invariant under a certain unipotent subgroup $u_{\mathbb{R}} \subset G$. This comes from the following observations:

LEMMA 20.11 [**Sh94**]. *Let $\theta : \mathbb{R} \to \mathrm{SL}(n, \mathbb{R})$ be a nontrivial polynomial. Then there exist $q \geq 0$ and a nontrivial unipotent subgroup $u : \mathbb{R} \to \mathrm{SL}(n, \mathbb{R})$ such that*

$$\lim_{t\to\infty} \theta(t + st^{-q})\theta(t)^{-1} = u(s) \quad \textit{for all } s \in \mathbb{R}.$$

For instance, if

$$\theta(t) = \begin{pmatrix} 1 & a_0 t^n + a_1 t^{n-1} + \dots \\ 0 & 1 \end{pmatrix},$$

where $a_0 \neq 0$, then $q = n - 1$ and

$$u(s) = \begin{pmatrix} 1 & a_0 n s \\ 0 & 1 \end{pmatrix}.$$

LEMMA 20.12 [**Sh94**].

$$\lim_{T\to\infty} \frac{1}{T} \int_1^T (f(t + st^{-q}) - f(t))dt = 0 \quad \textit{for any } f \in C_c(\mathbb{R}),\ q \geq 0,\ s \in \mathbb{R},$$

and the rate of convergence depends only on s, q and $\sup |f|$.

Now we prove that the limit measure μ is invariant under $u_{\mathbb{R}}$. Indeed,

$$\begin{aligned} \int_{G/\Gamma} f(u(s)x)d\mu_x &= \lim_{n\to\infty} \frac{1}{T_n} \int_0^{T_n} f(u(s)\theta(t)\Gamma)dt \\ &= \lim_{n\to\infty} \frac{1}{T_n} \int_0^{T_n} f(\theta(t + st^{-q})\Gamma)dt = \lim_{n\to\infty} \frac{1}{T_n} \int_0^{T_n} f(\theta(t)\Gamma)dt \\ &= \int_{G/\Gamma} f(x)d\mu_x, \end{aligned}$$

where the second equality follows from the choice of $u_{\mathbb{R}}$ and the uniform continuity of f, and the third follows from Lemma 20.12.

Thus, the limit measure μ is invariant under a unipotent subgroup, and hence each ergodic component of it has an algebraic origin. The further arguments follow the proof of the uniform distribution of unipotent trajectories in §19. More specifically, Theorem 12.2 (which says thata unipotent trajectory on a space of finite volume spends almost all its life inside a compact set) uses only the fact that the trajectory is polynomial, whence we derive that $\mu(G/\Gamma) = 1$. Then we use the countability Theorem 19.5, the linearization procedure, and Lemma 19.6 describing the behavior of polynomial trajectories near affine manifolds in a vector space.

The uniform distribution theorem holds also for polynomials of several variables. By a *parallelepiped* $B \subset \mathbb{R}^k$ we mean a set $[0, T_1] \times \dots \times [0, T_k]$. Let us consider the product map $\Theta : \mathbb{R}^k \to G \subset \mathrm{SL}(n, \mathbb{R})$ given by

$$\Theta(t_1, \dots, t_k) = \theta_1(t_1) \cdots \theta_k(t_k),$$

where $\theta_1, \dots, \theta_k$ is a family of k polynomials. Take a sequence of measures

$$\mu_{B_n}(f) = \frac{1}{l(B_n)} \int_{t\in B_n} f(\Theta(t)\Gamma)dl(t), \qquad f \in C_c(G/\Gamma),$$

given by a sequence of parallelepipeds in $\mathbb{R}^k$. As was shown in [**Sh94**], μ_{B_n} converges to an F-invariant measure ν_F on $F\Gamma$ for some subgroup $F \subset G$. Here $t = (t_1, \dots, t_k) \in \mathbb{R}^k$, l is the Lebesgue measure on $\mathbb{R}^k$, and $\{B_n\}$ is a sequence of k-dimensional parallelepipeds in $\mathbb{R}^k$ with lengths of their edges tending (independently) to infinity. In place of Θ one can take any polynomial map of $\mathbb{R}^k$ into $\mathrm{SL}(n,\mathbb{R})$, but then instead of B_n one should work with a sequence of balls in $\mathbb{R}^k$ with their radii growing to infinity.

These results imply the uniform distribution of orbits of multi-dimensional unipotent subgroups. This was proved by Ratner in the solvable case [**R90b**] and conjectured in the general case in [**R94**]. We say that a basis $B = \{b_1, \dots, b_k\}$ in a nilpotent Lie algebra $\mathfrak{n}$ is *triangular* if the structure constants satisfy the equality $c^l_{i,j} = 0$ whenever $l \le \max(i,j)$ (the subalgebra $\mathfrak{n}_l$ spanned by $b_l, \dots, b_k$, is an ideal and $[\mathfrak{n}, \mathfrak{n}_l] \subset \mathfrak{n}_{l+1}$; clearly, such a basis exists). A permutation of a triangular basis is called a *regular* basis in $\mathfrak{n}$.

THEOREM 20.13 [**Sh94**]. *Let Γ be a lattice in a Lie group G, and let $N \subset G$ be a simply connected unipotent subgroup of G. Choose a regular basis $\{b_1, \dots, b_k\}$ in the Lie algebra $\mathfrak{n}$ and consider a map $S : \mathbb{R}^k \to N$ given by $S(t_1, \dots, t_k) = \exp(t_1 b_1) \cdots \exp(t_k b_k)$. For any parallelepiped $B \subset \mathbb{R}^k$ we set*

$$\mu_B(f) = \frac{1}{\lambda(S(B))} \int_{t \in B} f(S(t_1, \dots, t_k)\Gamma)\, dt,$$

where λ stands for Haar measure on N. Then there exists a subgroup $F \subset G$ such that for any sequence of parallelepipeds $\{B_n\} \subset \mathbb{R}^k$ with lengths of all their edges tending to infinity, the sequence μ_{B_n} converges to an F-invariant measure ν_F on the orbit $F\Gamma$.

If G is a real-algebraic Lie group, the result obviously follows from the previous ones because for a regular basis the map S sends the Lebesgue measure on $\mathbb{R}^k$ to Haar measure on N (see [**R90b**]). The case of an arbitrary Lie group G reduces to the algebraic case.

Following [**Sh94**], we pose an open problem from this area. What happens if instead of a polynomial curve $\theta(\mathbb{R})\Gamma \subset G/\Gamma$ one considers a sequence $\theta(\mathbb{Z})\Gamma$? If $\theta : \mathbb{R} \to G$ is a unipotent subgroup, then this reduces to the uniform distribution of orbits of unipotent one-generator actions, which holds true by virtue of Theorem 19.1. On the other hand, if $\theta : \mathbb{R} \to \mathbb{R}^n$ is a polynomial map, then by Weyl's theorem (see §3), the uniform distribution on the torus $\mathbb{R}^n/\mathbb{Z}^n$ holds both for $\theta(\mathbb{R})x$ and for $\theta(\mathbb{Z})x$, $x \in \mathbb{R}^n/\mathbb{Z}^n$. However, if $G = \mathrm{SL}(2,\mathbb{R})$, $\Gamma = \mathrm{SL}(2,\mathbb{Z})$ and, say, $\theta(t) = \begin{pmatrix} 1 & t^2 \\ 0 & 1 \end{pmatrix}$, then the structure of countable "trajectories" $\theta(\mathbb{Z})g\Gamma \subset G/\Gamma$ is not well understood. Are their closures algebraic, and are they uniformly distributed in the closures?

CHAPTER 3

Dynamics of Nonunipotent Flows

The information on unipotent actions obtained in Chapter 2 considerably simplifies the study of arbitrary (not necessarily unipotent) flows. First of all, all the results can be rather easily extended to the class of quasi-unipotent flows. It turns out that the algebraicity of orbit closures and ergodic measures is to be replaced by "quasi-algebraicity". More specifically, orbit closures and supports of ergodic measures are diffeomorphic to infra-homogeneous spaces of the form $C\backslash F/\Gamma$, where Γ is a lattice in F, and C is a compact group of affine automorphisms of F/Γ (and the ergodic measure is obtained from an F-invariant measure on F/Γ). In particular, any orbit closure is a smooth manifold with a smooth ergodic measure on it.

In §21 we show that all orbit closures of a one-parameter flow are smooth if and only if the flow is quasi-unipotent (Starkov, [**St90**]). In addition, after proving a stronger form of Ratner's topological theorem we give a "classification" of minimal sets of homogeneous flows [**St95a**]. Then we comment on Zeghib's results [**Ze**] on the structure of rectifiable invariant sets.

Although measure rigidity need not hold for ergodic quasi-unipotent flows, it does hold for mixing quasi-unipotent flows (Witte [**Wi87**]). As in the ergodic case, there is a "fiber-wise" rigidity [**St95c**] (see §23).

In §22 we study the multiple mixing problem for homogeneous flows. This problem can be reduced to the unipotent case, which is then dealt with by combining the joining technique with Ratner's measure theorem. Following [**St93**], we prove Marcus' conjecture: if a homogeneous flow is mixing then it is mixing of all orders. Marcus himself proved this for flows on semisimple homogeneous spaces [**Mar78**]; subsequently this result was generalized by Mozes [**Moz92**] and Ryzhikov [**Ry91**]. Our exposition of the semisimple case follows [**Moz92**].

In §24 we study topological rigidity of homogeneous flows. Following Benardete [**Ben**] and Witte [**Wi90**], we will find conditions under which a topological equivalence implies the existence of an affine equivalence. Whenever this implication holds, the following question arises: Do all time changes reduce to linear time changes? This problem is much more intricate and less studied. In this regard we present results of Livshitz [**Liv**] and Katok and Spatzier [**KS94**] on the first cohomology group for Anosov actions. As regards time changes for minimal flows, the only known nontrivial results are due to Kolmogorov [**Ko**] (developed in [**An73, Ar, P72, Skl, Ste82**]) for a rectilinear flow on the torus, and due to Ratner [**R86,87b**] for the horocycle flow.

In §25 we turn to studying individual orbits of partially hyperbolic flows. The main question we are concerned with is: when do *divergent* and *bounded* orbits exist? Certainly, we have in mind flows on noncompact spaces, and pioneering results in both directions are due to Dani [**Da85,86c**]. The existence of divergent orbits in the general case is not yet established. As for bounded orbits, Dani has studied some examples of reductive flows and showed that bounded orbits not only exist but form a set of full Hausdorff dimension. In full generality the Hausdorff dimension for the family of bounded orbits was recently found by Kleinbock and Margulis [**KM96**]. In particular, they proved that bounded orbit always exist. It is worth noting that results of [**KM96**] are of more general nature: they describe the Hausdorff dimension of the family of points whose orbit closures are disjoint from a given closed invariant subset. In particular, in [**KM96**] and [**Kl98a**] the Hausdorff dimension of the family of *nondense* orbits was found. We also mention a related result of Dani [**Da88**] concerning the family of nondense orbits of a toral automorphism.

In §26 we consider the structure of ergodic measures and orbit closures for actions of multi-dimensional connected subgroups $F \subset G$. Here many problems are still open. Nevertheless, some classes of subgroups $F \subset G$ not generated by unipotent elements, whose ergodic measures and orbit closures are of algebraic origin, have been found. This was done by Ratner [**R94**], Mozes [**Moz95a**], Shah and Weiss [**Sh95, W98, SW**], and Margulis and Tomanov [**MT96**]. In the case where F is a Cartan subgroup of a semisimple group of $\mathbb{R}$-rank at least 2, partial results were recently obtained by Katok and Spatzier [**KS96**]. Apparently, understanding Cartan subgroup actions is the most interesting problem in the theory of homogeneous flows at present (especially due to applications in number theory—see Chapter 4).

Mozes' result implies that any uniquely ergodic homogeneous action $(G/\Gamma, F)$ is minimal whenever the space G/Γ is compact. An example of Raghunathan (see Example 26.4) shows that this is not the case if the space is noncompact. Some progress in the opposite direction (minimality $\Rightarrow$ unique ergodicity) was obtained by Mozes and Weiss [**MW**], Starkov [**St99b**], and Tomanov [**T**].

In §27 we discuss the phenomenon of exponential mixing of partially hyperbolic flows in classes of smooth and Hölder functions. Results of this kind have numerous applications (see §§24 and 25). This phenomenon is best seen for Anosov toral automorphisms; then we present a result of Lind [**Lin**] concerning partially hyperbolic automorphisms. In the semisimple case the proof of exponential mixing requires intensive use of representation theory. The exposition starts with the paper of Katok and Spatzier [**KS94**]. Then we turn to papers of Kleinbock and Margulis [**KM96,99**], where exponential mixing was proved in the class of smooth functions on noncompact semisimple spaces. Moreover, [**KM96**] sketches a method of proving similar results for Hölder functions. Earlier, corresponding results for Hölder functions on $\mathrm{SL}(2,\mathbb{R})/\Gamma$ were obtained by Moore [**Mo87**] and Ratner [**R87a**].

§21. Structure of orbits, invariant sets, and ergodic measures of homogeneous flows

This section contains a survey of the papers [**St90**], [**St95a**] by the author and [**Ze**] by Zeghib on the structure of invariant sets and finite ergodic measures of one-parameter homogeneous flows. We recall once again that all orbit closures and finite ergodic measures of a unipotent flow on a finite volume space are homogeneous. A somewhat more general class consists of quasi-unipotent flows. We will show that although the closure of an arbitrary orbit $\overline{g_{\mathbb{R}}x}$ of such a flow need not be homogeneous, it is still a smooth manifold. Moreover, it is smoothly diffeomorphic to an infra-homogeneous space $C\backslash F/\Gamma$, where C is a compact subgroup of the group

$\mathrm{Aff}(F/\Gamma)$ of affine maps of F/Γ, and the flow on $\overline{g_{\mathbb{R}}x}$ is compactly covered by a homogeneous flow on F/Γ. This easily implies the uniform distribution property for quasi-unipotent orbits and the fact that any finite ergodic measure is a smooth measure supported on a submanifold.

Sometimes one can ensure that an individual orbit of a partially hyperbolic flow has smooth (or even algebraic) closure. For instance, this applies to orbits which "do not exit to infinity inside the neutral fiber" (this generalizes Ratner's topological theorem). The result proves helpful for the classification of minimal sets of homogeneous flows.

At the end of the section we describe results of Zeghib on *rectifiable* invariant subsets for the action of an $\mathbb{R}$-diagonalizable subgroup $g_{\mathbb{R}} \subset G$. Ergodic components of Hausdorff measure (of the corresponding dimension) on such a subset turn out to be algebraic. Sometimes this enables one to establish the algebraicity of closed invariant sets under rather weak smoothness conditions.

1. In what follows, Γ is a discrete subgroup of a Lie group G. We recall that the dynamics of a homogeneous flow $(G/\Gamma, g_{\mathbb{R}})$ depends drastically on whether or not the subgroup $g_{\mathbb{R}}$ is quasi-unipotent. Relative to the metric induced by a right-invariant metric on G, the rate of divergence of trajectories of a quasi-unipotent flow $(G/\Gamma, g_{\mathbb{R}})$ is quasi-polynomial (possibly with a trigonometric component), whereas a nonquasiunipotent flow has exponential divergence of trajectories and is uniformly partially hyperbolic.

As was already mentioned, orbits of the geodesic flow can be "very bad." More specifically, an orbit can be nonrecurrent and not exiting to infinity or, on the contrary, it can be recurrent and have nowhere locally connected closure (as the Morse minimal set). The following observation of Margulis generalizes this fact.

LEMMA 21.1 [**St90**]. *Any uniformly partially hyperbolic flow on a manifold with finite smooth invariant measure has a nonrecurrent orbit which does not exit to infinity* (*and hence its closure is not a submanifold*).

Proof. Let a flow $(M, g_{\mathbb{R}})$ satisfy the assumptions of the lemma. By F^e we denote the nontrivial invariant expanding foliation on M. Let $k = \dim F^e$, and let $F^e_r m \subset F^e(m)$ be the k-dimensional closed disk centered at $m \in M$ of radius r.

We also select three closed $(n-1)$-dimensional, $n = \dim M$, disks $V_r m \subset V_{2r} m \subset V_{4r} m$ centered at $m \in M$ that are transversal to the flow and locally invariant relative to the foliation F^e. Without loss of generality we can assume that the following three conditions are satisfied:

1) Inside the time interval $\Delta t = 1$ the distance between any two points from $F^e_r m$ increases at least by a factor of 4.

2) The radius $r > 0$ is so small that there is a homeomorphism

$$g_{(-1,1)} V_{4r} m \simeq (-1, 1) \times V_{4r} m.$$

3) Almost all points of $F^e_r m$ return infinitely often to $V_{2r} m$ as $t \to +\infty$ (the Poincaré recurrence theorem).

We define on $V_{4r} m$ the function

$$\tau(x) = \min\{t > 0 : \ g_t x \in V_{2r} m\}, \ x \in V_{4r} m,$$

which is the first-return time to $V_{2r} m$. Let $\tau_0 = \inf\{\tau(x), \ x \in F^e_r m\}$. Then $1 < \tau_0 < \infty$. Since the diameter of the disk $g_{\tau_0}(F^e_r m)$ exceeds $8r$ and $g_{\tau_0}(F^e_r m) \cap$

$V_{2r}m \neq \varnothing$, it follows from geometric considerations that there exists a disk

$$F_r^e m_1 \subset g_{\tau_0}(F_r^e m) \cap (V_{4r}m - V_{2r}m).$$

We set

$$C_1 = g_{-\tau_0}(F_r^e m_1) \subset F_r^e m$$

and denote $\tau_1 = \inf\{\tau(x) : x \in C_1\}$. Then $2 < \tau_1 < \infty$, and there exists a disk

$$F_r^e m_2 \subset g_{\tau_1}(C_1) \cap (V_{4r}m - V_{2r}m).$$

We construct by induction a sequence of closed sets $F_r^e m \supset C_1 \supset C_2 \supset \dots$ such that $\tau(x) > k+1$ for all $x \in C_k$. Let $x_0 = \bigcap_k C_k \in V_r m$. Then the positive semiorbit $g_{\mathbb{R}^+}x_0$ returns infinitely often to $V_{4r}m$ but never to $V_{2r}m$. Thus, it is locally closed but not closed.

In a similar fashion we examine negative semiorbits of points from the contracting leaf of x_0. While doing so, we find a point x_1 with its negative semiorbit locally closed. It may be closed, because we cannot ensure that almost all points of the contracting leaf of x_0 are recurrent in the negative direction. Since the positive semiorbit $g_{\mathbb{R}^+}x_1$ asymptotically approaches $g_{\mathbb{R}^+}x_0$, it follows that the orbit $g_{\mathbb{R}}x_1$ is locally closed but not closed. Hence its closure is not a smooth manifold. □

Remark. It would be interesting to find an orbit whose closure is not locally connected. Orbits of this kind exist for Anosov flows (for instance, in a neighborhood of the closure of a homoclinic orbit; see [**AS**] for more detail).

As for quasi-unipotent flows, it is convenient to treat their orbits as "twisted" orbits of unipotent flows. Indeed, if G is a semisimple group with finite center, then for any quasi-unipotent subgroup $g_{\mathbb{R}} \subset G$ there exist a torus $T \subset G$ and a unipotent subgroup $u_{\mathbb{R}} \subset G$ such that $g_t = c_t \times u_t$, $t \in \mathbb{R}$, where $c_{\mathbb{R}}$ is a dense subgroup of the torus T (this is a Jordan decomposition of the subgroup $g_{\mathbb{R}}$ into reductive and unipotent parts). Thus, the subgroup $g_{\mathbb{R}}$ can be viewed as diagonal in the "cylinder" $T \times u_{\mathbb{R}}$ with vertical axis $u_{\mathbb{R}}$.

In general, G can fail to admit such a decomposition. However, as the solvable case shows, one can "slightly" extend the group G to a group G^* which does admit the required decomposition.

LEMMA 21.2 [**St90**]. *Let $g_{\mathbb{R}}$ be a quasi-unipotent subgroup in a Lie group G. Then there exist*

1. *a Lie group G^* that contains G as a normal subgroup, the factor-group G^*/G being a torus, and*
2. *a unipotent subgroup $h_{\mathbb{R}} \subset G^*$*

such that, given a discrete subgroup $\Gamma \subset G$, if $M' = \overline{h_{\mathbb{R}}x} \subset G^*/\Gamma$ is a homogeneous space of finite volume, then $M = \overline{g_{\mathbb{R}}x} \subset G/\Gamma$ is an analytic submanifold with finite smooth $g_{\mathbb{R}}$-ergodic measure, and the flow $(M, g_{\mathbb{R}})$ is compactly covered by the ergodic unipotent flow $(M', h_{\mathbb{R}})$.

Proof. Without loss of generality $x = \Gamma \in G/\Gamma$ and the group G is simply connected. Let $T \subset \operatorname{Aut} G$ be the compact component of the algebraic envelope of the quasi-unipotent subgroup $\operatorname{Ad}(g_{\mathbb{R}}) \subset \operatorname{Aut} G$, and let $G^* = T \cdot G$. Then G^* admits a Jordan decomposition $g_t = c_t \times h_t$, $t \in \mathbb{R}$, into semisimple and unipotent parts. Here the subgroup $c_{\mathbb{R}}$ is dense in the torus T, and $h_{\mathbb{R}} \subset G^*$. Let $\pi : G^* = T \cdot G \to G$ be the projection along T which induces a compact covering of flows

$$\pi^* : (G^*/\Gamma, h_{\mathbb{R}}) \to (G/\Gamma, g_{\mathbb{R}}).$$

Let
$$H = (\overline{h_{\mathbb{R}}\Gamma})_0 \subset G^* \quad \text{and} \quad \widetilde{M} = (\overline{g_{\mathbb{R}}\Gamma})_0 \subset G.$$
Then $\widetilde{M} = \pi(H) = TH \cap G$, where $\widetilde{M}$ is the orbit space for the left action of the compact subgroup $T' = H \cap T$ on H. We set $\Gamma' = \Gamma \cap \widetilde{M}$. By Lemma 7.2, Γ' is a subgroup of Γ, and $M = \overline{g_{\mathbb{R}}\Gamma} \subset G/\Gamma$ is the quotient of $\widetilde{M} \subset G$ by orbits of the right free Γ'-action. Obviously, $\Gamma \cap H$ is of finite index in Γ'. Since $T \cap G = 1$, the closed set M is diffeomorphic to $T' \backslash H/\Gamma'$. Hence M is a manifold. The flow $(H/\Gamma \cap H, h_{\mathbb{R}})$ is a topologically transitive flow on a finite volume space, and hence is ergodic. The H-invariant measure projects to a smooth $g_{\mathbb{R}}$-invariant measure on M. The rest is evident. □

It is obvious that $\operatorname{vol}(G/\Gamma) < \infty \iff \operatorname{vol}(G^*/\Gamma) < \infty$. Hence given a lattice $\Gamma \subset G$, all orbits of the unipotent flow $(G^*/\Gamma, h_{\mathbb{R}})$ have homogeneous closures and are uniformly distributed. This implies that any orbit $g_{\mathbb{R}}x$ of quasi-unipotent flow on a finite volume space is also uniformly distributed (relative to the smooth measure on the manifold $M = \overline{g_{\mathbb{R}}x} \subset G/\Gamma$ obtained from H-invariant measure on the space $Hx = \overline{h_{\mathbb{R}}x} \subset G^*/\Gamma$). We recall that the manifold M can fail to be a homogeneous subspace (see Example 4.3).

Now we can formulate a general criterion.

THEOREM 21.3 [**St90**]. *Let Γ be a lattice in G. Then all orbit closures of a flow $(G/\Gamma, g_{\mathbb{R}})$ are smooth $\iff$ the subgroup $g_{\mathbb{R}}$ is quasi-unipotent. All orbits of a quasi-unipotent flow are uniformly distributed in their closures.*

2. Now let μ be a finite ergodic measure for a quasi-unipotent flow $(G/\Gamma, g_{\mathbb{R}})$ and let $M \subset G/\Gamma$ be its support (we do not assume that $\operatorname{vol}(G/\Gamma) < \infty$). Then the flow $(M, g_{\mathbb{R}})$ is ergodic with respect to a strictly positive measure, and hence possesses a dense orbit. By Lemma 21.2, the flow $(M, g_{\mathbb{R}})$ is compactly covered by a topologically transitive flow $(M', h_{\mathbb{R}})$, where $M' \subset G^*/\Gamma$ and $h_{\mathbb{R}}$ is a unipotent subgroup of G^*. Weil's construction of the semidirect product of measures (see §2.3) provides us with a finite $h_{\mathbb{R}}$-invariant measure on M' that projects to the measure μ. Let μ'' be the ergodic component of μ' that projects to μ. Then μ'' is an H-invariant measure on a closed orbit $Hx \subset M' \subset G^*/\Gamma$. Hence μ is a smooth measure on the manifold $M \subset G/\Gamma$. We have proved the following statement.

COROLLARY 21.4. *Let μ be a finite ergodic measure for a quasi-unipotent flow $(G/\Gamma, g_{\mathbb{R}})$. Then there exists a smooth manifold $M \subset G/\Gamma$ such that μ is supported on M and the restriction of μ onto M is a smooth measure.*

We recall once again that according to Sinai [**Si72**] and Bowen [**Bo75**], any Anosov flow has uncountably many ergodic measures, those with their support not locally connected, and those that are strictly positive (but singular to smooth volume measure).

One can formulate the following

CONJECTURE. *Any partially hyperbolic flow on a finite volume homogeneous space has a finite ergodic measure whose support is not a smooth manifold. If the flow is ergodic, then there exists an ergodic strictly positive singular measure.*

3. Now we turn to orbits of partially hyperbolic homogeneous flows. If $\operatorname{vol}(G/\Gamma) < \infty$, then almost all orbits have smooth closures (a generic orbit of an ergodic flow is everywhere dense, and that of a nonergodic flow is dense in the corresponding ergodic submanifold; see Corollary 10.11). Among the remaining orbits one can find those with nonsmooth (see Lemma 21.1) and with smooth closures (for instance, periodic orbits of the geodesic flow). In turns out that under certain assumptions one can ensure that an individual orbit closure of the flow $(G/\Gamma, g_{\mathbb{R}})$ is smooth or even homogeneous.

To formulate the result, we consider the decomposition $\mathfrak{g} = \mathfrak{g}^- + \mathfrak{g}^0 + \mathfrak{g}^+$ of the Lie algebra $\mathfrak{g}$ of G into the contracting, neutral and expanding subalgebras with respect to the operator Ad_{g_1}. Let G^-, Q and G^+ be the corresponding analytic subgroups of G. We know that orbits of G^-, Q and G^+ on G/Γ form the contracting, neutral and expanding foliation for the flow $(G/\Gamma, g_{\mathbb{R}})$. Note that $g_{\mathbb{R}} \subset Q$ and $g_{\mathbb{R}}$ is always quasi-unipotent in the neutral subgroup Q.

We say that an orbit $g_{\mathbb{R}}x \subset G/\Gamma$ *does not exit to infinity inside the neutral leaf* $Qx \subset G/\Gamma$ if there exist a sequence $\{t_n\} \to \infty$ and a compact subset $K \subset Q$ such that $g_{t_n}x \in Kx$ for all n. One can prove the following result.

THEOREM 21.5 [**St95a**]. *Let* $\mathrm{vol}(G/\Gamma) < \infty$ *and let an orbit* $g_{\mathbb{R}}x \subset G/\Gamma$ *not exit to infinity inside the neutral leaf* $Qx \subset G/\Gamma$. *Then the closure* $\overline{g_{\mathbb{R}}x} \subset G/\Gamma$ *is a smooth manifold with finite smooth ergodic measure.*

The proof involves a statement similar to Lemma 21.2. Namely, there exist a Lie group G^* and a subgroup $h_{\mathbb{R}} \subset G^*$ such that G^*/G is a torus, all eigenvalues of the operator Ad_{h_t} on $\mathfrak{g}^*$ are purely real, and each closure $\overline{g_{\mathbb{R}}x} \subset G/\Gamma$ is compactly covered by the closure $\overline{h_{\mathbb{R}}x} \subset G^*/\Gamma$. If the second closure is homogeneous and has finite volume, then the first closure is a manifold and carries a finite smooth ergodic measure. Now Theorem 21.5 implies the following result.

THEOREM 21.6 [**St95a**]. *Let* $\mathrm{vol}(G/\Gamma) < \infty$, *and let a subgroup* $g_{\mathbb{R}}$ *be such that all eigenvalues of the operators* Ad_{g_t} *on* $\mathfrak{g}$ *are purely real. If an orbit* $g_{\mathbb{R}}x \subset G/\Gamma$ *does not exit to infinity inside the neutral leaf* $Qx \subset G/\Gamma$, *then the closure* $\overline{g_{\mathbb{R}}x} \subset G/\Gamma$ *is a homogeneous space of finite volume.*

Note that this result generalizes Ratner's topological Theorem 19.13 for one-parameter unipotent flows. Indeed, if $\mathrm{vol}(G/\Gamma) < \infty$, then by Margulis' Theorem 12.1, no semiorbit of a unipotent flow exits to infinity. Theorem 21.6 is not an immediate corollary of Ratner's theorem, because the neutral leaf $Qx \subset G/\Gamma$ can be nonclosed and as a homogeneous Q-space it may have Q-invariant measure infinite.

We prove the theorem in the arithmetic case.

LEMMA 21.7. *Let* G *be a closed connected subgroup of* $\mathrm{GL}(n, \mathbb{R})$, *and let* $G_{\mathbb{Z}} = G \cap \mathrm{GL}(n, \mathbb{Z})$ *be a lattice in* G. *Given a one-parameter subgroup* $g_{\mathbb{R}} \subset G$, *let* $Q \subset G$ *be the maximal connected subgroup of* G *such that the subgroup* $g_{\mathbb{R}}$ *is* Ad-*unipotent in* Q. *Assume that the orbit* $g_{\mathbb{R}}Q_{\mathbb{Z}}$ *does not exit to infinity in* $Q/Q_{\mathbb{Z}}$. *Then there exists a subgroup* $F \subset Q$ *such that* $\overline{g_{\mathbb{R}}G_{\mathbb{Z}}} = FG_{\mathbb{Z}}$ *and* $F_{\mathbb{Z}} = G_{\mathbb{Z}} \cap F$ *is a lattice in* F.

Proof. Let $\widehat{Q}$ be the maximal subgroup of $\mathrm{GL}(n, \mathbb{R})$ such that $g_{\mathbb{R}}$ is Ad-unipotent in $\widehat{Q}$. Then $\widehat{Q}$, being the centralizer of the semisimple part of $g_{\mathbb{R}}$, is an algebraic subgroup of $\mathrm{GL}(n, \mathbb{R})$. Let $\mathcal{A}(\widehat{Q}_{\mathbb{Z}}) \subset \widehat{Q}$ and $\mathcal{A}(g_{\mathbb{R}}) \subset \widehat{Q}$ stand for the algebraic closures of subgroups $\widehat{Q}_{\mathbb{Z}}$ and $g_{\mathbb{R}}$. Then $\mathcal{A}(\widehat{Q}_{\mathbb{Z}})$ is defined over $\mathbb{Q}$ and admits no nontrivial $\mathbb{Q}$-characters. Hence by the Borel–Harish-Chandra theorem, $\widehat{Q}_{\mathbb{Z}} = \mathcal{A}(\widehat{Q}_{\mathbb{Z}})_{\mathbb{Z}}$ is a lattice in $\mathcal{A}(\widehat{Q}_{\mathbb{Z}})$. Note that the set $\mathcal{A}(g_{\mathbb{R}})\mathcal{A}(\widehat{Q}_{\mathbb{Z}})$, being a product of two algebraic subgroups, is locally closed in $\widehat{Q}$. If the orbit $g_{\mathbb{R}}\mathcal{A}(\widehat{Q}_{\mathbb{Z}}) \subset \widehat{Q}/\mathcal{A}(\widehat{Q}_{\mathbb{Z}})$ is recurrent, then the quotient space $\mathcal{A}(g_{\mathbb{R}})/\mathcal{A}(g_{\mathbb{R}})\cap\mathcal{A}(\widehat{Q}_{\mathbb{Z}})$ is obviously compact. It follows that $\mathcal{A}(g_{\mathbb{R}})\mathcal{A}(\widehat{Q}_{\mathbb{Z}}) = T\mathcal{A}(\widehat{Q}_{\mathbb{Z}})$, where T is the maximal compact subgroup of $\mathcal{A}(g_{\mathbb{R}})$. We claim that this holds under the weaker assumption that the orbit does not exit to infinity inside the algebraic variety $\widehat{Q}/\mathcal{A}(\widehat{Q}_{\mathbb{Z}})$.

Since $\mathcal{A}(\widehat{Q}_{\mathbb{Z}})$ is defined over $\mathbb{Q}$, there exists a $\mathbb{Q}$-representation $\pi : \mathrm{GL}(n,\mathbb{R}) \to \mathrm{GL}(V)$ such that $\mathcal{A}(\widehat{Q}_{\mathbb{Z}})$ is the stabilizer of a line $\mathbb{R}v \subset V$ for some point $v \in V_{\mathbb{Q}}$. But $\mathcal{A}(\widehat{Q}_{\mathbb{Z}})$ has no nontrivial $\mathbb{Q}$-characters, and hence it stabilizes the point $v \in V_{\mathbb{Q}}$. Decompose the vector space V into the contracting, neutral and expanding subspaces with respect to the action of $\pi(g_t)$, $V = V^- + V^0 + V^+$, and let $v = v^- + v^0 + v^+$ be the corresponding decomposition for $v \in V$. If the orbit $g_{\mathbb{R}}\mathcal{A}(\widehat{Q}_{\mathbb{Z}})$ does not exit to infinity, say, in the positive direction, then $v^+ = 0$. But $\pi(g_t)v^- \to 0$, and hence $\pi(g_t)v \to V^0$. We recall that the reductive component of $\mathcal{A}(g_{\mathbb{R}})$ is central in $\widehat{Q}$, and hence $\widehat{Q}V^0 = V^0$. Now if $v \notin V^0$, then any limit point of the sequence $\pi(g_{t_n})v$, where $t_n \to +\infty$, sits inside V^0 and outside $\widehat{Q}v = \widehat{Q}/\mathcal{A}(\widehat{Q}_{\mathbb{Z}})$. This contradicts our assumption, and hence $v \in V^0$. Recall that orbits of a unipotent subgroup of $\mathrm{GL}(V)$ are closed in V. Hence the action of $\pi(g_t)$ on V^0 is semisimple and all its eigenvalues have absolute value 1. This exactly means that $g_{\mathbb{R}} \subset T\mathcal{A}(\widehat{Q}_{\mathbb{Z}})$.

By our construction, T is a central subgroup of $\widehat{Q}$, and hence $T\mathcal{A}(\widehat{Q}_{\mathbb{Z}})$ is a Lie group with the lattice $\widehat{Q}_{\mathbb{Z}}$. Now applying Ratner's topological theorem to the unipotent flow $(T\mathcal{A}(\widehat{Q}_{\mathbb{Z}})/\widehat{Q}_{\mathbb{Z}}, g_{\mathbb{R}})$, we conclude that $F = (\overline{g_{\mathbb{R}}\widehat{Q}_{\mathbb{Z}}})_0$ is a subgroup of $\widehat{Q}$ with the lattice $F_{\mathbb{Z}}$. But $g_{\mathbb{R}} \subset G$ and $G_{\mathbb{Z}}$ is a lattice in G. It follows that $F \subset (\widehat{Q} \cap G)_0 = Q$. □

Theorem 21.6 is proved using repeated applications of Ratner's theorem and reducing the semisimple, solvable and general cases to the arithmetic case. One might try to deduce it from Theorem 19.3 of Dani and Margulis as follows. Choose $x = \Gamma \in G/\Gamma$ and look at the unipotent flow $(Q/(\Gamma\cap Q), g_{\mathbb{R}})$. By the hypothesis, the orbit $g_{\mathbb{R}}(\Gamma\cap Q)$ does not exit to infinity and thus comes back infinitely often to some compact set. If the orbit comes back to the compact set, and the time it spends there has a positive density, we could apply Theorem 19.3 and immediately deduce our result. But an orbit of the horocycle flow on a surface of constant curvature -1 and of infinite area may not exit to infinity and at the same time not come back to a compact set and there spend time with a positive density (this concerns, for example, recurrent nonperiodic orbits of the horocycle flow on a surface of the second kind; see [**St95b**]).

On the other hand, for a reductive (that is, consisting of Ad-semisimple elements) subgroup $g_{\mathbb{R}} \subset G$, Theorem 21.6 is trivial, since $g_{\mathbb{R}}$ belongs to the center of its neutral subgroup Q. Hence the orbit $g_{\mathbb{R}}(\Gamma \cap Q)$ does not exit to infinity in $Q/(\Gamma \cap Q)$ only if its closure is a torus. Thus, this theorem is most important for "mixed" (neither reductive nor unipotent) subgroups $g_{\mathbb{R}} \subset G$.

4. Now we are ready to give a classification of minimal sets of homogeneous flows. First we introduce a new class of invariant sets.

Let $(X, g_{\mathbb{R}})$ be a continuous flow on a locally compact space X. Then a closed invariant set $M \subset X$ is called *birecurrent* if each orbit inside M is recurrent (comes back arbitrarily close to the original point) in both directions.

This class of invariant sets arises naturally in topological dynamics. It is clear that a compact minimal set is birecurrent. An example of Bebutov (see [**Ne**] or [**KW**]) shows that a compact birecurrent set need not be minimal even if it is topologically transitive. However, in this example the set is of infinite dimension. Now, having Ratner's theorem at our disposal, we can construct algebraic examples of such sets. In fact, a compact homogeneous space of finite volume is birecurrent relative to any unipotent flow, but an ergodic unipotent flow can fail to be minimal. (An example: $\Gamma \subset \mathrm{SL}(4,\mathbb{R})$ is a uniform lattice which intersects $\mathrm{SL}(2,\mathbb{R})$ in a lattice, and $g_{\mathbb{R}} \subset \mathrm{SL}(2,\mathbb{R})$ is a unipotent subgroup. Then the flow $(\mathrm{SL}(4,\mathbb{R})/\Gamma, g_{\mathbb{R}})$ is ergodic and not minimal).

The proof of the next result follows Margulis's Lemma 21.1.

LEMMA 21.8 [**St95a**]. *Let $(X, g_{\mathbb{R}})$ be a uniformly partially hyperbolic flow on a manifold X with birecurrent set $M \subset X$. Let $O(x) \subset X$ be a small enough neighborhood of a point $x \in M$. Then the connected component of the point x in the intersection $M \cap O(x)$ belongs to the neutral leaf of $x \in X$.*

Theorem 21.5 and Lemma 21.8 immediately imply the following result.

THEOREM 21.9 [**St95a**]. *Let M be a topologically transitive birecurrent set for a flow $(G/\Gamma, g_{\mathbb{R}})$ on a homogeneous space of finite volume. Then either M is not locally connected at any point $x \in M$, or M is a smooth submanifold in G/Γ.*

Now assume that M is a minimal set of a continuous flow $(X, g_{\mathbb{R}})$. Then either all semiorbits inside M do not exit to infinity, or one of them does. Proposition 12.8 asserts that in the first case M is a compact set. Now we can give a classification of minimal sets of homogeneous flows.

THEOREM 21.10 [**St95a**]. *Let M be a minimal set of a homogeneous flow $(G/\Gamma, g_{\mathbb{R}})$. Then one of the following statements holds:*

1) *$M = g_{\mathbb{R}}x$ is an orbit exiting to infinity in both directions.*

2) *There exists a point $x \in M$ such that one of its semiorbits exits to infinity, and the other semiorbit is recurrent.*

3) *M is a compact nowhere locally connected set.*

4) *M is a compact smooth submanifold in G/Γ such that the flow $(M, g_{\mathbb{R}})$ is compactly covered by a uniquely ergodic unipotent flow.*

Note that the theorem holds for arbitrary discrete subgroups $\Gamma \subset G$. If G/Γ fails to have finite volume, then instead of Ratner's theorem we can apply Theorem 19.3 of Dani and Margulis, which ensures that the compact closure of a unipotent orbit is always homogeneous.

Assume now that $\mathrm{vol}(G/\Gamma) < \infty$. It follows from Ratner's theorems that for a unipotent one-parameter flow, the properties of minimality and unique ergodicity are equivalent. The previous theorem implies that a quasi-unipotent flow can be minimal only if the space G/Γ is compact, in which case the flow is uniquely ergodic. On the other hand, it follows from Lemma 21.1 that a homogeneous flow can be minimal only if it is quasi-unipotent. Therefore minimality implies unique ergodicity[1]. Conversely, as was recently proved by Kleinbock and Margulis (see §25), a partially hyperbolic flow also has a compact invariant set. Obviously, such

[1]This gives a positive answer to Furstenberg's question raised in [**Fu73**].

a set carries an invariant measure, and hence only a flow on a compact space can be uniquely ergodic. But if a one-parameter flow on a compact space is uniquely ergodic with respect to a strictly positive measure, it must be minimal. We have proved the following statement.[2]

THEOREM 21.11 [**St95a**]. *For one-parameter flows on spaces of finite volume, minimality and unique ergodicity are equivalent.*

Orbits of the geodesic flow exiting to infinity in both directions demonstrate the first type of minimal sets in our classification. Morse minimal sets illustrate the third type, and periodic orbits the last type. It seems plausible that (if the surface has finite area and is noncompact) the geodesic flow has minimal sets of the second type as well.

Now let $(G/\Gamma, g_{\mathbb{R}})$ be an ergodic flow on a homogeneous space of finite volume. The existence of compact minimal sets follows from results of Kleinbock and Margulis (see §25). The question arises: Does an ergodic partially hyperbolic flow have minimal sets of all possible types? Namely, does it have minimal sets of the third and fourth types in the compact case and of all four types in the noncompact case (provided that the flow has the K-property)?

Note that the closure of a horocycle orbit on a surface of infinite area need not be homogeneous even if it is minimal. Indeed, following Hedlund, it is easy to prove (see [**St95b**]) that the nonwandering set Ω for the horocycle flow $(\mathrm{SL}(2,\mathbb{R})/\Gamma, u_{\mathbb{R}})$ is minimal if the Fuchsian group Γ is finitely generated and has no parabolic elements (the same follows from Burger's paper [**Bu**]). If Γ is not a uniform lattice, then the minimal set Ω is of the second type in our classification: otherwise it would be compact and hence consist of one periodic orbit (which is impossible because Γ contains no unipotent elements). We note also that in this case Ω is nowhere locally connected.

To conclude, we emphasize a contrast between the class of homogeneous flows and the more general class of smooth flows. As we mentioned, a homogeneous flow on a compact space can be minimal only if it has zero entropy (that is, is quasi-unipotent). On the other hand, Herman [**H**] has constructed a minimal diffeomorphism of a compact manifold with positive entropy. Further, according to Theorem 21.10, any compact minimal set of a homogeneous flow is locally connected either everywhere or nowhere. The first example of a homeomorphism of the plane with minimal set that is locally connected on a proper nonempty subset was given in [**Re**]. A smooth flow with a similar minimal set was constructed by Johnson [**Jo**] by making use of quasi-periodic systems of Millionshchikov [**Mi68**] and Vinograd [**Vi**].

5. Now we give several definitions from geometric measure theory [**Fed**]. Let (X, d) be a measure space and $n \in \mathbb{N}$. One says that a subset $Y \subset X$ is *n-rectifiable* if $Y = f(A)$, where $A \subset \mathbb{R}^n$ is a bounded subset and $f : A \to X$ is a Lipschitz map.[3] A countable union of n-rectifiable subsets of X is called *σ-n-rectifiable.*

---

[2] Note that a homogeneous action $(G/\Gamma, H)$ on a finite volume space can be uniquely ergodic but not minimal (see Example 26.4).

[3] A map $f : (A, d_A) \to (X, d_X)$ of two measure spaces is said to be *Lipschitz* if there exists a constant $C > 0$ such that $d_X(f(a), f(b)) \leq C d_A(a, b)$ for all $a, b \in A$.

As an illustration, we consider the case $n = 1$. It is easily seen that a 1-rectifiable curve in $\mathbb{R}^2$ has finite length. At the same time, one can construct a homeomorphism of the segment $[0, 1]$ to the image in $\mathbb{R}^2$ which is of infinite length (such a curve can be σ-1-rectifiable). The square $[0, 1] \times [0, 1]$ is not a σ-1-rectifiable set, but it is a continuous image of the segment $[0, 1]$ (the Peano curve).

Let $\mathcal{H}^n$ be the n-dimensional Hausdorff measure on X (see §25). A subset $Y \subset X$ is called *$\mathcal{H}^n$-rectifiable* if $\mathcal{H}^n(Y) < \infty$ and there exists a σ-n-rectifiable subset $Y' \subset X$ such that $\mathcal{H}^n(Y - Y') = 0$. If in addition $\mathcal{H}^n(Y) > 0$, then the Hausdorff dimension of Y is n (but not every subset $Y \subset X$ of Hausdorff dimension n is n-rectifiable).

A nontrivial example of an $\mathcal{H}^1$-rectifiable subset Y of the square $[0, 1]^2$ is given by a countable union $\bigcup_i \{a_i\} \times B_i$, where $a_i \in [0, 1]$ and $B_i \subset [0, 1]$ is a measurable subset with $\sum_i l(B_i) < \infty$.

Now let X be a Riemannian manifold. We have [**Fed**] the following criterion: A subset $Y \subset X$ is $\mathcal{H}^n$-rectifiable $\iff$ modulo a set of $\mathcal{H}^n$-measure 0 it belongs to a countable union of n-dimensional C^1-submanifolds in X .

Now we are in a position to formulate results of [**Ze**] on the structure of invariant rectifiable subsets for a flow $(K\backslash G/\Gamma, a_\mathbb{R})$, where $a_\mathbb{R}$ is an $\mathbb{R}$-diagonalizable subgroup that commutes with a compact subgroup $K \subset G$. Here Γ is an arbitrary discrete subgroup of G (not necessarily a lattice).

THEOREM 21.12 [**Ze**]. *Let $(K\backslash G/\Gamma, a_\mathbb{R})$ be an $\mathbb{R}$-diagonalizable flow and let $Y \subset K\backslash G/\Gamma$ be an $\mathcal{H}^n$-rectifiable invariant subset for some $n \le \dim G$. Then the flow $(Y, \mathcal{H}^n, a_\mathbb{R})$ is measure preserving, and all its ergodic components are algebraic.*

Having in mind Ratner's measure theorem (or, more precisely, Corollary 21.4), it is natural to make the following conjecture.

CONJECTURE. *Let $(K\backslash G/\Gamma, g_\mathbb{R})$ be a homogeneous flow and let $Y \subset K\backslash G/\Gamma$ be an $\mathcal{H}^n$-rectifiable invariant subset. Then the flow $(Y, \mathcal{H}^n, g_\mathbb{R})$ is measure-preserving, and all its ergodic components are smooth manifolds.*

Theorem 21.12 can be made more precise; more specifically, one can describe the subset Y modulo a zero measure set. Let $R(a_\mathbb{R}, \Gamma)$ be the collection of all closed subgroups $H \subset G$ such that $\Gamma \cap H$ is a lattice in H, and $g^{-1}a_\mathbb{R}g \subset H$ for some $g \in G$.

THEOREM 21.13 [**Ze**]. *Under the assumptions of Theorem 21.12 there exist countable collections of elements $g_i \in G$, subgroups $H_i \in R(a_\mathbb{R}, \Gamma)$ and subsets $S_i \subset G$ such that each S_i commutes with $a_\mathbb{R}$ and is $\mathcal{H}^k$-rectifiable for some $k = k(i)$, where $Y = \bigcup_i KS_ig_iH_i\Gamma$ modulo a set of zero $\mathcal{H}^n$-measure.*

It is clear that ergodic components in the theorem are (locally homogeneous) subspaces $Kyg_iH_i\Gamma$, $y \in S_i$, of finite volume.

Here is an example illustrating the theorem. Let $a_\mathbb{R} \subset H \subset G$, where $\Gamma \cap H$ is a lattice in H, and $\overline{a_{\mathbb{R}+}\Gamma} = H\Gamma \subset G/\Gamma$. Suppose that $\dim H = n > 1$ and that outside H there exists an element $u \in G$ for which $a_t u a_{-t} \to 1$, $t \to +\infty$ (such an example can be easily constructed). Then the limit set for the semiorbit $a_{\mathbb{R}+}u\Gamma$ is the homogeneous subspace $H\Gamma$. The closure $Y = \overline{a_{\mathbb{R}+}u\Gamma}$ is an $\mathcal{H}^n$-rectifiable set invariant under the semigroup $a_{\mathbb{R}+}$. The restriction of $\mathcal{H}^n$ onto Y is an ergodic measure supported on the homogeneous subspace $H\Gamma$. But the set Y is not homogeneous. In a similar fashion one can construct a full orbit $a_\mathbb{R}x$ having two distinct limit sets which are homogeneous subspaces of finite volume of dimension $n > 1$ (compare with construction of homoclinic orbits in §14). Then the closure $Y = \overline{a_{\mathbb{R}} x}$ is an $\mathcal{H}^n$-rectifiable set with two ergodic homogeneous components.

The results obtained have especially simple formulation in the case where $(K\backslash G/\Gamma, a_{\mathbb{R}})$ is an Anosov flow. For instance, let G be a simple group of $\mathbb{R}$-rank 1 and let $K \subset G$ be a maximal compact subgroup that commutes with $a_{\mathbb{R}}$ (the flow is just the geodesic flow on the unit tangent bundle T^1M of a locally symmetric manifold M of negative curvature). Since in this case the centralizer of $a_{\mathbb{R}}$ in G is $Ka_{\mathbb{R}}$, an $\mathcal{H}^n$-invariant subset Y is a countable union of closed subspaces of finite volume:

$$Y = \bigcup_i Y_i \pmod 0, \qquad \text{where } Y_i = Kg_iH_i\Gamma.$$

Moreover, it is proved in [**Ze**] that the decomposition has finitely many terms (this follows from the condition $\mathcal{H}^n(Y) < \infty$), and each component is the unit tangent bundle of some totally geodesic submanifold of finite volume $W_i \subset M$. Hence $Y = T^1W \pmod 0$, where $W = \bigcup_i W_i$ is a closed (not necessarily connected) totally geodesic submanifold in M.

§22. Multiple mixing

1. Here we are concerned with the mixing property of one-parameter flows on G/Γ with respect to the probability G-invariant measure ν. We already know that either $G = \overline{G_\infty \Gamma}$ and all ergodic flows are weak mixing, or $G \neq \overline{G_\infty \Gamma}$ and G/Γ admits no weak mixing homogeneous flows. In particular, the presence of a weak mixing flow implies that the radical of the group G is nilpotent.

Suppose that the group G is semisimple. Then every ergodic flow on G/Γ is weak mixing. Recall that measure mixing is a stronger property. We have already deduced the mixing property for the geodesic flow from the ergodicity of the horocycle flow. In a similar fashion one can prove that if a subgroup $g_{\mathbb{R}}$ is reductive and the flow $(G/\Gamma, g_{\mathbb{R}})$ is ergodic, then it is mixing. However, this method does not work for unipotent flows.

Suppose that G has no compact components. Without loss of generality we can assume that the center $Z(G)$ is finite (since the intersection $Z(G) \cap \Gamma$ is of finite index in $Z(G)$, we can replace G by $G/Z(G) \cap \Gamma$). Moreover, we assume that the lattice $\Gamma \subset G$ is irreducible. Then every unbounded one-parameter subgroup of G induces an ergodic flow on G/Γ. Let us prove that the G-action on G/Γ is mixing; that is, for any sequence $g_i \to \infty$ in G and any two measurable subsets $A, B \subset G/\Gamma$ one has $\nu(A \cap g_iB) \to \nu(A)\nu(B)$. This property follows from a theorem of Howe and Moore (see [**Z**]) on the decay of matrix coefficients of unitary representations of the group G. However, we will follow a more elementary approach of Mozes [**Moz92**] based on the joinings technique. It enables us to prove a stronger result, namely that the G-action on G/Γ is mixing of all orders.

First we formulate the definition of multiple mixing for an action of a locally compact group F on a probability space (X, μ).

An action of F on (X, μ) is called $(k-1)$-*mixing* (or of class $\mathrm{Mix}(k-1)$) if for any k measurable sets $A_1, \ldots, A_k \subset X$ and any k sequences $\{g_i(1)\}, \ldots \{g_i(k)\}$ of elements in F such that

$$g_i(l)(g_i(m))^{-1} \to \infty, \quad i \to \infty, \qquad \text{for all } l \neq m,$$

we have

$$\lim_{i\to\infty} \mu\Big(\bigcap_{l=1}^{k} g_i(l)A_l\Big) = \prod_{l=1}^{k} \mu(A_l).$$

Let an action of F on (X,μ) belong to the class Mix(k). It follows from the definition that the action of any closed noncompact subgroup $H \subset F$ on (X,μ) belongs to the same class.

An old problem of Rokhlin is whether mixing of a higher order for a group action on a probability space follows from mixing of the first order, that is, Mix$(1) \Rightarrow$ Mix(∞)? In the classical case $F = \mathbb{Z}$, $\mathbb{R}$ (that is, for one-generator actions and one-parameter flows) the problem is only solved under rather strong restrictions on the action. For instance, Kalikow [**Ka**] has proved that Mix$(1) \Rightarrow$ Mix(2) if the flow has rank of approximation 1. Recently this result was superseded by Ryzhikov [**Ry93**], who proved that Rokhlin's problem has a positive solution if the flow or the automorphism in question has finite rank of approximation. On the other hand, there is an example of Ledrappier [**L**] of a $\mathbb{Z}^2$-action of class Mix(1) but not Mix(2).

Now we return to our situation: Γ is an irreducible lattice in a semisimple Lie group G with finite center and without compact components. Let us prove that the action of G on G/Γ is $(k-1)$-mixing for every $k \geq 2$. Let $X = G/\Gamma$ and let ν_Δ be the *diagonal measure* on X^k, that is,

$$\nu_\Delta\Big(\prod_{1}^{k} A_l\Big) = \nu\Big(\bigcap_{1}^{k} A_l\Big).$$

Given $\overline{g} = (g(1), \dots, g(k)) \in G^k$, let $\nu_\Delta^{\overline{g}}$ be the translate of ν_Δ:

$$\nu_\Delta^{\overline{g}}\Big(\prod_{1}^{k} A_l\Big) = \nu_\Delta\Big(\prod_{1}^{k} g(l)A_l\Big) = \nu\Big(\bigcap_{1}^{k} g(l)A_l\Big).$$

Note that $\nu_\Delta^{\overline{g}\,\overline{g}'} = \nu_\Delta^{\overline{gg}'}$. The measure ν_Δ and all its translates are joinings; that is,

$$p_l(\nu_\Delta^{\overline{g}}) = \nu,$$

where $p_l : X^k \to X_l = X$, $l = 1, \dots, k$, is the corresponding projection. In these terms $(k-1)$-mixing is equivalent to the property that $\nu_\Delta^{\overline{g}_i} \to \nu^k$ whenever $g_i(l)(g_i(m))^{-1} \to \infty$, $i \to \infty$, for all $l \neq m$. Since the measure ν is G-invariant, we can assume without loss of generality that $g_i(1) \equiv 1 \in G$, and that the other $k-1$ sequences tend to infinity independently of each other.

It is well known that the convergence of joinings is equivalent to their weak* convergence. In addition, the space $\mathcal{P}(X^k)$ is compact if X is compact. In the noncompact case we consider the one-point compactification X_c; then $\mathcal{P}(X_c^k)$ is compact. Hence to prove the mixing property it suffices to show that from $\nu_\Delta^{\overline{g}_i} \to \mu \in \mathcal{P}(X_c^k)$ it follows that $\mu = \nu^k$. Since $p_l(\mu) = \nu$, $l = 1, \dots, k$, we deduce that $\mu \in \mathcal{P}(X^k)$.

To prove the equality $\mu = \nu^k$ we use the following elegant result obtained independently by Mozes and Ryzhikov.

LEMMA 22.1 [**Ry91**], [**Moz92**]. *Let (X',ν') and (X'',ν'') be two probability spaces and let a group G act ergodically on (X'',ν''). Assume that a measure μ on $X' \times X''$ possesses the following properties:*

1) $\mu(A \times X'') = \nu'(A)$ *for each measurable set* $A \subset X'$.
2) $\mu(X' \times B) = \nu''(B)$ *for each measurable set* $B \subset X''$.
3) μ *is invariant under the element* $\mathrm{Id} \times g$, $g \in G$.
Then $\mu = \nu' \times \nu''$.

Proof [**Ry91**]. Let $A \subset X'$ and $\nu'(A) > 0$. Define

$$\mu_A(B) = \mu(A \times B)/\nu'(A), \qquad B \subset X''.$$

It follows from 2) that μ_A is absolutely continuous with respect to ν''. According to 1) and 3), μ_A is a G-invariant probability measure on X''. Since the measure ν'' is G-ergodic, we deduce that $\mu_A = \nu''$ for each subset $A \subset X'$ of positive measure (otherwise μ_A and ν'' would be mutually singular as two distinct G-ergodic measures). Hence $\mu = \nu' \times \nu''$. □

Now we can prove our result.

THEOREM 22.2 [**Moz92**]. *Let* Γ *be an irreducible lattice in a semisimple Lie group* G *with finite center and with no compact components. Then the* G*-action on* G/Γ *is mixing of all orders.*

Proof. Assume that we have already established j-mixing for all $j < k-1$ (by 0-mixing we understand ergodicity). Let $\nu_\Delta^{\overline{g}_i} \to \mu \in \mathcal{P}(X^k)$. Since $g_i(2) \to \infty$ and the center $Z(G)$ is finite, there exists an element v of the Lie algebra $\mathfrak{g}$ such that either $\mathrm{Ad}_{g_i(2)}^{-1} v \to \infty$ or $\mathrm{Ad}_{g_i(2)} v \to \infty$. Hence we can find a sequence $a_i \to 1 \in G$ and an element

$$\overline{h} = (1, h(2), \dots, h(k)) \in G^k, \qquad \overline{h} \neq 1,$$

such that

$$\text{either } g_i^{-1}(l) a_i g_i(l) \to h(l),\ l = 1, \dots, k, \text{ or } g_i(l) a_i g_i^{-1}(l) \to h(l),\ l = 1, \dots, k$$

(passing to a subsequence if necessary). It follows from consideration of eigenvalues that the element $\overline{h} \in G^k$ is Ad-unipotent. In the first case every measure $\nu_\Delta^{\overline{g}_i}$ is invariant under

$$(a_i, g_i(2)^{-1} a_i g_i(2), \dots, g_i(k)^{-1} a_i g_i(k)),$$

and hence the limit measure μ is invariant under the limit element $\overline{h} \in G^k$. Let

$$X' = \prod_{h(l)=1} X_l,\ X'' = \prod_{h(l) \neq 1} X_l.$$

Then $X^k = X' \times X''$, and both the spaces X' and X'' are nontrivial. By the induction hypothesis, the projections of μ on X' and X'' coincide with the corresponding power of the measure ν. Since the element $\overline{h}$ induces an ergodic one-generator action on X'', we can use Lemma 22.1 to conclude that $\mu = \nu^k$. The second case is considered in a similar way. □

Remark. We used only the fact that any nontrivial unipotent element $h \in G$ induces an ergodic one-generator action on G/Γ. Hence the theorem holds true for a semisimple group $G = S \times K$ provided that the center $Z(G)$ is finite, $G = \overline{S\Gamma}$ and the lattice Γ is irreducible. In particular, any flow $(G/\Gamma, g_{\mathbb{R}})$ on the space G/Γ induced by an unbounded subgroup $g_{\mathbb{R}} \subset G$ is mixing of all orders. If (X', g'_t) and (X'', g''_t) are two k-mixing flows, then the diagonal flow (X, g_t), where $X = X' \times X''$ and $g_t = g'_t \times g''_t$, is also k-mixing. Hence an ergodic flow on G/Γ is k-mixing even if the lattice Γ is not assumed to be irreducible. Finally, $G = \overline{S\Gamma}$, and the finiteness of the center $Z(G)$ is not essential if G/Γ admits an ergodic homogeneous flow (see §8). We have deduced the following result, originally proved by Marcus:

THEOREM 22.3 [**Mar78**]. *If a group G is semisimple, then any ergodic flow $(G/\Gamma, g_{\mathbb{R}})$ is mixing of all orders.*

This settles Sinai's 1962 conjecture that the horocycle flow is k-mixing for all k (we recall that Rokhlin's problem for the geodesic flow is much easier because it is a K-flow).

2. We comment on Marcus' paper [**Mar78**] and then, using Ratner's measure theorem, solve Rokhlin's problem for any homogeneous flow.

First of all Marcus reduced the problem to the case of a unipotent subgroup $h_{\mathbb{R}} \subset G$. We say that a unipotent subgroup $h_{\mathbb{R}}$ of a Lie group G is *weakly horocyclic* if there exists a subgroup $a_{\mathbb{R}} \subset G$ such that

$$a_t h_1 a_{-t} = h_{e^t} \qquad \text{for all } t \in \mathbb{R}.$$

(this notion is a little weaker than the notion of horocyclicity introduced in §17 because we do not require here that $h_{\mathbb{R}}$ and $a_{\mathbb{R}}$ be embedded into a subgroup locally isomorphic to $\mathrm{SL}(2,\mathbb{R})$; for instance, a solvable Lie group may contain a weakly horocyclic subgroup but it never contains a horocyclic subgroup). Note that a weakly horocyclic subgroup is a one-parameter subgroup of a horospherical subgroup. By the Jacobson–Morozov lemma, any one-parameter unipotent subgroup of a semisimple Lie group is horocyclic. Now Theorem 22.3 follows from the main result of Marcus' paper.

THEOREM 22.4 [**Mar78**]. *Let $a_{\mathbb{R}}$ and $h_{\mathbb{R}}$ be two continuous flows on a manifold X with an invariant measure μ. Assume that the flow $h_{\mathbb{R}}$ is ergodic and that $a_t \circ h_1 = h_{e^t} \circ a_t$ for all $t \in \mathbb{R}$. Then $h_{\mathbb{R}}$ is mixing of all orders.*

However, a unipotent subgroup of a Lie group need not be weakly horocyclic even if it induces a mixing flow (see example 22.7). Nevertheless, Marcus conjectured that $\mathrm{Mix}(1) \Rightarrow \mathrm{Mix}(\infty)$ for any one-parameter flow on a homogeneous space of finite volume (see also [**Ma91b**]).

Almost simultaneously and by quite similar methods, Ryzhikov and Mozes proved important generalizations of Theorem 22.4. First, we note that its assumptions imply the existence of a measure-theoretic isomorphism $h_s \simeq h_1$ for all $s > 0$, induced by another flow $a_{\mathbb{R}}$. A more general result holds.

THEOREM 22.5 [**Ry91**]. *Let $(X, h_{\mathbb{R}})$ be a measure-preserving ergodic flow on a Lebesgue space and let $\mathcal{M} = \{s \in \mathbb{R} |\ h_s \simeq h_1\}$. Then $h_{\mathbb{R}}$ is mixing of all orders provided that $\mathcal{M}$ is of positive Lebesgue measure.*

A natural question arises: Can this result be applied to the class of homogeneous flows? Quite easily the multiple mixing problem reduces to the unipotent case. If it is true that for a mixing unipotent flow the set $\mathcal{M}$ is of positive measure, then Marcus' conjecture follows from Theorem 22.5. However, we will show below that if the set $\mathcal{M}$ for a homogeneous flow $(G/\Gamma, g_{\mathbb{R}})$ is uncountable, then the subgroup $g_{\mathbb{R}} \subset G$ is weakly horocyclic. Hence for homogeneous flows Theorems 22.4 and 22.5 are equivalent. Therefore Marcus' conjecture is not settled by Theorem 22.5.

On the other hand, the relation between $a_{\mathbb{R}}$ and $h_{\mathbb{R}}$ as in Theorem 22.4 means that a two-dimensional center-free solvable group acts on X. Mozes proved that $\mathrm{Mix}(1) \Rightarrow \mathrm{Mix}(\infty)$ for measurable actions of the so-called Ad-proper Lie groups. A Lie group F is called Ad-*proper* if the center $Z(F)$ is finite and the adjoint group $\mathrm{Ad}(F) = F/Z(F)$ is closed (in the Euclidean topology) in the algebraic group $\mathrm{Aut}(\mathfrak{f})$ of all automorphisms of the Lie algebra $\mathfrak{f}$.

THEOREM 22.6 [**Moz92,95b**]. *Let an* Ad-*proper Lie group* F *act by measure-preserving transformations on a probability space* (X, μ). *Then* $\mathrm{Mix}(1) \Rightarrow \mathrm{Mix}(\infty)$.

This result is proved similarly to Theorem 22.2.

Thus, Rokhlin's problem has a positive solution in the class of Ad-proper groups. But this does not help us in solving the problem for homogeneous flows.

Example 22.7. Let $G = \mathrm{SL}(2,\mathbb{R}) \cdot \mathbb{R}^2$, let Γ be a lattice in G, and let $n = \begin{pmatrix} 0 & 1 \\ 0 & 0 \end{pmatrix} \in \mathrm{sl}(2,\mathbb{R})$ with respect to some basis e_1, e_2 in $\mathbb{R}^2$. Then $[n, e_1] = 0$, $[n, e_2] = e_1$. It is easy to verify that the element $u = n + e_2$ is not weakly horocyclic. At the same time, the unipotent flow $(G/\Gamma, \exp(\mathbb{R}u))$ is mixing by Dani's criterion 9.7. There exist precisely two Ad-proper subgroups of G that contain $\exp(\mathbb{R}u)$. More specifically: $F_1 = B \cdot \mathbb{R}^2$, where B is the two-dimensional triangular subgroup of $\mathrm{SL}(2,\mathbb{R})$ containing $\exp(\mathbb{R}n)$, and $F_2 = G$. But the action of both subgroups is not mixing because the action of $\mathbb{R}^2$ on G/Γ is not even ergodic.

In full generality Marcus' conjecture was proved by the author.

THEOREM 22.8 [**St93**]. *Let* $(G/\Gamma, g_{\mathbb{R}})$ *be a mixing flow on a homogeneous space of finite volume. Then it is mixing of all orders.*

We briefly indicate the proof. First we reduce the problem to the unipotent case. According to Thouvenot (see [**Ka**]), a flow belongs to the class $\mathrm{Mix}(\infty)$ if and only if its maximal flow of zero entropy does. Hence to settle Marcus' conjecture it suffices to prove that $(G/\overline{A\Gamma}, g_{\mathbb{R}})$ is mixing of all orders if $(G/\Gamma, g_{\mathbb{R}})$ is a 1-mixing flow, $A \subset G$ being the Auslander normal subgroup for $g_{\mathbb{R}}$. By Witte's Theorem 9.9, if G/Γ admits a mixing flow, then the radical of G is nilpotent and the lattice Γ is Zariski dense in the Ad-representation of G.

Since $(\overline{A\Gamma})_0$ is normalized by the Zariski dense lattice Γ, it is normal in G. We have the isomorphism

$$(G/\overline{A\Gamma}, g_{\mathbb{R}}) \simeq (G'/\Gamma', g'_{\mathbb{R}}),$$

where $\Gamma' = \overline{A\Gamma}/(\overline{A\Gamma})_0$ is a lattice in $G' = G/(\overline{A\Gamma})_0$, and the subgroup $g'_{\mathbb{R}} \subset G'$ is quasi-unipotent. Hence from the beginning one can assume that $G = G'$ and the subgroup $g_{\mathbb{R}}$ is itself quasi-unipotent.

Since the group G is of the form $G = L \cdot N$, it follows that its Lie algebra is algebraic, and hence G admits a Jordan decomposition for any one-parameter subgroup. Let $g_t = c_t \times u_t$, $t \in \mathbb{R}$, where the group $\mathrm{Ad}(c_{\mathbb{R}})$ is relatively compact, and the group $\mathrm{Ad}(u_{\mathbb{R}})$ is unipotent in $\mathrm{Aut}(\mathfrak{g})$. Since the flow $(G/\Gamma, c_{\mathbb{R}})$ is uniformly continuous, following Marcus [**Mar78**] we conclude that it suffices to establish mixing of all orders for the unipotent flow $(G/\Gamma, u_{\mathbb{R}})$.

Now we apply the joinings technique. Henceforth $X = G/\Gamma$ and ν is the G-invariant probability measure on X relative to which the unipotent flow $(G/\Gamma, g_{\mathbb{R}})$ is mixing. Let $\{\bar{t}_i\} = \{t_i(1), \ldots, t_i(k)\}$ be a collection of k sequences such that $|t_i(l) - t_i(m)| \to \infty$ for all $l \neq m$. We define a translate of the diagonal measure

ν_Δ on X^k along the vector $\bar{t}_i$ as follows:

$$\nu_\Delta^{\bar{t}_i}\Big(\prod_{l=1}^{k} A_l\Big) = \nu\Big(\bigcap_{l=1}^{k} g_{t_i(l)} A_l\Big)$$

for measurable sets $A_1, \ldots, A_k \subset X$. To prove that the flow is $(k-1)$-mixing, it suffices to prove that $\mu = \nu^k$ whenever $\nu_\Delta^{\bar{t}_i} \to \mu$.

Let $(G/\Gamma, g_\mathbb{R})$ be a mixing unipotent flow. We know that $G = L \cdot N$, and by the mixing criterion, $G = \overline{G_\infty \Gamma}$, where G_∞ is the smallest subgroup normal in G that contains L. It is known [**Da77**] that $G_\infty = L \cdot M$, where $M \subset N$ is the group generated by the product $V \subset N$ of all nontrivial irreducible L-submodules in N (here we identify the nilradical N with its Lie algebra).

We denote by p_l the projection of G^k onto the l-th copy of G and by p_l^* the projection of G^k/Γ^k onto the l-th copy of G/Γ, $l = 1, \ldots k$. Note that $p_l^*(\nu_\Delta^{\bar{t}_i}) = \nu$ and hence $p_l^*(\mu) = \nu$. In particular, $\mu(G^k/\Gamma^k) = 1$. Clearly, all the measures $\nu_\Delta^{\bar{t}_i}$ are invariant under the diagonal subgroup $\{\bar{g}_t = (g_t, \ldots, g_t) \in G^k\}$, and hence the limit measure μ is $\bar{g}_\mathbb{R}$-invariant as well.

Let

$$q : G = L \cdot N \to L \quad \text{and} \quad q^* : G/\Gamma \to L/q(\Gamma).$$

It follows from Marcus' Theorem 22.3 applied to the mixing flow $(L/q(\Gamma), q(g_\mathbb{R}))$ that $(q^*)^k \mu = q^*(\nu)^k$. Let μ' be an ergodic component of μ with respect to the unipotent subgroup $\bar{g}_\mathbb{R} \subset G^k$ such that

$$(q^*)^k \mu' = q^*(\nu)^k \quad \text{and} \quad p_l^*(\mu') = \nu \quad \text{for all } l = 1, \ldots, k.$$

It suffices to prove that the measure μ' is G^k-invariant.

According to Ratner's measure theorem, there exist a subgroup $H \subset G^k$ and an element $g \in G^k$ such that μ' is an H-invariant measure supported on a closed orbit $Hg\Gamma^k \subset G^k/\Gamma^k$. Since $(q^*)^k \mu' = q^*(\nu)^k$, then $H = L^k \cdot N_H$, where $N_H \subset N^k$. On the other hand, $p_l^*(\mu') = \nu$, and hence $p_l(N_H) = N$, $l = 1, \ldots, k$.

It is easy to see that if $V' \subset V^k$ is an L^k-submodule such that $p_l(V') = V$, $l = 1, \ldots, k$, then $V' = V^k$. Therefore $V^k \subset M^k \subset N_H$, where $M \subset N$ is the normal subgroup of G generated by V. But from $\overline{LM\Gamma} = G$ it follows that $\overline{L^k M^k \Gamma^k} = G^k$, and since the orbit $Hg\Gamma^k$ closed, then $H = G^k$ and $\mu' = \nu^k$. □

Remark. Ryzhikov [**Ry97**] proved a general result asserting that Mix(1) $\Rightarrow$ Mix(∞) for a smooth flow ϕ_t on a manifold M with a finite smooth invariant measure μ whenever self-joinings of all orders are smooth measures on submanifolds. Thus, Marcus' conjecture can be deduced from Ratner's theorem without involving the theory of Lie groups.

CONJECTURE. *Rokhlin's problem has a positive solution for all homogeneous actions $(G/\Gamma, F)$, where F is a closed subgroup of G, and Γ is a lattice in G.*

§23. Partial rigidity of quasi-unipotent flows

1. Here we consider the measure rigidity property of homogeneous one-generator actions and flows. According to Theorem 18.3, ergodic unipotent flows are measure rigid: any measure-theoretical isomorphism is (almost everywhere) an affine map of homogeneous spaces. We already know that partially hyperbolic flows (for instance, the geodesic flow) do not possess this property. On the other hand, an ergodic quasi-unipotent flow on a solvable manifold is smoothly and measure-theoretically isomorphic to an ergodic nilflow. Hence measure rigidity does not hold in the quasi-unipotent case either.

However, as Witte proved, measure rigidity holds if the quasi-unipotent flows involved are mixing:

THEOREM 23.1 [**Wi87**]. *Any measure-theoretic isomorphism of mixing quasi-unipotent one-generator actions (or flows) is affine almost everywhere.*

In what follows, we need a generalization of the theorem to the class of ergodic quasi-unipotent one-generator actions. It turns out that although an isomorphism of such actions need not be an affine map globally, it is "fiber-wise" affine:

THEOREM 23.2 [**St95c**]. *If $f : (G/\Gamma, g_{\mathbb{Z}}) \to (G'/\Gamma', g'_{\mathbb{Z}})$ is a measure-theoretic isomorphism of ergodic quasi-unipotent one-generator actions, then for almost all $g\Gamma \in G/\Gamma$ we have $f(\overline{G_\infty g\Gamma}) = \overline{G'_\infty f(g\Gamma)}$, and the restriction of f onto almost every homogeneous subspace $\overline{G_\infty g\Gamma}$ is affine.*

This result can be also obtained by using methods from [**Wi87**] and [**Wi96**].

COROLLARY 23.3 [**St95c**]. 1) *Under the assumptions of the previous theorem, the isomorphism f induces an isomorphism of ergodic quotient actions*

$$\widehat{f} : (G/\overline{G_\infty \Gamma}, g_{\mathbb{Z}}) \to (G'/\overline{G'_\infty \Gamma'}, g'_{\mathbb{Z}})$$

on solvable homogeneous spaces,

2) *If the one-generator actions are mixing, then the isomorphism f is affine (almost everywhere).*

Note that the second statement is precisely Witte's theorem. It follows immediately from the first statement because solvable homogeneous spaces admit no mixing homogeneous translations.

We give the idea of the proof of Theorem 23.2. Since the actions are quasi-unipotent, it follows that the algebraic envelope of the subgroup $\mathrm{Ad}(g_{\mathbb{R}}) \subset \mathrm{Aut}(\mathfrak{g})$ is the direct product of a torus T and a unipotent subgroup. Hence there exists a unipotent subgroup $u_{\mathbb{R}} \subset T \cdot G$ that commutes with T and passes to $g_{\mathbb{R}}$ under the smooth projection $TG \to G$. Therefore we have constructed a compact cover of the original one-generator action by the unipotent one-generator action $(TG/\Gamma, u_{\mathbb{Z}})$. Likewise, we construct a cover of the one-generator action $(G'/\Gamma', g'_{\mathbb{Z}})$ by a unipotent one-generator action $(T'G'/\Gamma', u'_{\mathbb{Z}})$. The graph $\mathrm{Graph}(f) \subset G/\Gamma \times G'/\Gamma'$ is compactly covered by a subset $M \subset TG/\Gamma \times T'G'/\Gamma'$ invariant under the diagonal unipotent subgroup $(u \times u')_{\mathbb{Z}}$. A finite ergodic (relative to $(g \times g')_{\mathbb{Z}}$) measure on $\mathrm{Graph}(f)$ obtained from a G-invariant measure on G/Γ is covered by a finite $(u \times u')_{\mathbb{Z}}$-invariant measure on M. Each ergodic component of the measure has an algebraic origin and covers an ergodic measure on $\mathrm{Graph}(f)$. Thus, $\mathrm{Graph}(f)$ is compactly covered by a homogeneous space. Finally, the equalities $(TG)_\infty = G_\infty$ and $(T'G')_\infty = G'_\infty$ enable one to prove that although $\mathrm{Graph}(f)$ need not be a homogeneous subspace, it fibers into homogeneous subspaces

$$\mathrm{Graph}(f|_{\overline{G_\infty g\Gamma}}) \subset G/\Gamma, \qquad g \in G.$$

2. Ryzhikov's Theorem 22.5 relates the multiple mixing problem for a flow $(G/\Gamma, g_{\mathbb{R}})$ to the size of the set $\mathcal{M}$ of times $t > 0$ such that the translation g_t is measure-theoretically isomorphic to g_1. We will show that either $\mathcal{M}$ is at most countable, or $\mathcal{M}$ coincides with the group $\mathbb{R}^+$ and then the subgroup $g_\mathbb{R}$ is weakly horocyclic. It is clear that for a partially hyperbolic flow the set $\mathcal{M}$ is exhausted by $1 \in \mathbb{R}$ (this follows from the entropy formula $e(g_t) = te(g_1)$, $t > 0$). Thus, the set $\mathcal{M}$ can be nontrivial only for quasi-unipotent flows.

One easily constructs an example when $\mathcal{M} \neq \{1\}$. Let $\mathbb{T}^2 = \mathbb{R}^2/\mathbb{Z}^2$, $A \in \mathrm{SL}(2,\mathbb{Z})$ and $v \in \mathbb{R}^2$. Then the rectilinear flows $(\mathbb{T}^2, \mathbb{R}v)$ and $(\mathbb{T}^2, \mathbb{R}A(v))$ are isomorphic. In particular, if $v \in \mathbb{R}^2$ is an eigenvector, that is, if $A(v) = \lambda v$ for some $\lambda > 0$, then the translations $(\mathbb{T}^2, \mathbb{Z}v)$ and $(\mathbb{T}^2, \lambda\mathbb{Z}v)$ are isomorphic. It follows that the set $\mathcal{M}$ contains the multiplicative cyclic group $\lambda^\mathbb{Z} \subset \mathbb{R}^+$. Using the measure rigidity of such translations proved already by Parry [**P71**], it is not hard to show that in this case the set $\mathcal{M}$ is countable.

On the other hand, by the definition, for a weakly horocyclic subgroup $u_\mathbb{R} \subset G$ one has $\mathcal{M} = \mathbb{R}^+$. Actually, there are no other possibilities.

THEOREM 23.4 [**St95c**]. *Assume that the set $\mathcal{M} = \{t > 0 |\ u_t \simeq u_1\}$ for an ergodic homogeneous flow $(G/\Gamma, u_\mathbb{R})$ on a finite volume space is uncountable. Then the subgroup $u_\mathbb{R}$ is weakly horocyclic in G (and hence $\mathcal{M} = \mathbb{R}^+$).*

First we prove the theorem for unipotent flows. Without loss of generality we may assume that the lattice $\Gamma \subset G$ is faithful. Suppose that there exists an uncountable set $\mathcal{M} \subset \mathbb{R}^+$ such that for each $t \in \mathcal{M}$ there exists a measure-theoretic isomorphism of ergodic unipotent one-generator actions

$$\phi_t : (G/\Gamma, u_\mathbb{Z}) \to (G/\Gamma, u_{t\mathbb{Z}}).$$

By virtue of measure rigidity, for each $t \in \mathcal{M}$ there exist an element $c(t) \in G$ and an automorphism $\sigma_t \in \mathrm{Aut}(G,\Gamma)$ such that (almost everywhere on G/Γ) the isomorphism ϕ_t is as follows:

$$\phi_t(g\Gamma) = c(t)\sigma_t(g)\Gamma.$$

Since Γ is Zariski dense in the Ad-representation, it follows that the group $\mathrm{Aut}(G,\Gamma)$ is at most countable. Then for some uncountable subset $\mathcal{M}' \subset \mathcal{M}$, the automorphisms σ_t coincide with a single automorphism $\sigma \in \mathrm{Aut}(G,\Gamma)$. Since ϕ_t sends the left action of $u_\mathbb{Z}$ to the left action of $u_{t\mathbb{Z}}$, it follows that

$$\phi_t(u_1 g\Gamma) = u_t\phi_t(g\Gamma) \quad \text{almost everywhere on } G/\Gamma.$$

Using the explicit form of ϕ_t, $t \in \mathcal{M}'$, one proves that

$$c(t)\sigma(u_1)\sigma(g)\Gamma = u_t c(t)\sigma(g)\Gamma \quad \text{for almost all } g \in G.$$

But then the element $c(t)^{-1}u_{-t}c(t)\sigma(u_1)$ acts trivially on G/Γ and hence belongs to the intersection of the center $Z(G)$ and the lattice Γ. Since Γ is faithful, we see that $c(t)^{-1}u_t c(t) = \sigma(u_1)$ for every $t \in \mathcal{M}'$. We fix an element $s \in \mathcal{M}'$ and note that

$$c(t)c(s)^{-1}u_s c(s)c(t)^{-1} = u_t \quad \text{for each } t \in \mathcal{M}'.$$

Now our theorem is a corollary of the following observation.

LEMMA 23.5 [**St95c**]. *Assume that $u_{\mathbb{R}}$ is a subgroup of G such that for each element t of an uncountable set $\mathcal{M}' \subset \mathbb{R}^+$ there exists $h(t) \in G$ such that $h(t)^{-1}u_1 h(t) = u_t$. Then the subgroup $u_{\mathbb{R}}$ is weakly horocyclic.*

Finally, we consider an ergodic quasi-unipotent action $(G/\Gamma, u_{\mathbb{Z}})$ for which the set $\mathcal{M} = \{t > 0 : \; u_t \simeq u_1\}$ is uncountable. By Corollary 23.3, for each $t \in \mathcal{M}$ the quotient action $(G/\overline{G_\infty \Gamma}, u_{t\mathbb{Z}})$ is isomorphic to the quotient action $(G/\overline{G_\infty\Gamma}, u_{\mathbb{Z}})$. But an ergodic quasi-unipotent flow on a solvable space is smoothly and measure-theoretically isomorphic to a nilflow. We have already proved Theorem 23.4 in the unipotent case, and know that for an arbitrary nilflow the set $\mathcal{M}$ is at most countable, because a nilpotent Lie group contains no weakly horocyclic subgroups. Hence the space $G/\overline{G_\infty\Gamma}$ degenerates to a point, and, by Theorem 9.8, any ergodic translation on G/Γ is mixing. Therefore we can apply Witte's Theorem 23.1 to conclude that any measure-theoretic isomorphism $u_t \simeq u_1$, $t \in \mathcal{M}$, is given by an affine map. The rest of the arguments repeat literally the proof in the unipotent case. □

§24. Topological equivalence and time changes

1. We recall that $f : (G/\Gamma, g_{\mathbb{R}}) \to (G'/\Gamma', g'_{\mathbb{R}})$ is a *topological equivalence* of flows if f is a homeomorphism of G/Γ and G'/Γ' that sends orbits to orbits (preserving their orientations).

The simplest example of a topological equivalence is given by an affine invertible map $\sigma : G/\Gamma \to G'/\Gamma'$ that sends $g_{\mathbb{R}}$-orbits to $g'_{\mathbb{R}}$-orbits. It is easy to see that σ induces a smooth isomorphism $\sigma : (G/\Gamma, g_t) \to (G'/\Gamma', g'_{ct})$ for some constant $c > 0$. Such a map will be referred to as an *affine orbit equivalence.*

Suppose that $f : (G/\Gamma, g_{\mathbb{R}}) \to (G'/\Gamma', g'_{\mathbb{R}})$ is a topological equivalence of flows. We ask the following questions.

1) Under which conditions are the flows affinely orbit equivalent?

2) Does there exist an affine orbit equivalence σ such that

$$f(g_{\mathbb{R}}x) = \sigma(g_{\mathbb{R}}x), \qquad x \in G/\Gamma$$

(that is, f and σ are the same modulo a time change for the flow $(G'/\Gamma', g'_{\mathbb{R}})$)?

We know at least two counterexamples to the first question: 1) A flow on a Euclidean solvable manifold can be smoothly isomorphic to a rectilinear flow on the torus (see §4). 2) The geodesic flows on homeomorphic compact surfaces of constant curvature -1 are always topologically equivalent, but the surfaces involved may not be isometric (see §14). Nevertheless, in a broad class of Lie groups the answers to both questions are positive. At the core of this phenomenon, discovered by Benardete [**Ben**] and studied also by Witte [**Wi90**], lies a remarkably simple proof of the classical result [**Ne**] for flows on the torus.

LEMMA 24.1. *If flows $(\mathbb{R}^n/\mathbb{Z}^n, \mathbb{R}v)$ and $(\mathbb{R}^n/\mathbb{Z}^n, \mathbb{R}v')$ are topologically equivalent, then they are affinely orbit equivalent.*

Proof [**Ben**]. Let $f : \mathbb{R}^n/\mathbb{Z}^n \to \mathbb{R}^n/\mathbb{Z}^n$ be a homeomorphism establishing the topological equivalence. We can assume, taking a composite of f and a translation if necessary, that $f(\mathbb{Z}^n) = \mathbb{Z}^n$, and then lift f to a homeomorphism $\widetilde{f} : \mathbb{R}^n \to \mathbb{R}^n$ such that $\widetilde{f}(x + \mathbb{Z}^n) = \widetilde{f}(x) + \mathbb{Z}^n$. Since $\widetilde{f}|_{\mathbb{Z}^n} = \sigma \in \mathrm{SL}(n, \mathbb{Z})$, then $\sigma^{-1}\widetilde{f}$ is a homeomorphism of $\mathbb{R}^n$ identical on $\mathbb{Z}^n$ and sending $\mathbb{R}v$-orbits to $\sigma^{-1}(\mathbb{R}v')$-orbits. The torus $\mathbb{R}^n/\mathbb{Z}^n$ is compact, and hence there exists a constant $c > 0$ such that the distance between x and $\sigma^{-1}\widetilde{f}(x)$ is uniformly bounded. On the other hand, any two distinct one-dimensional subgroups of $\mathbb{R}^n$ diverge from each other, and hence $\mathbb{R}v = \sigma^{-1}(\mathbb{R}v')$. □

LEMMA 24.2. *If under the assumptions of the lemma above the flows are ergodic, then topological equivalence coincides with an affine orbit equivalence modulo a time change.*

Proof [**Wi90**]. In the same notation, for $x \in \mathbb{R}^n$ we have

$$\sigma^{-1}\widetilde{f}(x+\mathbb{Z}^n) = \sigma^{-1}\widetilde{f}(x) + \mathbb{Z}^n \quad \text{and} \quad \sigma^{-1}\widetilde{f}(\mathbb{R}v + x) = \mathbb{R}v + \sigma^{-1}\widetilde{f}(x).$$

It follows from the last condition that $\sigma^{-1}\widetilde{f}$ induces a homeomorphism of the quotient space $(\mathbb{R}v)\backslash\mathbb{R}^n$, and from the first condition that this homeomorphism is the identical map. □

Note that the proof of Lemma 24.1 goes through for a topological equivalence $f : (G/\Gamma, g_{\mathbb{R}}) \to (G'/\Gamma', g'_{\mathbb{R}})$ if the following conditions are satisfied: a) the lattices Γ and Γ' are uniform, b) the isomorphism $\psi : \Gamma \to \Gamma'$ extends to an isomorphism $\widetilde{\psi} : G \to G'$ of the Lie groups involved, and c) any two distinct one-parameter subgroups of G *diverge*, that is,

$$\sup_t d_G(g_t, h_{\mathbb{R}}) = \infty \qquad \text{if } g_{\mathbb{R}} \neq h_{\mathbb{R}}.$$

Actually, as was shown by Benardete [**Ben**], the uniformity condition can be omitted. The condition b) stands for rigidity of lattices, and holds if G and G' are class (R) solvable groups (Saito's rigidity theorem)[4] or if both the groups are semisimple and satisfy the assumptions of the Mostow–Margulis–Prasad rigidity theorem [**VGS**]. Finally, any two distinct subgroups of a class (R) solvable group obviously diverge. As regards divergence in a semisimple group, it can fail to hold (for instance, a quasi-unipotent subgroup stays at a bounded distance from its unipotent part). However, condition c) in the semisimple case can be replaced by ergodicity of the flows. As a result, we have the following theorem.

THEOREM 24.3 [**Ben**]. *Let $f : (G/\Gamma, g_{\mathbb{R}}) \to (G'/\Gamma', g'_{\mathbb{R}})$ be a topological equivalence of ergodic flows. Assume that either*

1) *G and G' are simply connected class (R) solvable groups, or*
2) *G and G' are center-free semisimple groups with no compact components, and G contains no normal subgroup S locally isomorphic to* $\mathrm{SL}(2,\mathbb{R})$ *such that the product $S\Gamma$ is closed.*

Then f coincides modulo a time change with an affine orbit equivalence of the flows.

2. Witte [**Wi90**] extended the theorem for multi-parameter actions. By topological (or affine orbit) equivalence for such actions one can understand a homeomorphism (an affine isomorphism) sending orbits to orbits (not caring about the orientation). In this setting Witte's result can be formulated as follows.

THEOREM 24.4 [**Wi90**]. *Let $f : (G/\Gamma, F) \to (G'/\Gamma', F')$ be a topological equivalence of ergodic actions, where F and F' are connected unimodular subgroups.*

[4]In general, this is not the case. For a rigidity criterion for lattices in solvable Lie groups, see [**St94**].

Assume that G and G' are as in Theorem 24.3. Then f is a composite of an affine orbit equivalence and a homeomorphism of G'/Γ' preserving F'-orbits.

Note that the theorem breaks down for $G = G' = \mathrm{SL}(2,\mathbb{R})$. Indeed, to within a conjugation, $\mathrm{SL}(2,\mathbb{R})$ contains four proper subgroups: the compact subgroup $K = \mathrm{SO}(2,\mathbb{R})$, the diagonal subgroup $A = \mathrm{D}(2,\mathbb{R})$, the unipotent subgroup $U = \mathrm{U}(2,\mathbb{R})$, and the two-dimensional triangular subgroup $B = \mathrm{T}(2,\mathbb{R})$. It is shown in [**Wi90**] that the actions of K on compact spaces G/Γ and G/Γ' are topologically equivalent, if the uniform lattices Γ and Γ' are isomorphic (but not necessarily conjugate). The same applies to the actions of the subgroups A and B.

On the other hand, according to Marcus [**Mar83**], a topological equivalence of the horocycle flows $(G/\Gamma, U)$ and $(G/\Gamma', U)$ on compact spaces is a composite of an affine orbit equivalence and a homeomorphism of G/Γ' preserving U-orbits (and possibly one can avoid the compactness condition). This result was proved simultaneously with Ratner's measure rigidity theorem. In view of Benardete's results, one can conjecture that a topological equivalence of unipotent ergodic flows is a composite of an affine orbit equivalence and a time change.

3. Thus, under favorable circumstances, a topological equivalence of one-parameter homogeneous flows is a composite of an affine equivalence and a continuous time change. The study of time changes for a measurable flow $(X, g_{\mathbb{R}})$ is closely related to studying the first cohomology group. Recall the general definition.

A measurable function $v : \mathbb{R} \times X \to \mathbb{R}$ is called a *cocycle* over a flow $(X, g_{\mathbb{R}})$ on a space X with invariant probability measure μ if

$$v(t_1 + t_2, x) = v(t_1, g_{t_2}x) + v(t_2, x) \qquad \text{for all } t_1, t_2 \in \mathbb{R} \text{ and } \mu\text{-almost all } x \in X.$$

The simplest example is a *constant cocycle* $v(t,x) = ct$. Given a cocycle $v(t,x)$, one can produce a constant cocycle via integration:

$$v^*(t,x) = \int_X v(t,x)\,d\mu_x.$$

A measurable function $P : X \to \mathbb{R}$ determines a cocycle

$$v(t,x) = P(g_t x) - P(x).$$

Such a cocycle is called a *cochain* or *cohomologous to zero*; this is written as $v \sim 0$.

The linear space of measurable cocycles modulo cochains is called the *first group of measurable cohomology* for the flow $(X, g_{\mathbb{R}})$. We denote it by $H^1(X, g_{\mathbb{R}})$ (here all our cocycles take values in $\mathbb{R}$). It is clear that the group contains $\mathrm{Hom}(\mathbb{R},\mathbb{R}) \simeq \mathbb{R}$ as a subgroup (two distinct constant cocycles cannot be cohomologous). We say that the first cohomology group *trivializes* if $H^1(X, g_{\mathbb{R}}) = \mathrm{Hom}(\mathbb{R},\mathbb{R})$, that is, if every cocycle is cohomologous to a constant cocycle.

If $(X, g_{\mathbb{R}})$ is a continuous flow on a topological space, it is natural to consider continuous cocycles $v : \mathbb{R} \times X \to \mathbb{R}$ and to require that the function $P : X \to \mathbb{R}$ involved into the definition of a cochain be continuous. The group obtained is called the *first group of continuous cohomology*. Assume that x_0 is a periodic point of a continuous flow $(X, g_{\mathbb{R}})$; that is, $g_a x_0 = x_0$ for some $a > 0$. If $v \sim 0$, then $v(a, x_0) = P(g_a x_0) - P(x_0) = 0$. So, for the condition $v \sim 0$ to hold it is necessary that $v(t,x) = 0$ whenever $g_t x = x$.

Thus, the first group of continuous cohomology does not trivialize if there are at least two distinct periodic orbits. Indeed, given any continuous function $f : X \to \mathbb{R}$, one can construct a continuous cocycle via

$$v_f(t,x) = \int_0^t f(g_s x)ds.$$

Now if x_1 and x_2 give rise to distinct periodic orbits with periods t_1 and t_2, let $f \equiv 1$ on $g_{\mathbb{R}}x_1$ and let $f \equiv a$ on $g_{\mathbb{R}}x_2$, where a is such that $t_1 \neq at_2$. One can extend f to a continuous function on X and consider the corresponding cocycle $v_f(t,x)$. It is clear that no cocycle of the form $v_f(t,x) - ct$ is a cochain. Therefore periodic orbits present obstacles for the first group of continuous cohomology to trivialize.

If $(X, g_{\mathbb{R}})$ is a smooth flow on a Riemannian manifold, one studies the groups of C^∞- and *Hölder cohomology*. We recall that $f : M \to \mathbb{R}$ is a *Hölder* function with exponent $\alpha > 0$ if

$$|f(x) - f(y)| < Cd(x,y)^\alpha \qquad \text{for all } x, y \in M \text{ and some constant } C.$$

Obviously, periodic orbits present obstacles for trivialization of groups of such cohomology as well. As a rule, if periodic orbits exist, there are many of them: they may form a countable dense subset of X. In this case the first cohomology group is infinitely dimensional. It is of interest that for an Anosov flow the family of periodic orbits describes all the obstacles.

THEOREM 24.5 (LIVSHITZ [**Liv**]). *Let $(X, g_{\mathbb{R}})$ be a topologically transitive Anosov flow on a compact Riemannian manifold. Assume that $v : X \times \mathbb{R} \to \mathbb{R}$ is a C^∞-cocycle such that $v(t,x) = 0$ whenever $g_t x = x$. Then $v \sim 0$ (in the class of C^∞-functions). The same holds for Hölder cocycles.*

An example of Anosov flow is given by the suspension over a hyperbolic toral automorphism (see §4). Veech [**V86**] extended Livshitz's theorem to the class of C^∞-cocycles over any ergodic toral automorphism $A \in \mathrm{Aut}(\mathbb{T}^n)$. He also discovered that Livshitz's theorem fails to hold for C^1-cocycles in the partially hyperbolic case.

Actually, all Anosov flows of algebraic origin arise as follows. Let a subgroup $g_{\mathbb{R}} \subset G$ be such that its neutral subgroup $Q \subset G$ decomposes into the product $Q = C \times g_{\mathbb{R}}$, where $C \subset G$ is a compact connected subgroup. Let Γ be a lattice in G. Since C and $g_{\mathbb{R}}$ commute, the flow $(C\backslash G/\Gamma, g_{\mathbb{R}})$ on an infra-homogeneous space is well defined. If the intersection $hCh^{-1} \cap \Gamma$ is independent of $h \in G$, the space $C\backslash G/\Gamma$ is a manifold. The neutral foliation of the flow is given by its orbits, and hence the flow is Anosov. For example, an $\mathbb{R}$-diagonalizable subgroup $g_{\mathbb{R}}$ of a semisimple group of $\mathbb{R}$-rank 1 induces an Anosov flow on an infra-homogeneous manifold $C\backslash G/\Gamma$, where $C \subset G$ stands for the maximal connected compact subgroup that commutes with $g_{\mathbb{R}}$. A classification of Anosov $\mathbb{R}^1$-flows on compact infra-homogeneous manifolds is given in [**Tom**].

The definition of a cocycle over a flow can easily be generalized. Assume, for instance, that an Abelian group A acts on a space X and that our cocycles take values in another Abelian group F. Then the cocycle identity is written as follows:

$$v(a+b,x) = v(a,bx) + v(b,x), \qquad a,b \in A, \quad x \in X.$$

We say that the first cohomology group trivializes if every cocycle is cohomologous to a constant cocycle $v \in \mathrm{Hom}(A,F)$.

An important class of such actions is given by Anosov $\mathbb{R}^k$-actions for $k \geq 2$. All known examples of the actions (except those obtained from natural operations with one-parameter Anosov flows) have an algebraic origin. More specifically, let $A \subset G$ be a subgroup isomorphic to $\mathbb{R}^k$, let $C \subset G$ be a maximal compact connected subgroup that commutes with A, and let $\Gamma \subset G$ be a uniform lattice. Then the action $(C\backslash G/\Gamma, A)$ is called *Anosov* if there exists a partially hyperbolic element $a \in A$ with its neutral foliation given by the A-orbits.

For example, let G be a semisimple Lie group, Γ a lattice and $A \subset G$ a maximal $\mathbb{R}$-diagonalizable subgroup.[5] The class of discrete Anosov $\mathbb{Z}^k$-actions arises if one takes k commuting automorphisms of the torus $\mathbb{T}^n$, one of them being hyperbolic (clearly, the suspension over the action $(\mathbb{T}^n, \mathbb{Z}^k)$ leads to an Anosov $\mathbb{R}^k$-action on a solvable manifold). Note that every Anosov action on a compact manifold has a dense countable family of compact orbits.

At present, Anosov $\mathbb{R}^k$-actions are one of the most interesting objects to study (see §§26 and 30). Katok and Spatzier [**KS94**] discovered surprising rigidity properties of cocycles over Anosov actions on compact manifolds. More specifically, they studied a series of examples called *standard Anosov actions.* In all the examples one imposes different irreducibility restrictions. We describe two very important classes of standard actions. Given a semisimple group of $\operatorname{rank}_{\mathbb{R}} G \geq 2$, one requires that the lattice $\Gamma \subset G$ be uniform and irreducible. Given a $\mathbb{Z}^k$-action on $\mathbb{T}^n$, one assumes that there is a subgroup $\mathbb{Z}^2 \subset \mathbb{Z}^k$ consisting of ergodic elements. The following theorem holds.

THEOREM 24.6 [**KS94**]. *Assume that (M, A), $A = \mathbb{R}^k$ or $\mathbb{Z}^k$, is a standard Anosov action on a compact infra-homogeneous manifold. Then the first C^∞-cohomology group with values in $\mathbb{R}^n$ trivializes. The same applies to Hölder cohomology.*

The standard Anosov actions studied in [**KS94**] have no quotient actions which degenerate to an $\mathbb{R}^1$- or $\mathbb{Z}^1$-action. One can conjecture that for Anosov actions of rank ≥ 2 admitting no rank 1 quotient actions of algebraic origin, the first C^∞ cohomology group trivializes. For $\mathbb{Z}^k$-actions on the torus[6] this was proved in [**St98**]. Progress in the proof of this conjecture and new classes of rigid Anosov actions were obtained in [**St99a**]. Note that the proof of cocycle rigidity in [**KS94**] and [**St99a**] crucially involves the phenomenon of exponential mixing (see §27).

4. We turn to time changes for a one-parameter flow $(X, g_{\mathbb{R}})$ with an invariant probability measure μ. Assume that a function $\tau \in L^1(X, \mu)$ is positive and let $\overline{\tau} = \int_X \tau d\mu$. Define

$$g_t^\tau(x) = g_{v(t,x)}(x), \qquad \text{where} \quad \int_0^{v(t,x)} \tau(g_s(x))ds = t.$$

Then the flow $(X, g_{\mathbb{R}}^\tau)$ preserves the measure μ^τ given as

$$d\mu_x^\tau = \frac{\tau(x)}{\overline{\tau}} d\mu_x,$$

and is called a *time change* of the flow $(X, g_{\mathbb{R}})$.

---

[5] In [**KS94**] this action is called "Weyl chamber flow".

[6] About $\mathbb{Z}^k$-actions on a compact Abelian group, see the book [**Sc**].

It is easy to verify that $v(t, x)$ is a cocycle over the new flow $(X, g_{\mathbb{R}}^{\tau})$. A linear time change corresponds to the constant cocycle $v(t, x) = t/c$ and is obtained from the constant function $\tau \equiv c$. We say that a time change *reduces to a linear time change* if the cocycle $v(t, x)$ is cohomologous to a constant cocycle. It is easily seen that if $\overline{\tau} = 1$ and the time change reduces to a linear time change, then the map $Q(x) = g_{P(x)}(x)$ establishes a conjugacy: $g_t \circ Q = Q \circ g_t^{\tau}$.

To understand whether there exists a nontrivial time change, one has to study the first cohomology group. If we prove that this group trivializes (in some class of functions on X), then every time change reduces to a linear time change (in the same class). We already know that for a continuous flow with periodic orbits, far from every time change reduces to a linear time change. At the same time, the countable family of periodic orbits of an Anosov flow presents the full collection of obstacles for a C^∞-time change to reduce to a linear time change. More specifically, Theorem 24.5 implies that a time change with a C^∞-function τ for an Anosov flow on a compact Riemannian manifold reduces to a linear time change (via a C^∞-function P) $\Longleftrightarrow$ the fraction of new and old periods is $\overline{\tau}$ for all periodic orbits. A similar result holds in the class of Hölder functions.

By a time change for an $\mathbb{R}^k$-action one understands another $\mathbb{R}^k$-action with the same orbits. The simplest example of a time change (an analogue of a linear time change) arises from an automorphism of the group $\mathbb{R}^k$. The question of whether any time change reduces to an automorphism involves the study of the first cohomology group with values in $\mathbb{R}^k$. Indeed, let $\mathbb{R}^k$ act smoothly and locally freely on X in two ways: $(a, x) \to ax$ and $(a, x) \to \widetilde{a}x$, and let the orbits of the actions be the same. It is easy to see that the formula $\widetilde{a}x = v(a, x)x$ determines a smooth cocycle $v : \mathbb{R}^k \times X \to \mathbb{R}^k$ (over the action $(a, x) \to \widetilde{a}x$). Hence Theorem 24.6 implies that any C^∞-time change for a standard Anosov $\mathbb{R}^k$-action with $k \geq 2$ reduces to an automorphism of $\mathbb{R}^k$. A similar statement holds in the class of Hölder time changes.

5. Now we return to the one-parameter case and assume that the flow under question has no periodic orbits at all (for instance, it is minimal). Then the question of whether there exists a nontrivial time change becomes more intricate. For example, there is a classical result of Kolmogorov for a rectilinear flow on the torus.

THEOREM 24.7 [**Ko**]. *For almost all irrational numbers $\alpha \in \mathbb{R}$, a positive function $\tau \in C^\infty(\mathbb{T}^2)$ determines a time change for the rectilinear flow $(\mathbb{T}^2, \mathbb{R}v_\alpha)$ with tangent vector $v_\alpha = (1, \alpha) \in \mathbb{R}^2$ that reduces to a linear time change (in the C^∞-class). On the other hand, there exist irrational numbers α with nontrivial smooth time changes for $(\mathbb{T}^2, \mathbb{R}v_\alpha)$.*

We outline the proof (see also [**KSF**] and [**Si96**]). *Without loss of generality we normalize τ by the condition*

$$\overline{\tau} = \int_{\mathbb{T}^2} \tau(x)dx = 1.$$

Let (u, v) be coordinates on the torus $\mathbb{T}^2 = \mathbb{R}^2/\mathbb{Z}^2$. It is easy to see that the flow T_t obtained by a time change is given by the system of differential equations

$$\frac{du}{dt} = \frac{1}{\tau(u, v)}, \qquad \frac{dv}{dt} = \frac{\alpha}{\tau(u, v)}.$$

In addition, the flow is smoothly isomorphic to the original flow if and only if there exists a smooth closed curve $\gamma \subset \mathbb{T}^2$ with no self-intersections that is transversal to the flow direction and satusfies the condition $T_1(\gamma) = \gamma$ (and the isomorphism is given by the map $Q(x) = T_{P(x)}(x)$ for a function $P : \mathbb{T}^2 \to \mathbb{R}$).

Let us find the condition for such a curve to exist. Let $\rho(v)$ be the time needed for $(0, v)$ to reach the curve γ along the flow T_t. It is clear that

$$t(v) = \int_0^1 \tau(u, v + \alpha u)du$$

is the time needed for the point $(0, v)$ to move to the point $(1, v + \alpha)$. Now the so-called *homology equation* for $\rho(v)$ takes the form

$$t(v) - \rho(v) + \rho(v + \alpha) \equiv 1.$$

It is clear that smoothness of the solution $\rho(v)$ (if it exists) determines smoothness of the isomorphism between T_t and the original rectilinear flow. It turns out that smoothness of $\rho(v)$ depends both on smoothness of $t(v)$ and on Diophantine properties of the number α. Expanding $t(v)$ and $\rho(v)$ into Fourier series, we obtain

$$\sum_{n \neq 0} (t_n - \rho_n + \rho_n e^{2\pi i n \alpha}) e^{2\pi i n v} \equiv 0,$$

whence

$$\rho_n = \frac{t_n}{1 - e^{2\pi i n \alpha}}, \qquad n \neq 0.$$

Assume that the number α is not extremely well approximable by rational numbers (see §29). This means that there exist $C > 0$ and $k \in \mathbb{N}$ such that

$$|n\alpha - m| > \frac{C}{|n|^k} \qquad \text{for all } n, m \in \mathbb{Z},\ n \neq 0.$$

Then $|1 - e^{2\pi i n \alpha}| > C_1/|n|^k$ for all $n \neq 0$, whence $|\rho_n| < |t_n||n|^k/C_1$. Since $t(v)$ is an infinitely smooth periodic function, it follows that its Fourier coefficients t_n decay faster than any negative power of n. Hence the ρ_n satisfy the same property; that is, $\rho(v)$ is an infinitely smooth periodic function as well. It follows from the homology equation that the closed curve γ has no self-intersections, and hence the flow T_t is conjugate to the original flow via a C^∞-smooth map.

It now suffices to note that numbers α with extremely fast approximation by rationals form a zero measure set (see §29). □

In §29 for any function $\psi : \mathbb{N} \to \mathbb{R}^+$ we will construct a number α and a sequence p_n/q_n of rational approximations such that

$$|\alpha - p_n/q_n| < \frac{1}{\psi(q_n)}, \qquad n \in \mathbb{N}.$$

Therefore even for an analytic function $t(v)$ (if it is not a trigonometric polynomial) one can find α such that $\sum \rho_n^2 = \infty$, and hence the homology equation has no solutions even in the class $L^2[0,1]$ (this was first observed in [**Win**]).

Moreover, Kolmogorov [**Ko**] noticed that for some α, an analytic time change can lead to a flow with continuous spectrum![7] Certainly, such a flow is not measure-theoretically isomorphic to a rectilinear flow (in other words, the homology equation for the function $\rho(v)$ has no measurable solutions). Anosov [**An73**] constructed (disproving a conjecture stated in [**Ko**]) an example of an analytic time change such that the homology equation has a measurable solution but has no solutions in $L^2[0,1]$.

Arnold [**Ar**] studied in detail how the smoothness of $\rho(v)$ depends on the smoothness of $t(v)$ and the choice of α. We give a corollary of his results formulated by Parry [**P72**]. Choose a positive function $\tau \in C^2(\mathbb{T}^2)$ normalized by $\overline{\tau} = 1$. Assume that α satisfies the condition

$$|n\alpha - m| > 1/|n|^{1+\varepsilon} \quad \text{for all } n, m \in \mathbb{Z},\ n \neq 0, \text{ and some } 0 < \varepsilon < 1/2.$$

Then the time change given by τ determines a flow which is topologically (that is, via a homeomorphism) and measure-theoretically isomorphic to the original flow $(\mathbb{T}^2, \mathbb{R}v_\alpha)$. Note that the set of such numbers $\alpha \in \mathbb{R}$ has full measure in $\mathbb{R}$ (see §29). This is in contrast with Stepin's result [**Ste82**] that numbers α such that the flow obtained is measure-theoretically isomorphic to the original flow form a first category set (provided that the C^2-smooth function $t(v) = \int_0^1 \tau(u, v + \alpha u)du$ does not degenerate to a trigonometric polynomial).

6. As for the horocycle flow on a compact surface, the structure of smooth time changes is not well understood (see [**R87b**]). Nevertheless, it is known that even a weak smoothness condition imposed on time changes for the horocycle flow preserves the measure rigidity property. More specifically, let Γ be a lattice in $G = \mathrm{SL}(2,\mathbb{R})$ and let the class $K(G/\Gamma) \subset L^2(G/\Gamma, \nu)$ consist of positive functions τ such that τ and $1/\tau$ are bounded and τ is a Hölder function in the direction of the rotation subgroup $\mathrm{SO}(2,\mathbb{R}) = \{c_t,\ t \in [-\pi, \pi]\} \subset \mathrm{SL}(2,\mathbb{R})$, that is, there exist constants $C, \alpha > 0$ such that

$$|\tau(x) - \tau(c_t x)| < C|t|^\alpha \quad \text{for all } x \in G/\Gamma,\ t \in [-\pi, \pi].$$

THEOREM 24.8 [**R86**]. *Let flows $(G/\Gamma_1, u_t^{\tau_1})$ and $(G/\Gamma_2, u_t^{\tau_2})$ be obtained from the horocycle flows by time changes with functions $\tau_i \in K(G/\Gamma_i)$, $i = 1, 2$, where $\overline{\tau}_1 = \overline{\tau}_2$. If the flows are measure-theoretically isomorphic, then the corresponding horocycle flows are isomorphic (and hence Γ_1 and Γ_2 are conjugate in G).*

We recall that according to [**R78**], any two horocycle flows $(G/\Gamma_1, u_t)$ and $(G/\Gamma_2, u_t)$ are Kakutani equivalent. In other words, there exists a measurable time change of the first flow that leads to a flow isomorphic to the second flow. But, as we have seen, this time change cannot belong to the class $K(G/\Gamma_1)$. Note that the proof is based on exponential mixing of the geodesic flow, which was established by Moore [**Mo87**] in the class of functions Hölder under rotations with an exponent $\alpha > 1/2$. Subsequently Ratner [**R87a**] removed the restriction on the exponent (see §27).

If one restricts the class of admissible time changes, then other rigidity properties are also preserved. More specifically, let the class $B^1(G/\Gamma)$ consist of positive C^1-functions τ bounded along with $1/\tau$. Then, as was shown in [**R87b**], time

[7] The full proof was given by Šklover [**Skl**]. Earlier, von Neumann constructed a similar example with a continuous time change.

changes within this class preserve the R-property of the horocycle flow and rigidity of quotients and joinings: they have an algebraic origin.

For other examples of unipotent flows (e.g., nilflows), similar results (even for C^∞-time changes) have not been proved. For instance (see [**R95**]), is it true that measure-theoretic isomorphism of two flows obtained by smooth time changes of ergodic unipotent flows implies that original flows are affinely equivalent?

§25. Divergent and bounded trajectories

1. First we find when there exist divergent (i.e., exiting to infinity) semiorbits of a homogeneous flow $(G/\Gamma, a_\mathbb{R})$ on a finite volume space. Surprisingly, for reductive subgroups $a_\mathbb{R}$ the solution is quite simple.

THEOREM 25.1 [**Da85**]. *If G/Γ is a noncompact space of finite volume and a subgroup $a_\mathbb{R} \subset G$ is reductive, then the flow $(G/\Gamma, a_\mathbb{R})$ has divergent semiorbits whenever it is ergodic.*

Before we prove the theorem, a few general remarks are in order. Let $G = (S \times K) \cdot R$ be a Levi decomposition of the group G. Since the space $KR/\Gamma \cap KR$ is compact, the problem reduces to the case where the group G is semisimple and has no compact components. But then the lattice $\Gamma \subset G$ can be assumed irreducible (and nonuniform). Certainly, the subgroup $a_\mathbb{R} \subset G$ is partially hyperbolic (otherwise the flow is nonergodic).

Proof of the theorem. Let U be the contracting horospherical subgroup for $a_\mathbb{R}$. By Theorem 13.1, the semiorbit $\{a_t g\Gamma,\ t \geq 0\}$ exits to infinity whenever the orbit $Ug\Gamma$ is not dense in G/Γ. But the unipotent flow $(G/\Gamma, U)$ on the noncompact space cannot be minimal (see §12)! □

The situation is more delicate if the unipotent part of the Jordan decomposition of the subgroup $a_\mathbb{R} \subset G$ is nontrivial. Then $\operatorname{rank}_\mathbb{R} G > 1$, and hence the lattice Γ is arithmetic. The existence of divergent semiorbits in this situation has not been established, although the case $G = \mathrm{SL}(n, \mathbb{R})$ was well studied by Dani [**Da85**]. First, the following holds.

THEOREM 25.2 [**Da85**]. *A flow $(\mathrm{SL}(n,\mathbb{R})/\mathrm{SL}(n,\mathbb{Z}), a_\mathbb{R})$ has no divergent semiorbits $\Longleftrightarrow$ the subgroup $a_\mathbb{R}$ is quasi-unipotent.*

Proof. Certainly, it suffices to find a divergent semiorbit for a partially hyperbolic subgroup $a_\mathbb{R}$. By Mahler's criterion F.1, it suffices to find a lattice $\Lambda = g(\mathbb{Z}^n)$ in $\mathbb{R}^n$ and a vector $e \in \Lambda$ such that $a_t(e) \to 0$. Since the subgroup $a_\mathbb{R}$ is partially hyperbolic, its contracting subspace

$$W = \{v \in \mathbb{R}^n : \ a_t(v) \to 0,\ t \to +\infty\}$$

is nontrivial. Hence it suffices to select a lattice $\Lambda = g(\mathbb{Z}^n)$ having nontrivial intersection with W. □

The divergent semiorbits found in the course of the proof of Theorem 25.2 can be viewed as "obvious". It is somewhat less evident that a semiorbit $\{a_t g\,\mathrm{SL}(n,\mathbb{Z}),\ t \geq 0\}$ exits to infinity if $\det(a_t g(\Sigma)) \to 0,\ t \to +\infty$, for a discrete subgroup $\Sigma \subset \mathbb{Z}^n$, $1 \leq \operatorname{rank}(\Sigma) \leq n-1$. In this case, by Minkowski's lemma (see §29), for any neighborhood $\Omega = O(0) \subset \mathbb{R}^n$, there exists $T > 0$ such that $a_t g(\Sigma) \cap \Omega \neq 0,\ t \geq T$. Hence by Mahler's criterion, $a_t g\,\mathrm{SL}(n,\mathbb{Z}) \to \infty,\ t \to +\infty$. Divergent semiorbits of this kind will be called *degenerate*. Roughly speaking, degenerate semiorbits exit to infinity very fast. One can show [**Da85**] that for $n = 2$ all divergent semiorbits are degenerate. On the other hand, for $n = 3$ and the simplest diagonal subgroup

$$a_t = \begin{pmatrix} e^t & 0 & 0 \\ 0 & e^t & 0 \\ 0 & 0 & e^{-2t} \end{pmatrix},$$

there exist nondegenerate divergent semiorbits (see §29).

2. Now we look for bounded orbits of a homogeneous flow $(G/\Gamma, g_{\mathbb{R}})$. We already know (see §12) that any unipotent flow has a compact invariant subspace. The same applies to reductive subgroups $g_{\mathbb{R}}$. Indeed, a standard argument leads us to the situation when Γ is an irreducible lattice in a semisimple group G with no compact components. The reductive subgroup $g_{\mathbb{R}}$ is contained in a Cartan subgroup $H \subset G$. But by the Prasad–Raghunathan Theorem E.5, the flow $(G/\Gamma, H)$ has invariant tori (which form an everywhere dense family in G/Γ).

For a "mixed" subgroup $g_{\mathbb{R}}$, the problem of finding bounded orbits turned out to be rather nontrivial, and their existence was only recently established by Kleinbock and Margulis [**KM96**]. The first informative results in this direction were obtained by Dani [**Da85,86c**], who studied the Hausdorff dimension of the family of bounded orbits for certain partially hyperbolic flows. Somewhat later (in Subsection 3) we will introduce all the definitions; right now we only note that the Hausdorff dimension of a manifold coincides with its usual dimension.

As we know, nondense orbits of an ergodic (quasi)unipotent flow form a singular set, which is a countable union of locally analytic sets. This implies that the singular set of such a flow has Hausdorff dimension strictly less than the dimension of the ambient space. In particular, the family of bounded orbits of an ergodic (quasi)unipotent flow on a noncompact space cannot have Hausdorff dimension equal to $\dim G$.

On the other hand, in §14 (following [**Da85**]) we have already found a criterion for trajectories of $g_t = \left(\begin{smallmatrix} e^t & 0 \\ 0 & e^{-t} \end{smallmatrix}\right)$ on $G/\Gamma = \mathrm{SL}(2,\mathbb{R})/\mathrm{SL}(2,\mathbb{Z})$ to be bounded. More specifically, if $p \in G$ is a lower-triangular matrix and $U_\alpha = \left(\begin{smallmatrix} 1 & \alpha \\ 0 & 1 \end{smallmatrix}\right)$, then the positive trajectory $\{g_t p U_\alpha \Gamma,\ t \geq 0\}$ is bounded $\iff$ the number α is badly approximable. It is well known that the set of badly approximable numbers is of Hausdorff dimension 1 (see [**Sch80**]) and hence the family of bounded positive semiorbits of the geodesic flow has full Hausdorff dimension (in this case 3). One can deduce the same for bounded (two-sided) orbits.

In [**Da85,86c**] Dani extended this result in two directions: either G is a semisimple group of $\mathbb{R}$-rank 1, or $G = \mathrm{SL}(n,\mathbb{R})$, $\Gamma = \mathrm{SL}(n,\mathbb{Z})$, and $\{g_t\}$ is a diagonal subgroup such that g_1 has only two distinct eigenvalues. This was substantially sharpened in [**KM96**], where the following criterion for the thickness (in the sense of Hausdorff dimension) of the family of bounded orbits was essentially found.

THEOREM 25.3. *The family of bounded orbits of an ergodic flow* $(G/\Gamma, g_{\mathbb{R}})$ *is of full Hausdorff dimension* $\iff$ *the maximal quotient space* $G/\overline{A\Gamma}$ *with zero entropy is compact (here* $A \subset G$ *is the Auslander normal subgroup for* $g_{\mathbb{R}}$*).*

We will prove this result in Subsection 3. Right now we make the following observation.

COROLLARY 25.4. *If* $\mathrm{vol}(G/\Gamma) < \infty$, *then every flow* $(G/\Gamma, g_{\mathbb{R}})$ *has bounded orbits.*

Proof. Without loss of generality we can assume that the flow in question is ergodic (see §10). Replacing G if necessary by $G' = C \cdot G$, where C is a torus in $\mathrm{Aut}(G)$, one can find a decomposition $g_t = c_t \times h_t$ such that $c_{\mathbb{R}} \subset C$ and all eigenvalues of Ad_{h_t} are purely real. Here C commutes with $h_{\mathbb{R}}$, and the projection $\alpha : G' = C \cdot G \to G$ along C induces a covering of flows $\alpha^* : (G'/\Gamma, h_{\mathbb{R}}) \to (G/\Gamma, g_{\mathbb{R}})$. If an orbit $h_{\mathbb{R}}x \subset G'/\Gamma$ is bounded, so is the orbit $g_{\mathbb{R}}\alpha^*(x) \subset G/\Gamma$. Hence one can assume that $g_{\mathbb{R}}$ has only purely real eigenvalues (with respect to the Ad-representation). Then the flow $(G/\overline{A\Gamma}, g_{\mathbb{R}})$ is unipotent, and by Theorem 12.10 it has a compact invariant subspace $Fx \subset G/\overline{A\Gamma}$. Let $\pi : G/\Gamma \to G/\overline{A\Gamma}$. Then by Theorem 25.3, the family of bounded orbits in the inverse image $\pi^{-1}(Fx) \subset G/\Gamma$ has full Hausdorff dimension. □

3. Here we follow the paper of Kleinbock and Margulis on the Hausdorff dimension of the families of bounded and nondense orbits. First we give the definition. Let E be a Borel subset of $\mathbb{R}^n$ and let $B(x, r) \subset \mathbb{R}^n$ be the ball of radius r centered at $x \in \mathbb{R}^n$. We fix $\alpha, \varepsilon > 0$ and define

$$\Lambda_\alpha^\varepsilon(E) = \inf\Big\{\sum_{j=1}^{\infty} c_j^\alpha : E \subset \bigcup_j B(x_j, c_j);\ c_j < \varepsilon\Big\}.$$

Clearly, $\Lambda_\alpha^\varepsilon$ increases as ε decreases, and hence the number

$$\Lambda_\alpha(E) = \lim_{\varepsilon \to 0} \Lambda_\alpha^\varepsilon(E) \le +\infty,$$

called the *α-dimensional Hausdorff measure* of E, is well defined. The *Hausdorff dimension* is given by

$$\dim_H E = \inf\{\alpha : \Lambda_\alpha(E) = 0\} = \sup\{\alpha : \Lambda_\alpha(E) = +\infty\}.$$

It is easy to check that $\dim_H E \le \dim_H \mathbb{R}^n = n$. In addition, $\dim_H \bigcup_i E_i = \sup_i \dim_H E_i$.

Similarly, one defines the Hausdorff dimension for subsets in metric spaces.

The simplest example of a set with fractional Hausdorff dimension is constructed as follows. Let $0 < \delta < 1$. We define a sequence of closed subsets $[0,1] = K_0(\delta) \supset K_1(\delta) \supset \dots$, where $K_{i+1}(\delta)$ is obtained from $K_i(\delta)$ by discarding the central interval of relative length δ in each segment appearing in $K_i(\delta)$. Then $K_i(\delta)$ is a union of 2^i segments of length $(\frac{1-\delta}{2})^i$. Consider the closed nowhere dense set $K(\delta) = \bigcap_i K_i(\delta)$. Since $l(K_i(\delta)) = (1-\delta)^i$, it follows that $K(\delta)$ is a null set. However, one can verify that

$$\dim_H K(\delta) = \frac{\ln 2}{\ln \frac{2}{1-\delta}}.$$

Taking $\bigcup_i K(1/i)$, one obtains a Borel set of measure 0 and of Hausdorff dimension 1.

A similar example can be easily constructed in higher dimensions. For instance, take a rectangle in $\mathbb{R}^2$ and at each step discard from every rectangle the central cross of relative area $0 < \delta < 1$.

One of useful properties of the Hausdorff dimension is that $\dim_H(A \times B) \ge \dim_H A + \dim_H B$. A stronger result also holds.

LEMMA 25.5 (Marstrand, see [**KM96**]). *Let $C \subset \mathbb{R}^n \times \mathbb{R}^m$, where*

$$\dim_H(C \cap (x \times \mathbb{R}^m)) \geq \beta$$

for all x in a Borel subset $A \subset \mathbb{R}^n$ of Hausdorff dimension n. Then $\dim_H C \geq n+\beta$.

Now we are ready to outline the proof of the main result from [**KM96**].

THEOREM 25.6 [**KM96**]. *Let G be a connected semisimple Lie group with no compact components, Γ an irreducible lattice in G, and F a partially hyperbolic one-parameter subgroup of G. Then the set of bounded orbits of the flow $(G/\Gamma, F)$ is of Hausdorff dimension* $\dim G$.

If $F = g_{\mathbb{R}}$, we let $F^+ = \{g_t,\ t \geq 0\}$, $F^- = \{g_t,\ t \leq 0\}$. Denote by $H, H^-, Q \subset G$ the expanding, contracting and neutral subgroups for $g_1 \in F$. Since the group G is semisimple, we have $0 < \dim H = \dim H^- = k < \dim G$. Define

$$\Phi_t(h) = g_t h g_{-t}.$$

Then Φ_t expands H with exponential speed in $t > 0$ (and contracts H^- exponentially fast).

Let $C, C^+, C^- \subset G/\Gamma$ stand for the sets of bounded orbits under the actions of F, F^+, F^- respectively. Then $C = C^+ \cap C^-$ and $C^+ = H^-C^+$, $C^- = HC^-$. Since locally $G = HQH^-$, Theorem 25.6 is easily deduced from Lemma 25.5 and the following result.

THEOREM 25.7 [**KM96**]. *For any $x \in G/\Gamma$ and any open subset $V \subset H$ one has*

$$\dim_H\{h \in V : \textit{the semiorbit } F^+hx \textit{ is bounded}\} = k.$$

We choose a point $x \in G/\Gamma$ and an open subset $V \subset H$. Under the action of $g_t \in F^+$, the area element $Vx \subset G/\Gamma$ is sent to the area element $g_tVx = \Phi_t(V)g_tx$ being expanded in all directions. To prove Theorem 25.7, we need an analogue of the uniform distribution property for the family of area elements g_tVx as $t \to +\infty$ uniformly in all x from a compact set $L \subset G/\Gamma$. If the subgroup F is reductive, then Φ_t acts isometrically on Q, and contracts V^- for $t > 0$. In this case the uniform distribution property does hold. This means that, given $f \in L^2(H)$ with compact support, a uniformly continuous function $\psi \in L^2(G/\Gamma)$, a point $x \in G/\Gamma$, and $\varepsilon > 0$, there exists $T_0 > 0$ such that

$$(*) \qquad \left| \int_H f(h)\psi(g_thx)dm_h - \int_H f dm \int_{G/\Gamma} \psi d\nu \right| < \varepsilon$$

for all $t \geq T_0$, where m and ν are the Haar measures on H and G/Γ. The proof involves the facts that the flow $(G/\Gamma, F)$ is mixing and the action of Φ_t on QV^- is nonexpanding for $t > 0$. Results of this kind were obtained already in [**Ma70**] and [**Mar75**]; in [**Bo76**] and [**EP**] they were used to prove that the horospherical flow on a compact space G/Γ is uniquely ergodic. We have to extend the property in different directions. First, it is easy to show that the parameter T_0 in (*) can be chosen the same for all x from a given compact set $L \subset G/\Gamma$. Second, instead of f and ψ one can take characteristic functions of *good* subsets in H and G/Γ (more precisely, of bounded sets with boundaries of zero measure). Third, we need a similar statement for a nonreductive F. The main difficulty here is that the action of Φ_t on the neutral subgroup is not isometric, because the unipotent part need not be trivial.

To avoid this obstacle, two possibilities are considered in [**KM96**]. Let $G' \subset G$ be the product of all simple components such that the projection of $g_{\mathbb{R}}$ to each of them is partially hyperbolic. We call G' the *essential* component. Assume that the projection of $g_{\mathbb{R}}$ to G' is a reductive subgroup; in this case F is called *essentially reductive*. If one applies the Dani–Margulis Homecoming Theorem 20.5 to the unipotent part of F and looks at the proof of inequality (*), then a weaker statement can be proved: (*) is satisfied not for all $t \geq T_0$ but for most values of the parameter $t \in [T, 2T]$ with $T \geq T_0$. This works, for instance, in the case when all essential simple components have $\mathbb{R}$-rank 1 (because a one-parameter subgroup of an $\mathbb{R}$-rank 1 group is either reductive or quasi-unipotent). On the contrary, if all essential simple components have $\mathbb{R}$-rank at least 2, one manages to prove exponential mixing of the flow $(G/\Gamma, F)$ (see §27), whence the inequality (*) follows for all $t \geq T_0$. Let us summarize the results.

LEMMA 25.8 [**KM96**]. *Let $V \subset H$ be a bounded open set and let $K \subset G/\Gamma$ be a compact set such that $m(\partial V) = \nu(\partial K) = 0$. Then for any compact subset $L \subset G/\Gamma$ and any $\varepsilon, \sigma > 0$ there exists $T_0 > 0$ such that for all $x \in L$ and all $T > T_0$ we have*

$$l\{t \in [T, 2T] : \ |m\{h \in V : \ g_t hx \in K\} - m(V)\nu(K)| \leq \varepsilon\} \geq (1-\sigma)T.$$

We proceed to the proof of Theorem 25.7. Let $K = L$ be a compact with nonempty interior and null boundary. We fix a point $x \in G/\Gamma$ and a good enough neighborhood of the identity element $V \subset H$. Then for every $y \in K$ there exists $T \leq t(y) \leq (1+\sigma)T$ such that

$$(**) \qquad |m\{h \in V : \ g_{t(y)} hy \in K\} - m(V)\nu(K)| \leq \varepsilon.$$

Let us prove that the Hausdorff dimension of points in Vx whose positive trajectories visit the compact set K almost periodically is equal to $\dim(H) = k$.

For this purpose we take an open set $V \subset H$ with null boundary whose translates form a *tesselation* of H. More specifically, there exists a countable subset $\Lambda \subset H$ such that

$$m(V\gamma \cap V\gamma') = 0, \quad \gamma, \gamma' \in \Lambda, \ \gamma \neq \gamma', \qquad \text{and} \qquad H = \bigcup_{\gamma \in \Lambda} \overline{V}\gamma.$$

An example is given by a fundamental region V for the right action of a lattice $\Lambda \subset H$. However, not every nilpotent Lie group contains a lattice. Nevertheless, by induction on dimension one can prove (see [**KM96**]) that a tesselation (V, Λ) for a nilpotent group always exists (and the open set V can be made arbitrarily small).

We fix a tessellation (V, Λ) of the group H and select a compact set $K = L$ such that $\overline{V}x \subset K$ and $\nu(\partial K) = 0$. The closure $\overline{V} \subset H$ will be called a *rectangle*. Let $t(x) \in [T, (1+\sigma)T]$ satisfy (**). The rectangle $\overline{V}$ (for large enough T) contains many small rectangles of the form

$$A = \Phi_{-t(x)}(\overline{V}\gamma), \qquad \gamma \in \Lambda.$$

We mark those of them for which $g_{t(x)}Ax \subset K$, and let $A_1(x) \subset \overline{V}$ be the union of all marked rectangles. Fix a small parameter $\varepsilon > 0$. By Lemma 25.8, one can prove (see [**KM96**]) that for large enough $T = T(\varepsilon)$, the relative measure of the marked subset $A_1(x)$ in $\overline{V}$ is at least $\nu(K) - 2\varepsilon$.

We perform the same procedure in each rectangle area

$$g_{t(x)}Ax = \Phi_{t(x)}Ag_{t(x)}x = \overline{V}\gamma g_{t(x)}x \subset K.$$

Thus, we break each rectangle $A \subset \overline{V}$ marked in the first step into small rectangles A' and mark those of them for which $g_{t(y)}A'y \subset K$, where $y = \gamma g_{t(x)}x$.

By induction, we obtain a sequence of closed sets

$$\overline{V} \supset A_1(x) \supset A_2(x) \supset \dots,$$

where the relative measure of $A_{i+1}(x)$ in $A_i(x)$ is at least $\nu(K) - 2\varepsilon$. We set $A_\infty(x) = \bigcap_i A_i(x)$. Using an estimate of McMullen and Urbansky for Hausdorff dimension, it is proved in [**KM96**] that

$$\dim_H(A_\infty(x)) > k - \frac{\log(\nu(K) - 2\varepsilon)}{\log c_T},$$

where $c_T < 1$ is the contraction bound for the right-invariant metric dist under the action of Φ_{-T}, that is,

$$c_T = \sup_{g,h\in V,\ g\neq h} \frac{\operatorname{dist}(\Phi_{-T}(g), \Phi_{-T}(h))}{\operatorname{dist}(g,h)}.$$

Note that $A_\infty(x)$ consists of points $h \in V$ such that the semiorbit F^+hx belongs to the compact set $g_{[0,(1+\sigma)T]}K$. If $T \to \infty$, then $c_T \to 0$ and hence $\dim_H(A_\infty(x)) \to k$. It follows that the Hausdorff dimension of the set of elements $h \in V$ with bounded semiorbit F^+hx is equal to k. □

Now we can prove Criterion 25.3. Without loss of generality we may assume that the group G is semisimple and has no compact components. Let $G' = G/(\overline{A\Gamma})_0$ and $\Gamma' = \overline{A\Gamma}/(\overline{A\Gamma})_0$. Assume that the space G'/Γ' is noncompact. Since the projection of $g_\mathbb{R}$ to G' is quasi-unipotent, the family of its bounded orbits cannot have full Hausdorff dimension. The rest is evident. □

Note that a similar criterion can be found in the nonergodic case, but the formulation is more involved (see [**KM96**]).

4. Actually, Theorem 25.6 was proved in [**KM96**] in a stronger form (which makes sense for compact spaces as well). More specifically, let $Z \subset G/\Gamma$ be a closed F-invariant subset of zero measure. Then, under the assumptions of the theorem (when $(G/\Gamma, F)$ is a K-flow on a semisimple homogeneous space), the family of points $x \in G/\Gamma$ such that the closure $\overline{Fx}$ is compact and disjoint from Z has full Hausdorff dimension. In particular, the family of *nondense* orbits of the flow $(G/\Gamma, F)$ is of full Hausdorff dimension (since a partially hyperbolic flow cannot be minimal, the set $Z \subset G/\Gamma$ in question always exists). This easily implies a similar result for a homogeneous flow $(G/\Gamma, g_\mathbb{R})$ provided that the projection of $g_\mathbb{R}$ to a Levi subgroup L of the group $G = L \cdot R$ is partially hyperbolic. However, this condition can be relaxed as well.

THEOREM 25.9 [**Kl98a**]. *Nondense orbits of a partially hyperbolic homogeneous flow on a finite volume space form a set of full Hausdorff dimension.*

Previously the following related result was known.

THEOREM 25.10 [**Da88**]. *There exists a subset $W \subset \mathbb{T}^n$ of full Hausdorff dimension such that for any point $x \in W$ and any semisimple automorphism $\sigma \in \mathrm{Aut}(\mathbb{T}^n)$, the orbit closure $\overline{\sigma^{\mathbb{Z}}x}$ does not contain periodic points.*

Recall (see §4) that for an ergodic toral automorphism, the family of periodic orbits is exactly $\mathbb{Q}^n/\mathbb{Z}^n$.

§26. Ergodic measures and orbit closures for actions of arbitrary subgroups

1. We know that, given a connected subgroup $H \subset G$ generated by unipotent elements, all finite ergodic measures and orbit closures on G/Γ (if $\mathrm{vol}(G/\Gamma) < \infty$) have an algebraic origin. If H is generated by quasi-unipotent elements, then as in Lemma 21.2 it is not difficult to construct a compact extension G^* of the group G such that our subgroup H belongs to a compact extension of a connected subgroup $U \subset G^*$ generated by unipotent elements. This easily implies that any finite H-ergodic measure is "good": it is supported on a smooth submanifold in G/Γ and is a smooth measure thereon. Likewise, if G/Γ is of finite volume, all closures of H-orbits are smooth (and even analytic) submanifolds in G/Γ.

Suppose now that a connected subgroup H contains no (quasi)unipotent elements at all. Certainly, if $\dim H = 1$ (that is, for a partially hyperbolic flow) one cannot ensure that all ergodic measures and orbit closures are to some extent good. But for a large enough subgroup $H \subset G$, the situation can be much better. For instance, let G be a semisimple Lie group and let H be its maximal $\mathbb{R}$-diagonalizable subgroup, where $\dim H = \mathrm{rank}_{\mathbb{R}} G \geq 2$. Then, given an irreducible lattice $\Gamma \subset G$, one may hope that ergodic measures and closures of H-orbits have an algebraic origin; see [**Ma91b**]. However, as an example of Rees [**Ree**] shows, one has to be careful. This example gives a uniform lattice $\Gamma \subset G = \mathrm{SL}(3,\mathbb{R})$ which intersects two commuting subgroups $L, D \subset G$ in lattices, where L is isomorphic to $\mathrm{SL}(2,\mathbb{R})$, and D is one-dimensional and $\mathbb{R}$-diagonalizable. It is clear that, taking the product of a diagonal subgroup $D_L \subset L$ with D, we obtain a maximal $\mathbb{R}$-diagonalizable subgroup of G. Now any "bad" D_L-ergodic measure for the geodesic flow $(L/\Gamma \cap L, D_L)$ produces a "bad" $D_L \times D$-ergodic measure for the two-dimensional flow on G/Γ. The same applies to orbit closures. For these reasons, Katok and Spatzier formulated in [**KS96**] the following conjecture. If H is a maximal $\mathbb{R}$-diagonalizable subgroup of a semisimple Lie group G, and Γ is a lattice in G, then a finite H-ergodic measure μ on G/Γ either is algebraic or has the following origin: H-action on an invariant closed homogeneous subspace projects to a one-parameter quotient flow and the measure μ is obtained by lifting an ergodic measure of this quotient flow. Using Ratner's measure theorem and some additional assumptions, Katok and Spatzier in [**KS96**] proved that H-ergodic measures of positive entropy are ergodic, but in the general case the problem is open.

Note that for the nonuniform lattice $\Gamma = \mathrm{SL}(3,\mathbb{Z}) \subset G = \mathrm{SL}(3,\mathbb{R})$, the conjecture about the algebraicity of orbit closures of the diagonal subgroup $H \subset G$ has not been disproved. Moreover, as Margulis observed (implicitly this was done already in [**CS**]), it has strong applications in number theory. More specifically, if one proves that all bounded H-orbits in G/Γ are compact, then a well-known conjecture of Littlewood will be proved (see §30). There is an impression that further progress in the theory of homogeneous flows will be provided by in-depth study of actions of n-dimensional $\mathbb{R}$-diagonalizable subgroups on homogeneous spaces of a semisimple Lie group of $\mathbb{R}$-rank $n > 1$. These are multi-parameter Anosov actions whose dynamics contrasts drastically from that of Anosov $\mathbb{R}^1$-flows (see [**KS94,96**]).

In this connection we note that recently Lindenstrauss and Weiss [**LW**] have proved that an orbit closure of the diagonal subgroup on $\mathrm{SL}(,\mathbb{R})/\mathrm{SL}(n,\mathbb{Z})$, $n \geq 3$, is algebraic whenever it contains a compact orbit. Moreover, for n prime, an orbit whose closure contains a compact orbit is either compact itself or everywhere dense.

2. Now we consider the "mixed" case where H is generated by unipotent and $\mathbb{R}$-diagonalizable elements (see also [**R94,95b**], [**B**]). The simplest case is $H = AU$, where a one-dimensional $\mathbb{R}$-diagonalizable subgroup A is "diagonal" relative to a one-dimensional unipotent subgroup U (see §17). Then, according to Ratner's results (see Theorem 17.5), any finite H-ergodic measure μ is invariant (and, of course, ergodic) under a subgroup $\mathrm{SL}_2(U, A) \subset G$ containing H and locally isomorphic to $\mathrm{SL}(2,\mathbb{R})$. Since $\mathrm{SL}_2(U, A)$ is generated by unipotent elements, we have no problems: the measure μ is algebraic.

This suggests the following generalization.

THEOREM 26.1 [**R94,97**]. *Assume that a connected subgroup $U \subset G$ is generated by unipotent elements and that subgroups $A_1, \dots, A_n \subset G$ are "diagonal" for unipotent subgroups $U_1, \dots, U_n \subset U$ respectively. Let $H \subset G$ be generated by U and all $A_1, \dots, A_n$. Then each finite H-ergodic measure on G/Γ is algebraic. If* $\mathrm{vol}(G/\Gamma) < \infty$, *then all closures of H-orbits on G/Γ are homogeneous subspaces of finite volume in G/Γ.*

It is clear that ergodic measures and orbit closures for H are those for the subgroup $H' \subset G$ generated by U and all $\mathrm{SL}_2(U_i, A_i)$.

In particular, the conditions of the theorem are satisfied if $H \subset G$ is a parabolic subgroup of a semisimple group $G_1 \subset G$ with no compact components.

The measure part of Theorem 26.1 for algebraic groups was generalized by Mozes as follows. Let $G \subset \mathrm{SL}(n,\mathbb{R})$ be an algebraic group. According to Bien and Borel [**BB**], a subgroup $H \subset G$ is called *epimorphic* in G if H-fixed vectors for any algebraic (that is, finite-dimensional polynomial) representation of G are also G-fixed. It is known [**Rag**] that a parabolic subgroup of a semisimple Lie group with no compact components is epimorphic. If the group G in Theorem 26.1 is algebraic, then its subgroups H and H' are "almost algebraic" (are of finite index in their Zariski closures). It follows that the group H is epimorphic relative to the algebraic group $H' \subset G$ generated by U and all $\mathrm{SL}_2(U_i, A_i)$. Hence the following result of Mozes actually generalizes the measure part of Theorem 26.1.

THEOREM 26.2 [**Moz95a**]. *Let G be an algebraic Lie group with a discrete subgroup Γ. Assume that $H \subset H' \subset G$ are algebraic subgroups, where H' is generated by unipotent elements and H is epimorphic relative to H'. Then each finite H-ergodic measure on the space G/Γ is H'-invariant and hence algebraic.*

It is of interest that according to [**BB**], every simple noncompact Lie group G, $\dim G > 3$, contains a 3-dimensional solvable epimorphic subgroup H. Clearly, this subgroup acts uniquely ergodically on any quotient space of G of finite volume. Moreover, the unipotent radical U of H, $\dim U = 2$, acts uniquely ergodically on any compact quotient space of G of finite volume. This follows from the paper [**W99**] of Weiss, where Theorem 26.2 is sharpened in the situation when Γ is a uniform lattice in G.

Under certain assumptions, Theorem 26.2 is valid for nonalgebraic groups $H \subset H' \subset G$ (if $\mathrm{Ad}_G H$ is epimorphic in $\mathrm{Ad}_G H'$). For a precise formulation of this generalization of Theorem 26.2 the reader is referred to the paper [**SW**] of Shah and Weiss. Among other things, this paper contains also the topological counterpart of Mozes' theorem.

THEOREM 26.3 [**SW**]. *Let G be an algebraic Lie group with a lattice Γ. Assume that $H \subset H' \subset G$ are algebraic subgroups, where H is epimorphic relative to H'. Then $\overline{Hx} = \overline{H'x} \subset G/\Gamma$ for each point $x \in G/\Gamma$. In particular, if H' is generated by unipotent elements, then all H-orbits on G/Γ have algebraic closures.*

Earlier, Weiss [**W98**] obtained a particular case of the theorem: if G is an algebraic $\mathbb{Q}$-group with no nontrivial $\mathbb{Q}$-characters and $H \subset G$ is an algebraic epimorphic subgroup, then the action $(G/G_{\mathbb{Z}}, H)$ is minimal. As Raghunathan noticed (see [**W98**]), this is not the case if the epimorphic subgroup $H \subset G$ fails to be algebraic.

Example 26.4. Let

$$G = \mathrm{SL}(2,\mathbb{R}) \times \mathrm{SL}(2,\mathbb{R}), \quad \Gamma = \mathrm{SL}(2,\mathbb{Z}) \times \mathrm{SL}(2,\mathbb{Z}), \quad H = g_{\mathbb{R}} \cdot (U_1 \times U_2),$$

where U_i is the unipotent subgroup of strictly upper-triangular matrices in the corresponding copy of $\mathrm{SL}(2,\mathbb{R})$, and let

$$g_t = \begin{pmatrix} e^t & 0 & 0 & 0 \\ 0 & e^{-t} & 0 & 0 \\ 0 & 0 & e^{-\alpha t} & 0 \\ 0 & 0 & 0 & e^{\alpha t} \end{pmatrix},$$

where $\alpha > 0$ is an irrational number. Note that the conjugation $x \to g_t x g_{-t}$ expands U_1 and contracts U_2, and the orbits $U_1\Gamma$ and $U_2\Gamma$ are compact. Hence the orbit $H\Gamma \subset G/\Gamma$ is closed (otherwise an orbit of some subgroup U_i would degenerate to a point). On the other hand, since the number α is irrational, the subgroup H is Zariski dense in the 4-dimensional parabolic subgroup $B \subset G$ and hence epimorphic in G. Note that in this example the homogeneous subspace $H\Gamma$ is of infinite volume. Therefore the action $(G/\Gamma, H)$ is not minimal. Nevertheless, as was noticed in [**W98**], it is *uniquely ergodic*. This follows from Theorem 26.2 of Mozes (the proof for a solvable subgroup H goes through if the subgroup splits and is not necessarily algebraic).

Assume now that a subgroup $H \subset G$ is generated by elements $g \in G$ such that all eigenvalues of the operator Ad_g are real and positive. Denote by $H_u \subset H$ the normal subgroup generated by all unipotent one-parameter subgroups of H. The following result of Margulis and Tomanov reduces the study of H-ergodic measures on G/Γ (where the subgroup $\Gamma \subset G$ is discrete) to the case when the subgroup H_u is trivial. Consider the decomposition of an H-ergodic invariant probability measure μ into algebraic H_u-ergodic components. The following theorem holds.

THEOREM 26.5 [**MT96**]. *In the above situation there exists a connected closed subgroup $L \subset G$ such that almost every H_u-ergodic component is an L-invariant measure on a closed L-orbit. If $P \subset G$ is the group of all elements in G that normalize L and preserve its Haar measure, then $H \subset P$ and $\mu(Px) = 1$ for some point $x \in G/\Gamma$.*

Since the subgroup L is normal in P, the space X of L-orbits in Px is a homogeneous P/L-space. The measure μ, being L-invariant, projects to an A-invariant ergodic measure μ' on X, where $A = HL/L$. It is easy to see that μ' is algebraic $\Longleftrightarrow$ μ is algebraic (note that the group A is Abelian whenever H is connected). It is also shown in [**MT96**] how to deduce Mozes' Theorem 26.2 from Theorem 26.5.

Let us describe a result of Shah [**Sh96**] which also deals with actions of arbitrary subgroups. Assume that G_1 is a semisimple subgroup with no compact components in a Lie group G, $a \in G_1$ is a semisimple element, and $U \subset G_1$ is the expanding subgroup of G_1 for the element a (that is, $U = G^+ \cap G_1$, where G^+ is the expanding horospherical subgroup of G). Let Γ be a lattice in G and let $\pi : G \to G/\Gamma$ be the natural projection. Without loss of generality we can assume that U is not contained in a proper normal subgroup of G_1 and that $\overline{G_1\Gamma} = G/\Gamma$. Now consider any rectangular domain $B \subset U$ of finite positive Haar measure m on U. If $n \to +\infty$, then the translation $B\Gamma \to a^n B\Gamma = (a^n B a^{-n}) a^n \Gamma$ expands the domain B. It turns out that the sequence of domains gets more and more uniformly distributed in the finite volume space G/Γ. More precisely, if $\lambda = m|_B$ is the restriction of Haar measure on U to B, then the sequence of finite measures $a^n \pi(\lambda)$ on G/Γ converges to G-invariant measure in the weak* topology. For $G = G_1$ this fact was well known earlier, but in the general case the result is nontrivial. Of course, it implies that the orbit of the point $\Gamma \in G/\Gamma$ under the subgroup $a^{\mathbb{Z}}U$ is dense (though the same follows from Theorem 26.1).

3. Note that the results described above enable one to examine the structure of ergodic measures and orbit closures for the action of a connected subgroup $H \subset G$ on a finite volume space G/Γ, where G is a semisimple Lie group of $\mathbb{R}$-rank 1. Denote by $A \subset G$ a maximal $\mathbb{R}$-diagonalizable subgroup, $\dim A = 1$, and by $U \subset G$ the expanding horospherical subgroup for A. Also let $K_1 \subset G$ be the compact part of the centralizer of A. Then $P = (K_1 \times A) \cdot U$ is a parabolic subgroup of G. If a closed connected subgroup $H \subset G$ is not generated by quasi-unipotent elements, then it is conjugate to a subgroup of P. It is clear that among closed noncompact subgroups of P only reductive subgroups can have (and actually do have) "bad" ergodic measures and orbit closures (this can be derived either from Ratner's Theorem 26.1 or from the result of Shah above). Therefore the criterion here is as follows: if $H \subset G$ is a closed noncompact connected subgroup, then G/Γ carries a "bad" H-ergodic measure $\Longleftrightarrow$ the group H is conjugate to a subgroup of $K_1 \times A$. The same holds for orbit closures as well.

A similar problem for solvable Lie groups is still open. One of the most interesting cases here is the structure of orbits and ergodic measures for a pair of commuting ergodic toral automorphisms, one of which is Anosov. If the action admits a one-generator group of ergodic toral automorphisms, then the original action has "bad" ergodic measures and orbit closures. If there are no such quotients, then all minimal sets are finite (Berend, [**Be**]) and hopefully all orbits are good. Interesting results on the algebraicity of ergodic measures for such actions have been obtained by Katok and Spatzier [**KS96**].

4. We know that unique ergodicity and minimality for a homogeneous action $(G/\Gamma, H)$ on a finite volume space are equivalent if the connected subgroup H is generated by unipotent elements (corollary of Ratner's results), or if $\dim H = 1$ (Theorem 21.11). On the other hand, the following result holds.

LEMMA 26.6 [**St99b**]. *If the group H is connected, then every H-invariant compact subset $M \subset G/\Gamma$ carries an H-invariant probability measure.*

Indeed, let $H = (K_H \times S_H) \cdot R_H$ be a Levi decomposition for H (here as usual $K_H \times S_H$ is the decomposition of a Levi subgroup into compact and noncompact parts, and R_H is the radical of H). Let $P \subset S_H$ be a parabolic subgroup of S_H. Then the group $A = (K_H \times P) \cdot R_H$ is amenable, and hence M carries an A-invariant probability measure. Note that the subgroup P is epimorphic in S_H, and S_H is generated by Ad_G-unipotent elements. By Mozes' Theorem 26.2, any P-invariant probability measure is S_H-invariant.[8] □

As a corollary, we deduce that the action $(G/\Gamma, H)$ is minimal whenever it is uniquely ergodic and the space G/Γ is compact. As Example 26.4 shows, this does not hold in the general case. However, apparently the converse statement holds. In this direction there is the following result of Mozes and Weiss.

THEOREM 26.7 [**MW**]. *Let G be a semisimple $\mathbb{R}$-algebraic group, let H be $\mathbb{R}$-algebraic subgroup of G, and let* $\mathrm{vol}(G/\Gamma) < \infty$. *Then the action $(G/\Gamma, H)$ is uniquely ergodic whenever it is minimal.*

Simultaneously and independently, this theorem was generalized by Starkov and Tomanov.

THEOREM 26.8 [**St99b, T**]. *Let G be a connected Lie group, let H be a connected subgroup of it, and let* $\mathrm{vol}(G/\Gamma) < \infty$. *Assume that* $\mathrm{Ad}(H) \subset \mathrm{Aut}(\mathfrak{g})$ *splits into the semidirect product of a reductive subgroup and the unipotent radical. Then the action $(G/\Gamma, H)$ is uniquely ergodic whenever it is minimal.*

In the course of the proof, the paper [**St99b**] suggests a reduction of the study of finite ergodic measures for any action $(G/\Gamma, H)$, where $H \subset G$ is connected and $\Gamma \subset G$ is discrete, to the case when H is an Abelian subgroup that contains no unipotent elements (thus improving Theorem 26.5).

§27. Exponential mixing

Let $(G/\Gamma, a_{\mathbb{R}})$ be a mixing flow. Recall that, given functions $\phi, \psi \in L^2(G/\Gamma)$ with zero average (that is, orthogonal to constants), the correlation coefficient

$$(a_t\phi, \psi) = \int_{G/\Gamma} \phi(a_{-t}x)\overline{\psi(x)}d\nu_x$$

converges to zero as $t \to \infty$. In the class $L^2(G/\Gamma)$ (and even in the class of continuous functions) the rate of convergence can be arbitrarily slow. But for sufficiently smooth functions the correlation coefficient can decay exponentially fast. In other words, for such functions ϕ, ψ there exist constants $C > 0$ and $\lambda > 1$ such that

$$|(a_t\phi, \psi)| < C\lambda^{-|t|} \quad \text{for all } t \in \mathbb{R}.$$

As a rule, for applications it suffices to establish the fact in the class $C^\infty(G/\Gamma)$, although sometimes one needs to prove it in the class of Hölder functions.

---

[8]Rigorously speaking, Mozes' theorem can be applied only if G embeds into an algebraic group (in this case S_H and P are also algebraic). In the general case one can use results of [**SW**].

1. Consider, for example, an Anosov toral automorphism $A \in \mathrm{SL}(2, \mathbb{Z})$. We recall that its suspension is a solvable flow $(G/\Gamma, a_{\mathbb{R}})$, where $G = a_{\mathbb{R}} \cdot \mathbb{R}^2$, $\Gamma = A^{\mathbb{Z}} \cdot \mathbb{Z}^2$ and $a_1 = A$. Since this flow is not mixing, one estimates decay of correlations in the class of functions orthogonal to constant functions along invariant tori $\mathbb{R}^2 h\Gamma \subset G/\Gamma$. This problem reduces to the study of the A-action on the space of functions on $\mathbb{T}^2$ with zero average.

Let ϕ and ψ be two functions with zero average on the torus $\mathbb{T}^2$ of the class $C^{1+\varepsilon}$ (that is, their partial derivatives of the first order are Hölder functions with exponent $\varepsilon > 0$). Let

$$\phi(\overline{x}) = \sum_{I \in \mathbb{Z}^2} \phi_I e^{2\pi i (I, x)}$$

be the Fourier expansion (we recall that $\phi_0 = 0$, and $\overline{x} \in \mathbb{T}^2$ stands for the image of $x \in \mathbb{R}^2$ in the torus). This series converges absolutely, i.e., $|\phi|_1 = \sum_I |\phi_I| < \infty$, and there exist constants $c, \alpha > 0$ such that $|\phi_I| < c/\|I\|^\alpha$ for all $I \in \mathbb{Z}^2 - \{0\}$, where $\|x\|$ denotes any suitable norm on $\mathbb{R}^2$. This applies to the function ψ as well, and for simplicity we assume that the constants c and α are the same as those for ϕ. Recall that

$$A\phi(\overline{x}) = \phi(A^{-1}\overline{x}) = \sum_I \phi_{\mathcal{A}I} e^{2\pi i (I, x)},$$

where $\mathcal{A} = A^\intercal \in \mathrm{SL}(2, \mathbb{Z})$ stands for the transpose matrix.

With respect to the hyperbolic automorphism $\mathcal{A}$, the plane $\mathbb{R}^2$ decomposes into the sum $\mathbb{R}^2 = V^+ + V^-$ of expanding and contracting lines; that is, $x = x^+ + x^-$, where $x^+ \in V^+$, $x^- \in V^-$. Let $\|x\| = |x^+| + |x^-|$. The action of $\mathcal{A}$ on the open complement $\mathbb{R} - (V^+ \cup V^-)$ breaks the latter into (two-sided) sectors such that $\mathcal{A}$ maps each sector to an adjacent sector (say, in the clockwise direction). We select the sector S containing the bisector of the angle between V^+ and V^-. Then there exist $b > 0$ and $\lambda > 1$ such that $\|\mathcal{A}^i x\| \geq b\lambda^{|i|}\|x\|$ for all $x \in S$ and all $i \in \mathbb{Z}$ (here $\lambda > 1$ is the absolute value of the eigenvalue of $\mathcal{A}$ out of the unit circle). In addition, every $\mathcal{A}$-orbit contains precisely one point from S.

Note that

$$(A^n\phi, \psi) = \sum_{I \in \mathbb{Z}^2 - 0} \phi_{\mathcal{A}^n I} \psi_I = \sum_{J \in S \cap \mathbb{Z}^2} \sum_{l \in \mathbb{Z}} \phi_{\mathcal{A}^{n+l} J} \psi_{\mathcal{A}^l J}.$$

Decompose the sum over $l \in \mathbb{Z}$ into sums over $l > -n/2$ and $l \leq -n/2$. For $l > -n/2$ we have

$$|\phi_{\mathcal{A}^{n+l} J}| < \frac{c}{\|\mathcal{A}^{n+l} J\|^\alpha} < \frac{c}{(\lambda^{n+l} b \|J\|)^\alpha} < C\lambda^{-n\alpha/2}.$$

Likewise, for $l \leq -n/2$ we have $|\psi_{\mathcal{A}^l J}| < C\lambda^{-n\alpha/2}$, and hence

$$|(A^n\phi, \psi)| < C\lambda^{-n\alpha/2}(|\phi|_1 + |\psi|_1).$$

It follows that the rate of mixing is exponential within the class $C^{1+\varepsilon}(\mathbb{T}^2)$. In a similar fashion one proves that exponential mixing holds in the class $C^{n/2+\varepsilon}(\mathbb{T}^n)$ for $n > 2$. More delicate arguments show that exponential mixing holds in the class of Hölder functions as well. This was first established by means of Markov partitions (see [**Bo75**]).

On the other hand, if $\phi \in C^\infty(\mathbb{T}^n)$, then $|\phi_I| < c(\alpha)/\|I\|^\alpha$ with arbitrarily large α. Therefore the rate of mixing in the class of infinitely smooth functions is *super-exponential.*

Considerable difficulties caused by the neutral component V^0 arise while considering partially hyperbolic automorphisms. Here Diophantine properties of the subspaces V^+, V^-, V^0 come into play. For ergodic automorphisms the subspaces are determined by linear equations with algebraic coefficients. It is known (see §29) that algebraic numbers are badly approximable by rationals. This implies that an integer point $I \in \mathbb{Z}^n$ cannot occur very close to the subspace $V^- + V^0$: as was proved by Katznelson [**Kat**],

$$\|I^+\| > C\|I\|^{-n} \qquad \text{for all } I \in \mathbb{Z}^n - \{0\}.$$

Using this estimate, it was proved in [**Kat**] that any ergodic toral automorphism is Bernoulli (although a Markov partition may not exist). Also Veech [**V86**] proved Livshitz's theorem in the class $C^\infty(\mathbb{T}^n)$ (see §24), and Lind [**Lin**] found the rate of mixing.

THEOREM 27.1 [**Lin**]. *An ergodic toral automorphism is exponentially mixing in the class of Hölder functions and super-exponentially mixing in the class of infinitely smooth function.*

2. We proceed to the semisimple case. In what follows, G is a connected semisimple Lie group *with finite center and with no compact components*, $\mathfrak{g}$ is its Lie algebra, and $K \subset G$ is a maximal compact subgroup.

To establish exponential mixing in the semisimple case, one needs some information from the theory of representations which falls outside the framework of the book. The results expounded below can be found in [**Kir**] and [**How**].

Recall that any unitary representation of G decomposes into the direct integral of irreducible unitary representations. Let a unitary representation π of the group G on a Hilbert space $\mathbb{H}$ be irreducible. A vector $v \in \mathbb{H}$ is said to be *K-finite* if the linear hull of the orbit $\pi(K)v \subset \mathbb{H}$ is finite dimensional.

Let $l = \dim K$, let $Y_1, \dots, Y_l$ be an orthogonal basis in the Lie algebra $\mathfrak{k}$ of the group K, and let $Y = 1 - \sum_1^l Y_i^2$. Then Y belongs to the center of the universal enveloping algebra of $\mathfrak{k}$.

Vectors $v, w \in \mathbb{H}$ determine the *matrix coefficient*, that is, the function $m_{v,w} : G \to \mathbb{C}$ given by $m_{v,w}(g) = (\pi(g)v, w)$. If the function $m_{v,w}$ is infinitely smooth on G for all $w \in \mathbb{H}$, then the vector $v \in \mathbb{H}$ is called a *C^∞-vector.* The subspace of all C^∞-vectors in $\mathbb{H}$ is called the *Gårding space* of the representation π. It is dense in $\mathbb{H}$ (the Gelfand–Gårding Theorem) and carries the action of the universal enveloping algebra of $\mathfrak{g}$.

The first estimates of decay of matrix coefficients $m_{v,w}(g)$ as $g \to \infty$ for C^∞-vectors v, w appeared in papers of Harish-Chandra. Later they were sharpened for some series of irreducible representations in different papers (see references in [**KS94**]).

π is said to be a *strongly L^p-representation* if $\mathbb{H}$ contains a dense subspace $\mathbb{H}'$ such that for any $v, w \in \mathbb{H}'$ we have $m_{v,w} \in L^p(G)$. Cowling [**Cow**] proved that an irreducible unitary representation π with discrete kernel is a strongly L^p-representation for some $p > 0$. Moreover, p is independent of π if G contains no simple components locally isomorphic to $\mathrm{SO}(n, 1)$ or $\mathrm{SU}(n, 1)$.

As we will see, estimating decay of matrix coefficients $m_{v,w}(g)$ reduces to studying the restriction of this function to a maximal $\mathbb{R}$-diagonalizable subgroup $A \subset G$. We fix an arbitrary norm[9] on the Lie algebra $\mathfrak{a}$ of A. Then elements of A are of the form $\exp(tx)$, where $x \in \mathfrak{a}$, $\|x\| = 1$. Using Howe's estimates [**How**] for matrix coefficients of K-finite vectors in strongly L^p-representations, Katok and Spatzier have proved the following results.

THEOREM 27.2 [**KS94**]. *Let v, w be C^∞-vectors in an irreducible unitary representation π of the group G with discrete kernel. Then there exist constants $C, \alpha > 0$, $M \in \mathbb{N}$ and a natural integer $p = p(\pi)$ such that for all* $\exp(tx) \in A$ *and all $m \geq M$,*

$$|(\exp(tx)v, w)| \leq Ce^{-\alpha|t|/p}\|Y^m(v)\| \ \|Y^m(w)\|.$$

If G has no simple components locally isomorphic to $\mathrm{SO}(n,1)$ *or* $\mathrm{SU}(n,1)$, *then p is independent of π.*

Let ρ be the regular unitary representation of G on the space $L^2_0(G/\Gamma)$ of functions with zero average. Let the lattice $\Gamma \subset G$ be irreducible. Then in the decomposition $\rho = \int \rho_\mu$ into the direct integral of irreducible representations, almost every ρ_μ has discrete kernel (otherwise there would exist a connected subgroup normal in G, whose action on G/Γ is nonergodic; in view of Moore's ergodicity criterion this is impossible as long as G has no compact components and Γ is irreducible).

Let $C^\infty_{c0}(G/\Gamma)$ stand for the space of infinitely smooth functions on G/Γ *with compact support* and zero average (for functions with compact support all partial derivatives belong to $L^2(G/\Gamma)$). In the decomposition $\phi = \int \phi_\mu$ of a function $\phi \in C^\infty_{c0}(G/\Gamma)$ almost all elements $\phi_\mu \in \mathbb{H}_\mu$ are C^∞-vectors. This implies the following result.

COROLLARY 27.3 [**KS94**]. *Let Γ be an irreducible lattice in a group G with no simple components locally isomorphic to* $\mathrm{SO}(n,1)$ *or* $\mathrm{SU}(n,1)$. *Then there exist $C, \alpha > 0$ and $m \in \mathbb{N}$ such that, given $\phi, \psi \in C^\infty_{c0}(G/\Gamma)$ and an element* $\exp(tx) \in A$,

$$|(\exp(tx)\phi, \psi)| \leq Ce^{-\alpha|t|}\|Y^m\phi\| \ \|Y^m\psi\|.$$

Remark. The paper [**KS94**] deals with uniform lattices Γ in G. But all the arguments go through in the nonuniform case as well.

The results of [**KS94**] were extended in [**KM96**] by Kleinbock and Margulis. First, Corollary 27.3 provides an estimate of the correlation coefficient $(g\phi, \psi)$ *for all $g \in G$*. Indeed, a semisimple group with finite center admits the *Cartan decomposition $G = KAK$*. Note that given $g = kak'$, where $k, k' \in K$ and $a \in A$, we have $(kak'\phi, \psi) = (ak'\phi, k^{-1}\psi)$. The differential operator Y on $C^\infty_{c0}(G/\Gamma)$ commutes with the action of K, and hence $\|Y^m k'\phi\| = \|Y^m\phi\|$. On the other hand, given a K-bi-invariant metric dist on G, one has $\mathrm{dist}(1,g) \sim \mathrm{dist}(1,a)$ as $g \to \infty$. Hence Corollary 27.3 can be generalized as follows:

THEOREM 27.4 [**KS94**], [**KM96**]. *Let Γ be an irreducible lattice in a group G without simple components locally isomorphic to* $\mathrm{SO}(n,1)$ *or* $\mathrm{SU}(n,1)$. *Then there*

[9] In [**KS94**] one considered the norm on $\mathfrak{a}$ obtained by taking half the sum of the positive roots in the positive Weyl chamber.

exist $C, \alpha > 0$ *and* $m \in \mathbb{N}$ *such that for any* $\phi, \psi \in C^\infty_{c0}(G/\Gamma)$ *and any element* $g \in G$,

$$|(g\phi, \psi)| \leq Ce^{-\alpha \operatorname{dist}(1,g)} \|Y^m \phi\| \ \|Y^m \psi\|.$$

Second, the paper [**KM96**] provides some results on exponential mixing for a flow $(G/\Gamma, g_\mathbb{R})$, where G can have "bad" components locally isomorphic to $\mathrm{SO}(n,1)$ or $\mathrm{SU}(n,1)$. More precisely, following §25 a simple component $S \subset G$ is said to be *essential* for $g_\mathbb{R}$ if the projection of $g_\mathbb{R}$ to S is partially hyperbolic. Then exponential mixing takes place whenever at least one essential component S is "good". The argument is like this. Let $G = S \times G'$ and $g_t = s_t \times g'_t$, where $s_\mathbb{R} \subset S$, $g'_\mathbb{R} \subset G'$. Note that $\operatorname{dist}(1, s_t) \geq \beta t$ for some $\beta > 0$. By Theorem 27.3, for every irreducible representation with discrete kernel of the group S,

$$(*) \qquad |(s_t v, w)| \leq Ce^{-\alpha\beta|t|} \|Y^m(v)\| \ \|Y^m(w)\| \quad \text{for all } C^\infty\text{-vectors } v, w \text{ and all } t \in \mathbb{R}.$$

Now we consider the representation of $S \times g'_\mathbb{R}$ on $L^2_0(G/\Gamma)$ and decompose it into the direct integral over irreducible representations. Every irreducible representation of $S \times g'_\mathbb{R}$ is a tensor product of an irreducible representation of S and a one-dimensional representation of $g'_\mathbb{R}$. Integrating the estimate (*) over all such representations, we obtain the following result.

THEOREM 27.5 [**KM96**]. *Assume that at least one essential component for* $g_\mathbb{R} \subset G$ *is not isomorphic locally to* $\mathrm{SO}(n,1)$ *or* $\mathrm{SU}(n,1)$, *and that the lattice* $\Gamma \subset G$ *is irreducible. Then there exist constants* $C, \gamma > 0$ *and* $m \in \mathbb{N}$ *such that for all* $\phi, \psi \in C^\infty_{c0}(G/\Gamma)$,

$$|(g_t\phi, \psi)| \leq Ce^{-\gamma|t|} \|Y^m \phi\| \ \|Y^m \psi\|, \qquad t \in \mathbb{R}.$$

This result plays an important role in estimating the Hausdorff dimension of the family of bounded orbits of the flow $(G/\Gamma, g_\mathbb{R})$ (see §25).

Third, the proof of exponential mixing by Kleinbock and Margulis involves the isolation property of unitary representations of G from the trivial representation in the Fell topology (under some restrictions on G). It is well known that a simple Lie group S, if it is not locally isomorphic to $\mathrm{SO}(n,1)$ or $\mathrm{SU}(n,1)$, possesses the T-property of Kazhdan (see [**Ma91c**]), and hence all its irreducible unitary representations are outside a fixed neighborhood of the trivial representation.

On the other hand, as was proved in [**Cow**], all irreducible representations of a simple group S outside a fixed neighborhood of the trivial representation are strong L^p-representations for some $p \in \mathbb{N}$. Assume now that Π is a family of unitary representations of a semisimple group G with discrete kernel, and that the restriction of Π to every simple component $G_i \subset G$ is isolated from the trivial representation. Then the conclusion of Theorem 27.2 holds for every representation $\pi \in \Pi$ with constants C, α, p independent of π. For this reason, the following conjecture was formulated in [**KM96**]: the regular representation of G on $L^2_0(G/\Gamma)$ (viewed as a family of irreducible representations) always possesses this property (even if G has "bad" components).

In [**KM99**] this conjecture was proved for nonuniform lattices Γ. Therefore the following holds.

THEOREM 27.6 [**KM99**]. *Let Γ be an irreducible nonuniform lattice in a semisimple group G (with finite center and with no compact components). Then there exist $C, \alpha > 0$ and $m \in \mathbb{N}$ such that for any $\phi, \psi \in C_{c0}^{\infty}(G/\Gamma)$ and any element $g \in G$,*

$$|(g\phi, \psi)| \leq Ce^{-\alpha \operatorname{dist}(1,g)} \|Y^m \phi\| \, \|Y^m \psi\|.$$

Thus, the exponential mixing phenomenon in the class of smooth functions was established in a very important class of semisimple homogeneous spaces. The general conjecture is that a similar estimate is valid for any lattice Γ in a semisimple group G.

Finally, in [**KM96**] exponential mixing was proved in the class of Hölder functions by obtaining analogues of Theorems 27.4 and 27.5. Apparently, the method of [**KM96**] is strong enough to establish exponential mixing for Hölder functions on a semisimple space, as soon as it is verified for C^∞-functions. Results of this kind are used, for instance, to prove rigidity of Hölder cocycles over standard Anosov actions (see §24).

It is worth noting that exponential mixing in the class of Hölder functions for $G = \mathrm{SL}(2, \mathbb{R})$ was first proved in papers of Moore [**Mo87**] and Ratner [**R87a**]. In both papers one only used that the functions involved are Hölder in the direction of the subgroup $K = \mathrm{SO}(2) \subset G$ (see §24). Moore proved exponential mixing for Hölder functions with exponent $\alpha > 1/2$, whereas Ratner did the same for arbitrary exponent $\alpha > 0$. Moreover, in [**R87a**] an estimate for the rate of mixing of the horocycle flow was found. It is based on the Cartan decomposition $G = KAK$. Namely, already Hedlund used the relation

$$u_t = r_1(t) a_{\ln t} r_2(t) \text{ for } t > 1, \qquad \text{where } u_t = \begin{pmatrix} 1 & t \\ 0 & 1 \end{pmatrix}, \; a_t = \begin{pmatrix} e^t & 0 \\ 0 & e^{-t} \end{pmatrix}$$

and $r_i : \mathbb{R} \to \mathrm{SO}(2)$ is a continuous function. Now it is clear that from

$$|(a_t\phi, \psi)| < Ce^{-\alpha|t|},$$

where $C = C(\phi, \psi) > 0$ and $0 < \alpha = \alpha(\Gamma) < 1$, it follows that

$$|(u_t\phi, \psi)| < D|t|^{-\alpha}, \qquad |t| > 1.$$

CHAPTER 4

Applications to Number Theory

We have already exhibited certain connections of the theory of homogeneous flows with number theory. More specifically, Weyl's theorem on distribution of the fractional parts of values of a polynomial, was proved in §3 by studying the structure of orbits of nilflows. In §14, following Dani, we were able to relate the behavior of geodesic orbits on $\mathrm{SL}(2,\mathbb{R})/\mathrm{SL}(2,\mathbb{Z})$ with Diophantine properties of real numbers. In §20 we mentioned a connection between the well-known Riemann hypothesis and the distribution of periodic orbits of the horocycle flow on $\mathrm{SL}(2,\mathbb{R})/\mathrm{SL}(2,\mathbb{Z})$. Finally, the result of Kolmogorov presented in §24 revealed a connection between Diophantine properties of the rotation number of a rectilinear flow on $\mathbb{T}^2$ and the existence of nontrivial time changes.

As was already noted in the Introduction, progress in the study of unipotent flows was very much motivated by the Oppenheim–Davenport conjecture on density of the set of values $Q(\mathbb{Z}^n)$ of an indefinite irrational quadratic form Q in $n \geq 3$ variables. This conjecture was proved by Margulis. In §28, following [**DM90b**] and [**Ma90**], we give the full proof by reducing the problem to the case $n = 3$ and studying a particular case ($G = \mathrm{SL}(3,\mathbb{R})$) of the Raghunathan topological conjecture. One can make this result more precise by proving that the set of values of Q at primitive points of $\mathbb{Z}^n$ is dense in $\mathbb{R}$ (Dani and Margulis [**DM89,90a**], Borel and Prasad [**BP, B**]). The results also provide information on the Markov spectrum of quadratic forms in $n \geq 3$ variables.

In more detail the questions are considered in the surveys [**B**] and [**Ma97**]. The problem of finding the asymptotics of the number of integer solutions of the inequality $|Q(x)| < \varepsilon$ in the ball $|x| < R$ as $R \to \infty$ for indefinite forms Q was recently solved in papers of Dani and Margulis [**DM93**] (the lower estimate), and of Eskin, Margulis and Mozes [**EMM**] (the upper estimate). It is of interest that the upper estimate does not coincide with the lower estimate for certain indefinite forms of signatures $(2,1)$ and $(2,2)$.

Section 29 is devoted to preliminaries of the theory of Diophantine approximation. Here we introduce a classification of real numbers, vectors, and matrices based on the rate of approximation by corresponding rational objects. Incidentally, we prove some results of metric theory of Diophantine approximation which go back to Khintchine.

As will be shown in §30, the basic notions of the theory of Diophantine approximation are well interpreted by the structure of corresponding orbits of a certain homogeneous flow. Here we follow [**Da85**] and [**KM98**]. In particular, a conjecture of Littlewood reduces to studying orbit closures of the two-dimensional diagonal subgroup on $\mathrm{SL}(3,\mathbb{R})/\mathrm{SL}(3,\mathbb{Z})$. Then we discuss the recent proof by Kleinbock and Margulis [**KM98**] of conjectures due to Baker and Sprindzhuk from the theory of Diophantine approximation on manifolds; earlier progress in this direction (in particular, the proof of Mahler's conjecture) was mostly provided by Sprindzhuk's results (see the book [**Sp77**] and the survey [**Sp80**]).

Skriganov [**Skr98**] used the ergodic theory of the diagonal action on the space $\mathrm{SL}(n,\mathbb{R})/\mathrm{SL}(n,\mathbb{Z})$ to find the asymptotics of the number of lattice points inside polyhedra; the two-dimensional version of his results [**Skr89,93**] goes back to the twenties (Hardy and Littlewood [**HL**] and Khintchine [**Kh23**]). These problems are discussed in §31. Finally, we consider the paper of Eskin, Mozes and Shah [**EMS96**] on the asymptotics of the number of integer points on homogeneous $\mathbb{Q}$-manifolds in $\mathbb{R}^n$, which generalizes results of Duke, Rudnick and Sarnak [**DRS**] (see also [**EM**]) for the affine-symmetric case.

§28. The Oppenheim–Davenport conjecture

As was already mentioned, the most powerful stimulus to studying unipotent flows was provided by the Oppenheim–Davenport conjecture in number theory. A quadratic form on $\mathbb{R}^n$ is said to be *rational* if it is proportional to a form with rational coefficients, and *irrational* otherwise.

The Oppenheim–Davenport conjecture. Let $B(x_1,\dots,x_n)$ be an indefinite nondegenerate irrational quadratic form in $n \geq 3$ variables. Then for every $\varepsilon > 0$ there exists an integer nonzero vector $x \in \mathbb{Z}^n - \{0\}$ such that $|B(x)| < \varepsilon$.

By analytic number theory methods this conjecture was proved in the fifties for $n \geq 21$; the history of the problem is well described in [**Ma97**]. (Recently Bentkus and Gotze [**BG**] did this for $n \geq 9$. Also for $n \geq 9$ they proved a conjecture of Davenport and Lewis on values of *definite* irrational quadratic forms at integer points.)

First we deduce the conjecture from a certain result on homogeneous flows and discuss different generalizations and corollaries of it. Then we pass from a purely qualitative statement about the existence of integer solutions of the inequality $|B(x)| < \varepsilon$ to counting the asymptotics of the number of solutions in the ball $|x| < T$, $T \to \infty$. The conjecture itself can be settled by Ratner's topological Theorem 19.13 (see [**R95b**], [**B**]), but for completeness we give the original and relatively short proof due to Margulis.

1. Let us prove by induction that it suffices to examine the case $n = 3$ (following [**SM89**]). In fact, for $n > 3$ we select a rational basis $v_1,\dots,v_n \in \mathbb{Q}^n$ in $\mathbb{R}^n$ such that the restriction of B to the linear hull $L = \mathrm{Span}(v_1,\dots,v_{n-1})$ is an indefinite nondegenerate form and B is nontrivial on $\mathrm{Span}(v_1, v_2)$ (this is easy to do by reducing B to principal axes, and then perturbing the basis to make it rational). Assume that the form is rational on L. We set

$$L_t = \mathrm{Span}(v_1,\dots,v_{n-2}, v_{n-1} + tv_n).$$

Then for a sufficiently small $|t| < \delta$, the restriction of B to L_t is an indefinite nondegenerate form as well. Assume that for some rational $0 < t < \delta$, the form $B|_{L_t}$ is rational. Since the restriction of B to $L \cap L_t = \mathrm{Span}(v_1,\dots,v_{n-2})$ is nontrivial for $n > 3$, it follows that there exists $\alpha > 0$ such that the form αB has rational coefficients on L and L_t. But then αB has rational coefficients on all of $\mathbb{R}^n$. This contradiction shows that the restriction of B to some rational hyperplane L_t, $t \in \mathbb{Q}$, is a nondegenerate indefinite irrational form, and one can lower the dimension to $n - 1$.

Note that the conjecture is incorrect for $n = 2$. For instance, if

$$B(x_1, x_2) = x_1^2 - \lambda^2 x_2^2, \qquad \text{where } \lambda = 1 + \sqrt{2},$$

then 0 is an isolated point of the set $B(\mathbb{Z}^2)$. The point is that $\sqrt{2}$ (and hence λ) is badly approximated by rationals: it is straightforward to verify that

$$\left|\sqrt{2} - \frac{n}{m}\right| > \frac{1}{4m^2} \qquad \text{for all } n \in \mathbb{Z},\ m \in \mathbb{N}$$

(see §29).

We proceed to the case $n = 3$. By definition, $B(x) = x^{\mathsf{T}} \sigma_B x$, where σ_B is the symmetric matrix of the quadratic form B. Let

$$G = \mathrm{SL}(3, \mathbb{R}), \qquad \Gamma = \mathrm{SL}(3, \mathbb{Z}), \qquad S(x) = 2x_1 x_3 - x_2^2.$$

Since the form B is indefinite and nondegenerate, there exists a change of variables $x \to gx$, $g \in G$, such that $B(x) = \lambda S(gx)$ for some $\lambda \in \mathbb{R}$. Let

$$\mathrm{SO}(B) = \{h \in G:\ B(hx) = B(x), x \in \mathbb{R}^n\} = \{h \in G:\ h^{\mathsf{T}} \sigma_B h = \sigma_B\}$$

be the stabilizer of the form B, and let $H_B = \mathrm{SO}(B)_0$ be the identity component of it. Likewise, we introduce $H = \mathrm{SO}(S)_0$. Then $H = g H_B g^{-1}$. In the group H one distinguishes three one-parameter subgroups:

$$\left\{V_t = \begin{pmatrix} 1 & t & t^2/2 \\ 0 & 1 & t \\ 0 & 0 & 1 \end{pmatrix}\right\}, \left\{A_t = \begin{pmatrix} e^t & 0 & 0 \\ 0 & 1 & 0 \\ 0 & 0 & e^{-t} \end{pmatrix}\right\}, \left\{V_t^{\mathsf{T}} = \begin{pmatrix} 1 & 0 & 0 \\ t & 1 & 0 \\ t^2/2 & t & 1 \end{pmatrix}\right\}.$$

It is easy to see that $H \simeq \mathrm{SO}(2,1)_0 \simeq \mathrm{PSL}(2, \mathbb{R})$ and H is generated by two unipotent subgroups $V_{\mathbb{R}}$ and $V_{\mathbb{R}}^{\mathsf{T}}$ (by the way, for $n = 2$ the stabilizer of an indefinite form is a reductive one-parameter subgroup of $\mathrm{SL}(2, \mathbb{R})$ inducing the geodesic flow. This explains the drastic contrast between the cases $n = 2$ and $n = 3$). Raghunathan noticed[1] that the Oppenheim–Davenport conjecture reduces to the study of orbits of the action $(G/\Gamma, H)$.

LEMMA 28.1 (see [**Ma87**], [**B**]). *If each relatively compact orbit $Hg\Gamma$, $g \in G$, is compact, then the Oppenheim–Davenport conjecture holds.*

Proof. Let $B(x) = \lambda S(gx)$ for all $x \in \mathbb{R}^3$. It suffices to prove that the set $S(g\mathbb{Z}^3)$ has 0 as an accumulation point provided that the form B is irrational. Assume the opposite: 0 is an isolated point of the set. Then the orbit $Hg\Gamma \subset G/\Gamma$ is relatively compact. Indeed, suppose that $h_k g\Gamma \to \infty$ for some sequence $h_k \in H$. Then by Mahler's criterion, $0 \in \mathbb{R}^3$ is a limit point for the family of lattices $h_k g \mathbb{Z}^3$, $k \in \mathbb{N}$. But $S(h_k gx) = S(gx)$, and we arrive at a contradiction.

Hence the orbit $Hg\Gamma \subset G/\Gamma$ is relatively compact. By the hypothesis, it must be compact. Then $\mathrm{SL}(3, \mathbb{Z}) \cap g^{-1}Hg$ is a lattice in $g^{-1}Hg$ which is Zariski dense by Borel's Theorem E.4. It follows that $g^{-1}Hg$ is defined over $\mathbb{Q}$. But $g^{-1}Hg$ is the identity component of the stabilizer of B. Let us prove (after [**B**]) that the form B is rational.

Indeed, every symmetric matrix C preserved by the subgroup H is proportional to the matrix S (a direct computation shows that from $V_1 C V_1^{\mathsf{T}} = C$ and $V_1^{\mathsf{T}} C V_1 = C$ it follows that $C = \lambda S$). Therefore $\mathbb{R}B$ consists of all symmetric matrices preserved

[1] As was mentioned in [**Ma97**], this reduction was implicitly contained already in [**CS**].

by the $\mathbb{Q}$-defined subgroup $g^{-1}Hg$. Since the lattice $\mathrm{SL}(n,\mathbb{Z}) \cap g^{-1}Hg$ is Zariski dense, it follows that $\mathbb{R}B$ consists of all symmetric matrices preserved by $\mathrm{SL}(n,\mathbb{Z}) \cap g^{-1}Hg$. Hence $\mathbb{R}B$ is determined by a system of linear equations with integer coefficients. But such system has a rational solution, and hence for some $\lambda \in \mathbb{R}$ the matrix λB has all its entries rational. □

So, the Oppenheim–Davenport conjecture can be reduced to the proof of compactness of relatively compact orbits of the flow $(G/\Gamma, H)$. Essentially, it suffices to examine relatively compact orbits of the unipotent flow $(G/\Gamma, V_{\mathbb{R}})$ (it was this observation of Raghunathan which raised the interest in unipotent flows). This problem was solved completely by Margulis by using the structure of the group $\mathrm{SL}(3,\mathbb{R})$ [**Ma87**]; we give the proof at the end of the section. In [**Ma89**] Margulis has proved a stronger version of the conjecture: for any $\varepsilon > 0$ there exists $x \in \mathbb{Z}^n$ such that $0 < |B(x)| < \varepsilon$.

It is known [**Op**] that if 0 is the right limit point for the set $B(\mathbb{Z}^n)$, then it is the left one as well. Moreover, $B(\mathbb{Z}^n)$ is invariant under multiplication by the square of an integer. Hence the set $B(\mathbb{Z}^n)$ is dense in $\mathbb{R}$ for any form B as specified (the family of intervals $(\frac{a-\varepsilon}{n^2}, \frac{a+\varepsilon}{n^2})$, $n \in \mathbb{N}$, contains an interval $(0,\delta)$, and hence the set $B(\mathbb{Z}^n)$ intersects any ε-neighborhood of an arbitrary point $a > 0$). Thus, the following theorem holds.

THEOREM 28.2 [**Ma89**]. *Let $B(x)$ be an indefinite nondegenerate irrational quadratic form on $\mathbb{R}^n$, where $n \geq 3$. Then the set $B(\mathbb{Z}^n)$ is everywhere dense in $\mathbb{R}$.*

Note that the main problem in studying the flow $(G/\Gamma, V_{\mathbb{R}})$ is that *the subgroup $V_{\mathbb{R}} \subset G$ is not horospherical.* Thus, it was Margulis who first managed to study orbit structures of a unipotent flow that is not horospherical (for a semisimple group G).

A deep study of all (not only relatively compact) orbits of the action $(G/\Gamma, H)$ enabled Dani and Margulis [**DM89,90a**] to substantially refine the result on values of the form B at integer points. First we formulate the definitive result on the structure of orbits of the action $(G/\Gamma, H)$.

THEOREM 28.3 [**DM89,90a**]. *An orbit $Hg\Gamma \subset G/\Gamma$ is either closed, or everywhere dense.*

From the point of view of Ratner's topological Theorem 19.13 this is not surprising: there are no other closed connected subgroups between H and G.

One says that an integer vector $x \in \mathbb{Z}^n$ is *primitive* if there are no vectors $y \in \mathbb{Z}^n$ and integers $k \geq 2$ such that $x = ky$. We denote by $P(\mathbb{Z}^n)$ the set of all primitive vectors. Note that the set $P(\mathbb{Z}^n)$ is invariant under the action of $\mathrm{SL}(n,\mathbb{Z})$. The following statement holds.

THEOREM 28.4 [**SM89**]. *Let $B(x)$ be an indefinite nondegenerate irrational quadratic form on $\mathbb{R}^n$, where $n \geq 3$. Then the set $B(P(\mathbb{Z}^n))$ is everywhere dense in $\mathbb{R}$.*

We deduce this result from the statement on orbit structures. As in the proof of Theorem 28.2, it suffices to consider the case $n = 3$ (the only remark to be added is that $P(L \cap \mathbb{Z}^n) \subset P(\mathbb{Z}^n)$ for every rational hyperplane $L \subset \mathbb{R}^n$). We have to prove that $\overline{B(E)} = \mathbb{R}$, where $E = P(\mathbb{Z}^3)$ and B is an indefinite nondegenerate irrational quadratic form. But $B(E) = B(H_B E) = B(H_B \Gamma E)$, where $\Gamma = \mathrm{SL}(3,\mathbb{Z})$ and H_B

is the stabilizer of B. We know that the orbit $H_B\Gamma$ is either closed or everywhere dense in G/Γ. In the first case the Margulis Theorem 12.6 implies that the group $\Gamma \cap H_B$ is a lattice in H_B, and hence $H_B \subset G = \mathrm{SL}(3,\mathbb{R})$ is defined over $\mathbb{Q}$. The proof of Lemma 28.1 shows that the form B is rational. It follows that $\overline{H_B\Gamma} = G$, and hence $\overline{B(E)} = \overline{B(GE)} = \overline{B(\mathbb{R}^3)} = \mathbb{R}$. □

Using Ratner's general results, Theorem 28.4 was sharpened in an elegant way by Borel and Prasad.

THEOREM 28.5 [**BP**]. *Let B be as before. Then for any $c_1,\dots,c_{n-1} \in \mathbb{R}$ and $\varepsilon > 0$ there exist elements $x_1,\dots,x_{n-1} \in \mathbb{Z}^n$ that form a part of a basis in $\mathbb{Z}^n$ (and hence they are primitive elements) such that $|B(x_i) - c_i| < \varepsilon$, $i = 1,\dots,n-1$.*

Proof. It is clear that the level surface $\{x \in \mathbb{R}^n : B(x) = c\}$ is not contained in a hyperplane. Hence there exists a unimodular lattice $\Lambda = \mathbb{Z}y_1 + \cdots + \mathbb{Z}y_n$ such that $B(y_i) = c_i$, $i = 1,\dots,n-1$. Let $\mathrm{SO}(B) \subset \mathrm{SL}(n,\mathbb{R})$ be the stabilizer of B. Then $\mathrm{SO}(B)$ is conjugate to a group $\mathrm{SO}(p,q)$, where $1 \le p,q < n = p+q$. It follows that $\mathrm{SO}(B)$ is generated by unipotent elements. In addition, $\mathrm{SO}(B)$ is a maximal subgroup: there are no intermediate closed subgroups between it and $\mathrm{SL}(n,\mathbb{R})$. Now by Ratner's Theorem 19.13, the orbit $\mathrm{SO}(B)\,\mathrm{SL}(n,\mathbb{Z})$ is either closed or dense in the space $\mathrm{SL}(n,\mathbb{R})/\mathrm{SL}(n,\mathbb{Z})$ of unimodular lattices in $\mathbb{R}^n$. As in the proof of Lemma 28.1, the first possibility cannot occur for the irrational form B. Hence there exist $h \in \mathrm{SO}(B)$ and $z_1,\dots,z_n \in h(\mathbb{Z}^n)$ close to $y_1,\dots,y_n$ such that $|B(z_i) - B(y_i)| < \varepsilon$, $i = 1,\dots,n$. But $z_i = h(x_i)$ and $B(z_i) = B(x_i)$ for some $x_i \in \mathbb{Z}^n$. □

2. We denote by $\mathcal{O}_n$ the family of all nondegenerate indefinite quadratic forms in n variables. Let $d(P)$ be the determinant of the matrix of $P \in \mathcal{O}_n$, and let

$$m(P) = \inf\{|P(x)|,\ x \in \mathbb{Z}^n,\ x \neq 0\}, \qquad \mu(P) = \frac{m(P)^n}{|d(P)|}.$$

Clearly, $\mu(P)$ depends only on the equivalence class of P (here forms $P, P' \in \mathcal{O}_n$ are equivalent if $P(x) = \lambda P'(gx)$, $x \in \mathbb{R}^n$, for some $g \in \mathrm{SL}(n,\mathbb{Z})$ and $\lambda \in \mathbb{R}$). The set $M_n = \mu(\mathcal{O}_n)$ is called the *n-dimensional Markov spectrum*. It follows easily from Mahler's criterion that M_n is compact in $\mathbb{R}$.

As Markov proved (see [**Ca57**]), $M_2 \subset [0, 4/5]$, and the maximal value $\mu(P) = 4/5$ is attained on the form $P(x_1,x_2) = x_1^2 + x_1x_2 - x_2^2$. In addition, the intersection $M_2 \cap (4/9+\varepsilon, \infty)$ is finite for any $\varepsilon > 0$. At the same time, the set $M_2 \cap [0, 4/9]$ is uncountable and has a complicated topological structure.

By the celebrated theorem of Meyer (see [**Ca78**] or [**Se82**]), a rational form $P \in \mathcal{O}_n$, $n \ge 5$, has $x \in \mathbb{Z}^n$, $x \ne 0$, such that $P(x) = 0$. This and Theorem 28.2 imply that $M_n = \{0\}$ for $n \ge 5$. On the other hand, it was noticed in [**CS**] that if the Oppenheim–Davenport conjecture holds then the intersection $M_n \cap (\varepsilon, \infty)$ is finite for any $\varepsilon > 0$ and $n = 3, 4$. Thus, it follows from Margulis' Theorem 28.2 that the set $M_n \cap (0, \infty)$ is discrete for $n = 3, 4$. Moreover, the number of equivalence classes of forms $P \in \mathcal{O}_n$ such that $\mu(P) > \varepsilon$ is finite. Note that the 11 largest values of the set M_3 were found by Venkov [**Ven**]; see [**Ma97**] for more on the Markov spectrum.

3. Given a quadratic form P on $\mathbb{R}^n$, we define

$$V^P_{(a,b)}(T) = \{|x| < T :\ a < P(x) < b\}.$$

Here are results of [**DM93**] and [**EMM**] on the asymptotics of the number $N^P_{(a,b)}(T)$ of integer points in $V^P_{(a,b)}(T)$. Let $n \geq 3$, $1 \leq p \leq n/2$, and let $Q(p,n)$ be the space of symmetric $n \times n$ matrices with determinant ± 1 and signature $(p, n-p)$. Every matrix $P \in Q(p,n)$ specifies a quadratic form $P(x) = x^\intercal P x$, $x \in \mathbb{R}^n$, and the space $Q(p,n)$ is supplied with the natural locally compact topology of pointwise convergence in $\mathbb{R}^n$.

The first statement of the following theorem gives the lower asymptotic estimate of $N^P_{(a,b)}(T)$ uniformly on compact subsets $C \subset Q(p,n)$ in a compact set $K \subset Q(p,n)$ after discarding a finite collection of rational forms in K. Having in mind Meyer's theorem (see Subsection 2), it is natural to conjecture that for $n \geq 5$ one can find the lower asymptotic estimate of $N^P_{(-\varepsilon,\varepsilon)}(T)$ for all forms from an arbitrary compact set $K \subset Q(p,n)$. The second statement of the theorem supports this conjecture, although the estimate obtained is a little bit weaker.

THEOREM 28.6 [**DM93**]. *Let K be an arbitrary compact set in $Q(p,n)$.*

1. *For any $a < b$ and any $\delta > 0$ there exists a finite collection $N \subset K$ of rational forms such that for any compact subset $C \subset K - N$ there is $T_0 \geq 0$ for which*

$$N^P_{(a,b)}(T) \geq (1-\delta)\operatorname{Vol}(V^P_{(a,b)}(T)), \qquad P \in C, \quad T \geq T_0.$$

2. *If $n \geq 5$, then for any $\varepsilon > 0$ there are $c > 0$ and $T_0 \geq 0$ such that*

$$N^P_{(-\varepsilon,\varepsilon)}(T) \geq c\operatorname{Vol}(V^P_{(-\varepsilon,\varepsilon)}(T)), \qquad P \in K, \quad T \geq T_0.$$

For a single irrational form $P \in Q(p,n)$ the estimate (1) (but with a constant $0 < c < 1$ in place of $1-\delta$) was found by Dani and Mozes, and also independently by Ratner (both unpublished). One can show that

$$\operatorname{Vol}(V^P_{(a,b)}(T)) \sim \lambda_P (b-a) T^{n-2}, \qquad T \to \infty.$$

We give the idea of the proof of Theorem 28.6. Define a form $S \in Q(p,n)$ on $\mathbb{R}^n$ relative to the basis $e_1, \ldots, e_n$ by the formula

$$S(x_1, \ldots, x_n) = 2x_1x_n + \sum_2^p x_i^2 - \sum_{p+1}^{n-1} x_i^2.$$

Let

$$G = \operatorname{SL}(n,\mathbb{R}), \quad \Gamma = \operatorname{SL}(n,\mathbb{Z}), \quad H = \operatorname{SO}(S).$$

Since every form $B \in Q(p,n)$ reduces to S by changing variables, it follows that $Q(p,n)$ is homeomorphic to G/H. Let $U \subset H$ fix $e_1, \ldots, e_{n-2}$ and preserve the invariant subspaces spanned by e_1, e_{n-1} and e_1, e_{n-1}, e_n. Then U is a one-parameter unipotent subgroup (in the case $(p,n) = (1,3)$ it is just $V_\mathbb{R}$). By $M \simeq \operatorname{SO}(p-1,\mathbb{R})$ we denote the stabilizer in H of the subspace $\mathbb{R}(e_1+e_n)+\mathbb{R}e_2+\cdots+\mathbb{R}e_p$, on which the form S is positively defined.

The left action of G on $\mathbb{R}^n$ induces a homeomorphism between G/Γ and the space of unimodular lattices in $\mathbb{R}^n$. Any continuous function ψ on $\mathbb{R}^n - \{0\}$ with compact support determines a function $\widetilde{\psi}$ on G/Γ by the formula

$$\widetilde{\psi}(g\Gamma) = \sum_{x \in g\mathbb{Z}^n} \psi(x).$$

By Siegel's theorem, $\int_{\mathbb{R}^n} \psi dx = \int_{G/\Gamma} \widetilde{\psi} d\nu$. The proof of Theorem 28.6 relies upon the following obvious identity:

$$(*) \qquad \int_T^{kT} \int_A \sum_{v \in g\mathbb{Z}^n} \psi(U_t m v)\, dm\, dt = \int_T^{kT} \int_A \widetilde{\psi}(U_t m g \Gamma)\, dm\, dt$$

for any continuous function ψ on $\mathbb{R}^n - \{0\}$ with compact support, any Borel subset $A \subset M$, any $k > 1$, and any $T > 0$. The idea is that for a suitable choice of ψ and $g \in G$, the left side of the identity (with $A = M$) gives a good approximation of $N^P_{(a,b)}(T)$, whereas the asymptotics of the right side is estimated by the uniform distribution Theorem 19.1. However one has to keep in mind that the function $\widetilde{\psi}$ is not bounded if $\psi \neq 0$. Hence Theorem 19.1 (more precisely, its uniform analogue 20.1) should be applied to a bounded continuous function $f \leq \widetilde{\psi}$.

Clearly, this method does not provide us with an upper asymptotic estimate for $N^P_{(a,b)}(T)$. This obstacle was avoided by Eskin, Margulis and Mozes [**EMM**]. It turns out that the upper estimate of the right side of (*) can be obtained by using the uniform distribution theorem if one proves that, for some $\delta > 0$,

$$(**) \qquad \limsup_{T \to \infty} \frac{1}{T} \int_0^T \int_A \widetilde{\psi}(U_t m g \Gamma)^{1+\delta}\, dm\, dt < \infty.$$

In essence, in [**EMM**] this inequality is proved for any $0 \leq \delta < 1$ provided that $(p,n) \neq (1,3)$ and $(2,4)$. For exceptional values of (p,n) and $\delta = 0$, the inequality (**) becomes valid if one replaces $1/T$ by $1/(T \log T)$. This leads to the following important result.

THEOREM 28.7 [**EMM**]. *Let K be an arbitrary compact set in $Q(p,n)$, and let $a < b$.*

1. *If $(p,n) \neq (1,3)$ and $(p,n) \neq (2,4)$, then there exists a constant $c = c(K,a,b)$ such that*

$$N^P_{(a,b)}(T) < cT^{n-2} \qquad \text{for all } P \in K,\ T > 1.$$

Moreover, for any $\delta > 0$ there exists a finite collection $N \subset K$ of rational forms such that for any compact set $C \subset K - N$ there exists $T_0 \geq 0$ for which

$$(1-\delta)\lambda_P(b-a)T^{n-2} \leq N^P_{(a,b)}(T) \leq (1+\delta)\lambda_P(b-a)T^{n-2}, \quad P \in C, \quad T \geq T_0.$$

2. *If $(p,n) = (1,3)$ or $(p,n) = (2,4)$, then there exists a constant $c = c(K,a,b)$ such that*

$$N^P_{(a,b)}(T) < cT^{n-2} \log T, \qquad P \in K,\ T > 2.$$

Here for exceptional values of (p,n) the asymptotics $N^P_{(a,b)}(T) \sim T^{n-2}$ holds for almost all forms $P \in Q(p,n)$, but breaks down for the forms

$$P_\alpha(x) = x_1^2 + x_2^2 - \alpha^2 x_3^2 \quad \text{and} \quad \widehat{P}_\alpha(x) = x_1^2 + x_2^2 - \alpha^2(x_3^2 + x_4^2),$$

if α is extremely well approximable by rationals ([**EMM**]). The first examples of quadratic forms P for which the asymptotics $N^P_{(a,b)}(T) \sim T^{n-2}$ fails to hold were found by Sarnak and worked out in detail in [**Bre**].

Let us show (following [**EMM**]) that for some irrational parameters α, the number $N^{P_\alpha}_{(a,b)}(T)$ cannot be estimated from above better than $T \log T$. More precisely, for irrational parameters α from a second category set (of zero measure) in $\mathbb{R}$

there exists a sequence $T_i \to \infty$ such that $N^{P_\alpha}_{(a,b)}(T_i) > T_i(\log T_i)^{1-\varepsilon}$ with arbitrary small $\varepsilon > 0$. (A similar argument shows that for many irrational parameters $\alpha \in \mathbb{R}$ one cannot improve the estimate $N^{\widehat{P}_\alpha}_{(a,b)}(T) < cT^2 \log T$).

In fact, it is known from divisor theory that for every $\alpha \in \mathbb{Q}$ there exists $c_\alpha \in \mathbb{R}$ such that

$$\operatorname{card}\{x \in \mathbb{Z}^3 : \ P_\alpha(x) = 0, \ |x| < T\} \sim c_a T \log T.$$

Now let

$$L(\alpha, T) = \{x \in \mathbb{Z}^3 : \ P_\alpha(x) = 0, \ T/2 \le |x| \le T\}.$$

Then, given $x \in L(\alpha, T)$, we have

$$\frac{T^2}{4(1+\alpha^2)} \le x_3^2 \le \frac{T^2}{1+\alpha^2}.$$

Hence, taking $\beta^2 = \alpha^2 + (1+\alpha^2)/T^2$, we obtain $1/4 \le P_\beta(x) \le 1$ for any $x \in L(\alpha, T)$. Therefore

$$N^{P_\beta}_{(1/4,1)}(T) \ge \frac{c_\alpha}{4} T \log T > T(\log T)^{1-\varepsilon} \qquad \text{for large enough } T.$$

Since the set $\{\alpha^2, \ \alpha \in \mathbb{Q}\}$ is dense in $\mathbb{R}$, it follows that, for all $\varepsilon > 0$ and $S > 0$, the set of $\beta \in \mathbb{R}$ such that

$$N^{P_\beta}_{(1/4,1)}(T) > T(\log T)^{1-\varepsilon} \qquad \text{for some } T > S$$

is also dense in $\mathbb{R}$.

We set

$$\begin{aligned} U_S = \{\gamma \in \mathbb{R} : \ & \text{there exist } \beta \in \mathbb{R} \text{ and } T > S \text{ such that} \\ & N^{P_\beta}_{(1/4,1)}(T) \ge T(\log T)^{1-\varepsilon} \text{ and } |\beta^2 - \gamma^2| < 1/T^3\}. \end{aligned}$$

Clearly, the set U_S is open and dense in $\mathbb{R}$. Hence the intersection $U = \bigcap_j U_j$ is a second category (uncountable) set. Choose an irrational number $\gamma \in U$. Then there exist $\beta_j \to \gamma$ and $T_j \to \infty$ such that

$$|\beta_j^2 - \gamma^2| < 1/T_j^3 \quad \text{and} \quad N^{P_{\beta_j}}_{(1/4,1)}(T_j) \ge T_j(\log T_j)^{1-\varepsilon}.$$

It is clear that $|P_{\beta_j}(x) - P_\gamma(x)| < 1/T_j < 1/8$ for all $|x| < T_j$ and large enough T_j. Hence

$$N^{P_\gamma}_{(1/8,2)}(T_j) \ge T_j(\log T_j)^{1-\varepsilon}.$$

A similar estimate can be proved for the interval $(-2, -1/8)$, and by rescaling for any interval $(a, b) \subset \mathbb{R}$.

4. Here we prove the Oppenheim–Davenport conjecture following [**DM90b**] and [**Ma90**]. Let $V_\mathbb{R}, A_\mathbb{R}, \Gamma \subset G$ be as in Subsection 1, and let

$$U_t = \begin{pmatrix} 1 & 0 & t \\ 0 & 1 & 0 \\ 0 & 0 & 1 \end{pmatrix}, \qquad \sigma_0 = \begin{pmatrix} 0 & 0 & 1 \\ 0 & -1 & 0 \\ 1 & 0 & 0 \end{pmatrix}, \qquad q_0 = \begin{pmatrix} 0 & 0 & 0 \\ 0 & 0 & 0 \\ 0 & 0 & 1 \end{pmatrix}.$$

We use the quadratic form $S(y) = y^\intercal \sigma_0 y$; that is, $S(y_1, y_2, y_3) = 2y_1y_3 - y_2^2$. Let $H \subset G$ be the identity component of the stabilizer of the form S. Our ultimate goal is the following result.

THEOREM 28.8. *If the closure $\overline{Hx} \subset G/\Gamma$ is compact, then the orbit Hx is compact.*

We make use of the following notations:

$$A = A_{\mathbb{R}},\ U = U_{\mathbb{R}},\ V = V_{\mathbb{R}},\ B = A \cdot (U \times V),\ F = A \cdot V,\ U^+ = U_{\mathbb{R}^+},\ U^- = U_{\mathbb{R}^-}.$$

It is clear that Theorem 28.8 is an immediate corollary of the following two statements.

LEMMA 28.9. *The closures $\overline{HU^+x}$ and $\overline{HU^-x}$ are noncompact for any point $x \in G/\Gamma$.*

Proof. The set

$$\{y \in \mathbb{R}^3 :\ y_3 > 0,\ S(y) < 0\}$$

contains a ball of arbitrarily large diameter. Let $x = g\Gamma$. Then there exists a point $y \in g(\mathbb{Z}^3)$ such that $y_3 > 0$ and $S(y) < 0$. But $S(U_t y) = S(y) + 2ty_3^2$. Therefore $0 \in S(U^+y)$, and hence the intersection of the orbit U^+y with the zero cone of S is nonempty. The group H acts transitively on each half of the cone, and it follows that $0 \in \overline{HU^+y}$. By Mahler's criterion, the closure $\overline{HU^+x}$ is noncompact. □

LEMMA 28.10. *Let a noncompact orbit $Hx \subset G/\Gamma$ have compact closure. Then there exists a point $y \in \overline{Hx}$ such that either U^+y or U^-y is contained in $\overline{Hx}$.*

We proceed to the proof of Lemma 28.10. Basically, it relies upon the following simple observation, whose proof is left to the reader.

PROPOSITION 28.11. *Let a Lie group G act continuously on a topological space Ω. Suppose that closed subsets $Z, W \subset \Omega$ are invariant under closed subgroups $V, H \subset G$ respectively, and Z is a compact V-minimal set. Finally, let $mZ \cap W \neq \varnothing$ for all $m \in M \subset G$. Then $hZ \subset W$ for any $h \in N_G(V) \cap \overline{HMV}$.*

Let us prove that Lemma 28.10 is a corollary of the following two results.

LEMMA 28.12. *Let $Y \subset X \subset \overline{Hx} \subset G/\Gamma$, where the set Y is compact and minimal under $F = AV$, and the set X is minimal under H. Suppose that $\overline{Hx} \neq Hx$ and $y \in Y$. Then the closure of the set*

$$\{g \in G - HU :\ gy \in \overline{Hx}\}$$

contains the identity element $1 \in G$.

LEMMA 28.13. *Let $M \subset G - HU$ and $1 \in \overline{M}$. Then $\overline{HMV}$ contains either U^+ or U^-.*

Indeed, the compact closure $\overline{Hx}$ contains a compact H-invariant minimal set X. In turn, X contains an F-minimal set Y, and the latter contains a V-minimal set Z. Given $z \in Z$, let

$$M = \{g \in G - HU :\ gz \in \overline{Hx}\}.$$

Then $1 \in \overline{M}$ (Lemma 28.12) and $\overline{HMV}$ contains either U^+ or U^- (Lemma 28.13). But by virtue of Proposition 28.11, $(N_G(V) \cap \overline{HMV})Z \subset \overline{Hx}$. Since $U \subset N_G(V)$, Lemma 28.10 is proved. □

Let us prove Lemma 28.12. We start with an elementary statement.

PROPOSITION 28.14. *Let a Lie group G act continuously and transitively on a topological space Ω, and let a closed subset $Y \subset \Omega$ be invariant under a closed subgroup $F \subset G$. Let G_y be the stabilizer of a point $y \in Y$. Then:*

a) *If the closure of the subset $\{g \in G - FG_y : \ gy \in Y\}$ does not contain $1 \in G$, then the orbit Fy is open in Y.*

b) *Let a subgroup $B \subset G$ be closed and contain F, let $B_y = B \cap G_y$, and let $\pi : B \to B/B_y$ be the natural map. Assume that the set Y is compact and F-minimal, and the closure $\overline{\pi(F)}$ is noncompact. Then the closure of the subset $\{g \in G - B : \ gy \in Fy\}$ contains $1 \in G$.*

It is easy to check that a discrete subgroup $\Lambda \subset B = A \cdot (U \times V)$ either belongs to the nilradical $U \times V$ or is conjugate to a subgroup of A. In both cases the closure $\overline{F\Lambda} \subset B/\Lambda$ is noncompact. Now 28.14b) immediately implies the following result.

PROPOSITION 28.15. *Let Y be a compact F-minimal set, and let $y \in Y$. Then the closure of the set $\{g \in G - B : \ gy \in Fy\}$ contains $1 \in G$.*

The vector space E of symmetric 3×3 matrices carries the right action of the group G, defined as

$$g(\sigma) = (g^{-1})^{\mathsf{T}} \sigma g^{-1}, \qquad g \in G, \ \sigma \in E.$$

The next result is straightforward.

PROPOSITION 28.16. a) *$U_t(\sigma_0) = \sigma_0 - 2tq_0$, and hence $(U_tH)(\sigma_0) = \sigma_0 - 2tq_0$ for all $t \in \mathbb{R}$.*

b) *If $g \in G$ and $g(\sigma_0) = \sigma_0 - tq_0$, then $g \in U_tH$.*

c) *The map $(h, U_t) \to hU_t$ establishes a homeomorphism $H \times U \to HU$.*

This easily implies the following result.

PROPOSITION 28.17. a) *Let $h \in H$ and $h(q_0) = tq_0$ for some $t \in \mathbb{R}$. Then $h \in F$.*

b) *If $h \in H, t \neq 0$ and $hU_t \in UH$, then $h \in F$.*

Now we prove Lemma 28.12. Assume the opposite: there exists a neighborhood $O(1) \subset G$ such that

$$(*) \qquad O(1) \cap \{g \in G : \ gy \in \overline{Hx}\} \subset HU.$$

If the orbit Hy is open in $\overline{Hx} \supset X$, then since the set $X = \overline{Hy}$ is H-minimal, the orbit Hy must be compact. Hence $Hy = \overline{Hx}$, which contradicts the condition $Hx \neq \overline{Hx}$. Therefore the orbit Hy is not open in $\overline{Hx}$. By virtue of Proposition 28.14a), the closure of the set

$$\{g \in G - H : \ gy \in \overline{Hx}\}$$

contains $1 \in G$. In view of (*), there exist sequences $h_i \in H$ and $t_i \neq 0$ such that $h_i U_{t_i} y \in \overline{Hx}$ and $h_i U_{t_i} \to 1$. Then $U_{t_i} y \in \overline{Hx}$, and, in view of 28.16c), $t_i \to 0$ as $i \to \infty$.

Proposition 28.15 ensures that the closure of the set

$$\{g \in G - B : \ gy \in Fy\}$$

contains $1 \in G$. Hence there exist $u \in O(1) \cap (G - B)$ and $p \in F$ such that $uy = py$. But in view of (*), $u \in HU - B = (H - F)U$ and hence $u = hU_s$ for some $h \in H - F$

and $s \in \mathbb{R}$. Since $t_i \to 0$, it follows that for some k we have $pU_{t_k}p^{-1}u \in O(1)$. On the other hand, since $U_{t_k}y \in \overline{Hx}$, it follows that

$$pU_{t_k}p^{-1}uy = pU_{t_k}p^{-1}py = pU_{t_k}y \in p\overline{Hx} = \overline{Hx}.$$

Now, by virtue of (*), $pU_{t_k}p^{-1}hU_s \in HU$, and hence $h^{-1}pU_{-t_k}p^{-1} \in UH$. But $h^{-1} \in H$ and $pU_{-t_k}p^{-1} \in U - \{1\}$. Now it follows from Proposition 28.17b) that $h \in F$, and we arrive at a contradiction. □

Now it suffices to prove Lemma 28.13. By induction on the dimension of a vector space, it is easy to prove the following property of unipotent linear maps.

PROPOSITION 28.18. *Let E be a vector space and let $E_0 \subset E$ be the subspace of fixed points for a unipotent subgroup $V_{\mathbb{R}} \subset \mathrm{GL}(E)$. Let $M_0 \subset E - E_0$ and $p_0 \in \overline{M_0} \cap E_0$. Then there exist a nonconstant polynomial map $\psi : \mathbb{R} \to E_0$ and sequences $m_i \in M_0$ and $t_i \in \mathbb{R}$ such that $\psi(0) = p_0$ and $V_{st_i}(m_i) \to \psi(s)$ as $i \to \infty$ for any $s \in \mathbb{R}$.*

It is not hard to see that the orbit $G(\sigma_0)$ is closed in E and hence homeomorphic to the homogeneous space G/H. Moreover, the action of V on E is unipotent and the subspace $E_0 = \mathbb{R}\sigma_0 + \mathbb{R}q_0 \subset E$ consists of all V-fixed points. It is straightforward to verify that $E_0 \cap G(\sigma_0) = \sigma_0 + \mathbb{R}q_0 = U(\sigma_0)$. Let $M_0 = M^{-1}(\sigma_0)$. Now we conclude that $M_0 \cap E_0 = \varnothing$ because $M \subset G - HU$.

Proposition 28.18 applied to the point $\sigma_0 \in \overline{M_0} \cap E_0$ provides us with a polynomial map $\psi : \mathbb{R} \to E_0$ and sequences $g_i \in M$ and $t_i \in \mathbb{R}$ such that

$$\psi(0) = \sigma_0 \quad \text{and} \quad (V_{st_i}g_i^{-1})(\sigma_0) \to \psi(s), \ s \in \mathbb{R}.$$

Since $\psi(\mathbb{R}) \subset E_0 \cap G(\sigma_0)$, there exists a polynomial $\phi : \mathbb{R} \to \mathbb{R}$ such that

$$\psi(s) = U_{\phi(s)}(\sigma_0), \qquad s \in \mathbb{R}.$$

Therefore $V_{st_i}g_i^{-1}H \to U_{\phi(s)}H$, and hence $\overline{HMV} \supset U_{\phi(\mathbb{R})}$. But since $\phi(0) = 0$ and $\phi \neq \text{const}$, the set $\phi(\mathbb{R})$ contains either $\mathbb{R}^+$ or $\mathbb{R}^-$. This yields Lemma 28.13. □

§29. Foundations of the theory of Diophantine approximation

1. Diophantine approximation of real numbers. We extract basic definitions, examples, and results from [**GNS**], [**Kh78**], [**KSF**], [**BSh**].

We will study the rate of approximation of a number $x \in \mathbb{R}$ by rational fractions p/q.

One says that x is approximated by rationals to within order $\alpha > 0$ if there exists $c = c(x)$ such that the inequality

$$\left| x - \frac{p}{q} \right| < \frac{c}{q^\alpha}$$

has infinitely many solutions $p, q \in \mathbb{Z}$.

LEMMA 29.1. *Let $x = k/l \in \mathbb{Q}$ be a noncancellable fraction. Then*
1) *there exist infinitely many solutions of the equality $|x - p/q| = \frac{1}{lq}$, and*
2) *the inequality $|x - p/q| < \frac{1}{lq}$ has only trivial solution $p/q = x$.*

Proof. 1) Let $p_0, q_0 \in \mathbb{Z}$ be such that $kq_0 - lp_0 = 1$. Then $x - p_0/q_0 = \frac{1}{lq_0}$. Clearly, there exist infinitely many pairs p, q of the form in question (take, for instance, $p = p_0 + km, q = q_0 + lm$).

2) If $|k/l - p/q| < \frac{1}{lq}$, then $|kq - lp| < 1$, and we are done. □

THEOREM 29.2 (Dirichlet). *For any $x \in \mathbb{R}$ and $n \in \mathbb{N}$ there exist $p \in \mathbb{Z}, q \in \mathbb{N}$ such that $1 \le q \le n$ and $|x - p/q| < \frac{1}{nq} \le \frac{1}{q^2}$.*

Proof. Let, say, $x > 0$. We consider $n+1$ numbers $\{ix\} = ix - [ix]$, $i = 0, \dots, n$ in the segment [0,1]. Then there exist $0 \le i < j \le n$ such that $|\{jx\} - \{ix\}| < 1/n$ (the Dirichlet principle!). It follows that

$$\left| x - \frac{[jx] - [ix]}{j - i} \right| < \frac{1}{n(j-i)},$$

and hence it suffices to take $p = [jx] - [ix]$, $q = j - i$, where, obviously, $0 < q \le n$. □

Note that there exist (transcendental) numbers admitting any preassigned order of approximation.

Example 29.3. Let $\psi : \mathbb{N} \to \mathbb{R}^+$ be a monotone increasing unbounded function, where $\psi(n) \ge n$, $n \in \mathbb{N}$. We define a sequence $\{k_n\}$ inductively as follows: $k_1 = 1$, $2^{k_{n+1}} > \psi(2^{k_n})$. Let

$$\alpha = \sum_{i=1}^{\infty} \frac{1}{2^{k_i}} \quad \text{and} \quad \frac{p_n}{q_n} = \sum_{i=1}^{n} \frac{1}{2^{k_i}}.$$

Then

$$\left| \alpha - \frac{p_n}{q_n} \right| \le \frac{2}{2^{k_{n+1}}} < \frac{2}{\psi(2^{k_n})} = \frac{2}{\psi(q_n)}.$$

On the other hand, there exist (algebraic) numbers admitting no order of approximation better than 2.

Example 29.4. We have

$$\left| \sqrt{2} - \frac{p}{q} \right| \left| \sqrt{2} + \frac{p}{q} \right| = \left| 2 - \frac{p^2}{q^2} \right| > \frac{1}{q^2},$$

because $\sqrt{2}$ is an irrational number. If $|\sqrt{2} - p/q| < 1$, then $|\sqrt{2} + p/q| < 4$, and hence $|\sqrt{2} - p/q| > \frac{1}{4q^2}$ (whereas for $|\sqrt{2} - p/q| \ge 1$ this is obviously so).

This example is generalized as follows.

THEOREM 29.5 (Liouville). *If x is an algebraic number of degree n, then there exists $c = c(x)$ such that $|x - p/q| > \frac{c}{q^n}$ for all $p, q \in \mathbb{Z}$.*

Proof. Let

$$f(t) = a_0 t^n + a_1 t^{n-1} + \dots + a_n$$

be a polynomial irreducible over $\mathbb{Q}$ with coefficients in $\mathbb{Z}$ and roots $x = x_1, \dots, x_n$. With no loss of generality we can assume that $n \ge 2$, that is, $x_i \notin \mathbb{Q}$, $i = 1, \dots, n$. Then $f(t) = a_0 \prod_1^n (t - x_i)$, and hence

$$\left| f\left(\frac{p}{q}\right) \right| = |a_0| \left| x - \frac{p}{q} \right| \prod_2^n \left| x_i - \frac{p}{q} \right| \le c \left| x - \frac{p}{q} \right|$$

for all $p/q \in \mathbb{Q}$ such that $|x - p/q| < 1$, and some constant $c > 0$. On the other hand, $q^n f(p/q) \in \mathbb{Z}$ and hence $|q^n f(p/q)| \ge 1$. It follows that $|x - p/q| > \frac{c^{-1}}{q^n}$ for all $p \in \mathbb{Z}$, $q \in \mathbb{N}$. □

Using this result, Liouville established the existence of transcendental numbers (subsequently Cantor also derived this from the countability of the set of algebraic numbers).

The Liouville theorem has been generalized many times. The strongest result in this direction is as follows.

THEOREM 29.6 (Roth, see [**Ca57**], [**Sch80**]). *If x is an algebraic number of degree $n \geq 2$ and $\varepsilon > 0$, then the inequality $|x - p/q| < \frac{1}{q^{2+\varepsilon}}$ has only finitely many solutions.*

We say that $x \in VWA$ (*x is very well approximable*) if x has order of approximation $\alpha > 2$.

Likewise, $x \in BA$ (*x is badly approximable*) if there exists $c > 0$ such that

$$\left| x - \frac{p}{q} \right| > \frac{c}{q^2} \qquad \text{for all } p, q \in \mathbb{Z}.$$

It follows from the definition and Roth's theorem that $x \notin VWA$ for any algebraic irrational number x. Note that for a quadratic irrational x (that is, for an algebraic number of degree 2), Liouville's theorem gives a better estimate: $x \in BA$. So far it is not known whether an algebraic number of degree $n \geq 3$ can be badly approximable.

We turn to the metric theory of Diophantine approximation. Many results to follow will be based on the next, well known statement.

LEMMA 29.7 (Borel–Cantelli). *Let (X, μ) be a measure space and let $\{A_n, n \in \mathbb{N}\}$ be a sequence of subsets in X such that $\sum_1^\infty \mu(A_n) < \infty$. Then the family $B \subset X$ of points $x \in X$ belonging to infinitely many sets A_n has μ-measure 0.*

Proof. $B = \bigcap_{n=1}^\infty B_n$, where $B_n = \bigcup_{i=n}^\infty A_i$. Since $\sum_1^\infty \mu(A_n) < \infty$, it follows that $\mu(B_n) \to 0$ and hence $\mu(B) = 0$. □

Henceforth l stands for the usual Lebesgue measure on $\mathbb{R}$.

LEMMA 29.8. $l(VWA) = 0$.

Proof. We introduce some notation: for a given $\varepsilon > 0$, let

$$A_\varepsilon = \left\{ x \in \mathbb{R} : \left| x - \frac{p}{q} \right| < \frac{1}{q^{2+\varepsilon}} \text{ for infinitely many } p, q \in \mathbb{Z} \right\}.$$

$$A_\varepsilon^q = \left\{ x \in [0,1] : \min_{p=0,\dots,q} \left| x - \frac{p}{q} \right| \leq \frac{1}{q^{2+\varepsilon}} \right\}.$$

Then $l(A_\varepsilon^q) \leq \frac{2}{q^{1+\varepsilon}}$. Hence $\sum_{q=1}^\infty l(A_\varepsilon^q) < \infty$ and, by the Borel–Cantelli lemma, $l(A_\varepsilon) = 0$. It now suffices to note that $VWA = \bigcup_j A_{1/j}$. □

Clearly, $VWA \cap BA = \varnothing$. In the twenties Khintchine used continued fractions (see Subsection 2) to prove that the set BA has measure 0 as well, but this result is much less trivial. However, by methods of ergodic theory it can be obtained almost for nothing. More specifically, as we have seen in §14, there is a natural mapping of $\mathbb{R}$ to the set of orbit classes of an ergodic flow on a finite-volume noncompact space. Moreover, BA corresponds to the family of bounded orbits (which is a null set due to ergodicity).

Note that although the sets BA and VWA are of measure 0, they are rather thick from the topological viewpoint: the Hausdorff dimension of each of them is equal to 1 (Jarnik, Besikovitch; see [**Sch80**]).

The basic theorem in the metric theory of Diophantine approximation sharpens the equalities $l(BA) = l(VWA) = 0$.

THEOREM 29.9 (Khintchine). *Let $\psi(q)$ be a positive nonincreasing function on the set $\mathbb{N}$ of natural numbers. Then the set*

$$A_\psi = \{x \in \mathbb{R} : \text{ the inequality } |x - p/q| < \psi(q)/q \text{ has infinitely many solutions } p \in \mathbb{Z},\ q \in \mathbb{N}\}$$

has measure 0 $\iff \sum_q \psi(q) < \infty$, *and has full measure* $\iff \sum_q \psi(q) = \infty$.

Example. If $\psi(q) = 1/(q \ln^\alpha q)$, then $l(A_\psi) = 0$ for $\alpha > 1$; and, conversely, $l(\mathbb{R} - A_\psi) = 0$ for $\alpha \le 1$. Note that the first statement of Khintchine's Theorem is rather elementary and is deduced along the same lines as Lemma 29.8. The proof of the second statement is not that easy.

To conclude, we note that the sets BA and VWA are invariant under the group Γ of integer matrices $\gamma \in \mathrm{GL}(2, \mathbb{R})$ with determinant ± 1. The action is given by the formula

$$\gamma(x) = \frac{kx + l}{mx + n}, \qquad \text{where } \gamma = \begin{pmatrix} k & l \\ m & n \end{pmatrix} \in \Gamma.$$

Moreover, x and $\gamma(x)$ have equal orders of approximation by rationals, because this is true for numbers x, $1/x$, $x + 1$, and nx, where $n \in \mathbb{N}$. This enables one to introduce an equivalence relation as follows: $x \sim y$ if $y = \gamma(x)$ for some $\gamma \in \Gamma$. The Lebesgue measure on $\mathbb{R}$ is quasi-invariant under the action of Γ, and the action is ergodic (see §14). Hence it is not surprising that any Γ-invariant family of numbers is either null or conull set.

2. Continued fractions. We give some preliminaries from the theory of continued fractions (see [**Kh78**] or [**Vin**]).

Each number $x \in (0, 1)$ can be written uniquely in the form

$$x = \frac{1}{k_1 + \frac{1}{k_2 + \dots}}, \qquad \text{where } k_i \in \mathbb{N}.$$

An expression like this will be denoted by $[k_1, k_2, \dots]$. If x is rational, then the expansion is finite: $x = [k_1, k_2, \dots, k_n]$, whereas for irrational x it is infinite; moreover, $x = \lim [k_1, k_2, \dots, k_n]$ and the limit always exists.

Example. Given $\alpha = \sqrt{2} - 1$, one has $\alpha = \frac{1}{2+\alpha}$ and hence $\alpha = [2, 2, \dots]$. More generally, any quadratic irrational corresponds to an eventually periodic sequence (Euler, Lagrange). At the same time, no explicit continued-fraction decomposition is known for algebraic numbers of degree $n \ge 3$.

Let $p_n/q_n = [k_1, k_2, \dots, k_n]$. The following recurrent relations hold:

$$p_i = k_i p_{i-1} + p_{i-2}, \quad q_i = k_i q_{i-1} + q_{i-2}, \qquad i \ge 2,$$

where $p_0 = 0, p_{-1} = 1, q_0 = 1, q_{-1} = 0$.

It follows that the sequences $\{p_i\}$ and $\{q_i\}$ increase exponentially in i: $p_i \ge 2^{(i-2)/2}$, $q_i \ge 2^{(i-1)/2}$.

Approximations $\{p_i/q_i\}$ for an irrational number $x \in (0,1)$ are the best ones in the following sense:

$$|qx - p| \geq |q_i x - p_i| \quad \text{for all } p, q \in \mathbb{N},\ 0 \leq q < q_{i+1}.$$

The rate of approximation is estimated as follows:

$$\frac{1}{2q_i q_{i+1}} < \left| x - \frac{p_i}{q_i} \right| \leq \frac{1}{q_i q_{i+1}}.$$

Using the recurrent relation between q_{i+1}, q_i and q_{i-1}, one easily constructs numbers with preassigned order of approximation by rationals.

In addition, it follows from the estimates above that an irrational number $x = [k_1, k_2, \dots]$ is badly approximable if and only if the natural numbers k_i are uniformly bounded. Note that given $i, n \in \mathbb{N}$, numbers x with $k_i = n$ form a family of intervals. Now estimating the total measure of the intervals, one easily proves that the set of badly approximable numbers is of measure 0.

Given a number $\alpha = [k_1, k_2, \dots]$, we introduce an arithmetic function

$$Q_\alpha(t) = \sum_{1 \leq j \leq t} k_j.$$

This function proves to be helpful in finding the asymptotics of the remainder term in the Birkhoff–Khintchine ergodic theorem applied to an irrational rotation of the circle. More precisely, let $T_\alpha : [0,1) \to [0,1)$ be a map given as $T_\alpha(x) = \{x + \alpha\} = (x + \alpha) \pmod 1$, and let $f(x) = x,\ x \in [0,1)$. If α is an irrational number, then, by the ergodic theorem,

$$\sum_{j=1}^{m} \{j\alpha + x\} = \frac{m}{2} + r_\alpha(m), \qquad \text{where } r_\alpha(m) = o(m),\ m \to \infty.$$

(Moreover, the relation holds for *all* x because T_α defines a uniquely ergodic map of the circle and f is "almost" continuous.) As was noticed in [**Skr98**], the solution of Exercise 4 in Chapter 3 of Vinogradov's book [**Vin**] gives the following estimate, uniform in x:

$$|r_\alpha(m)| < \frac{3}{2} Q_\alpha(5 \lg m + 1).$$

Clearly, $Q_\alpha(t) \leq Ct$ if the number α is badly approximable. Hence

$$r_\alpha(m) = O(\ln m) \quad \text{as } m \to \infty \quad \text{for any } \alpha \in BA.$$

Actually, the estimate was first found by Hardy and Littlewood [**HL**]. Moreover, they proved that the estimate $r_\alpha(m) = o(m)$ cannot be improved in the class of all irrational numbers α. In other words, the function $r_\alpha(m)/m$ can decrease arbitrarily slowly.

On the other hand, as was proved by Khintchine [**Kh23**], for almost all $\alpha \in (0,1)$,

$$Q_\alpha(t) = O(t\phi(\ln t)),\ t \to \infty,$$

where ϕ is any positive nondecreasing function such that $\sum_1^\infty \frac{1}{\phi(j)} < \infty$. In particular, given $\phi(t) = t^{1+\varepsilon}$, we obtain

$$r_\alpha(m) = O(\ln m (\ln \ln m)^{1+\varepsilon}), \qquad m \to \infty,$$

for almost all $\alpha \in (0,1)$ and each $\varepsilon > 0$ (this will be used in §31).

The continued fraction for a real number $x \in \mathbb{R}$ is usually written as $x = [k_0; k_1, k_2, \dots]$, where $k_0 = [x]$ and $\{x\} = x - [x] = [k_1, k_2, \dots]$. It is easy to see that the *Gauss transform* of the interval $(0,1)$, given as $x \to \{1/x\}$, reduces to the shift of the continued fraction on the left. This implies that $x \sim y \Longleftrightarrow$ there exist integers $i, j \geq 0$ such that $k_{i+l} = m_{j+l}$, $l \in \mathbb{N}$, where $x = [k_0; k_1, k_2, \dots]$ and $y = [m_0; m_1, m_2, \dots]$. Note that the Gauss transform of $(0,1)$ is mixing relative to an invariant smooth measure [**KSF**].

3. Diophantine approximation in $\mathbb{R}^n$. We follow [**Ca57,59**]. To proceed to Diophantine approximation in n-dimensional space, we need one of the fundamental results in the geometry of numbers:

LEMMA 29.10 (Minkowski, see [**Ca59**]). *Let O be a convex centrally symmetric domain in $\mathbb{R}^n$, and let* $\mathrm{vol}(O) > 2^n$. *Then $\mathbb{Z}^n \cap O \neq \{0\}$.*

If in addition O is closed, it can be easily seen that the result holds true if $\mathrm{vol}(O) \geq 2^n$. Note that 2^n is the volume of the largest open cube that contains no points of $\mathbb{Z}^n - \{0\}$. As a rule, for applications one takes a (skew) parallelopiped P given by a system of inequalities

$$\begin{cases} |a_{11}x_1 + \cdots + a_{1n}x_n| \leq c_1, \\ |a_{21}x_1 + \cdots + a_{2n}x_n| \leq c_2, \\ \dots\dots \\ |a_{n1}x_1 + \cdots + a_{nn}x_n| \leq c_n, \end{cases}$$

where c_i, $i = 1, \dots, n$, are positive numbers and the matrix $A = \{a_{ij}\}$ is nondegenerate. A straightforward computation shows that

$$\mathrm{vol}(P) = \frac{2^n \prod_i c_i}{\det(A)}.$$

Let $\mathbf{x} = (x_1, \dots, x_n)$ be a vector in $\mathbb{R}^n$. We will estimate the rate of approximation of $\mathbf{x}$ by rational vectors with equal denominators. Define the norm $\|\mathbf{x}\| = \max_i |x_i|$. We are concerned with $\|\mathbf{x} - \mathbf{p}/q\|$, where $\mathbf{p} \in \mathbb{Z}^n$ and $q \in \mathbb{N}$.

The following result is a counterpart of Dirichlet's Theorem 29.2 in the multidimensional case.

THEOREM 29.11. *Given $\mathbf{a} \in \mathbb{R}^n$ and $R > 1$, there exist $\mathbf{p} \in \mathbb{Z}^n$ and $q \in \mathbb{N}$, $q \leq R$, such that $\|q\mathbf{a} - \mathbf{p}\| < \frac{1}{R^{1/n}}$, $q \leq R$. In particular, the inequality $\|q\mathbf{a} - \mathbf{p}\|^n < \frac{1}{q}$ has infinitely many solutions $\mathbf{p} \in \mathbb{Z}^n$, $q \in \mathbb{N}$.*

Proof. Apply Minkowski's lemma to a parallelepiped in $\mathbb{R}^{n+1}$ given by the system of inequalities

$$\begin{cases} |y_1 + a_1 y_{n+1}| \leq R^{-1/n}, \\ \dots\dots \\ |y_n + a_n y_{n+1}| \leq R^{-1/n}, \\ |y_{n+1}| \leq R. \end{cases}$$

□

By analogy with the case $n = 1$, we say that a vector $\mathbf{a} \in \mathbb{R}^n$ is *very well approximable* ($\mathbf{a} \in VWA$), if for some $\varepsilon > 0$ the inequality

$$\|q\mathbf{a} - \mathbf{p}\|^n < \frac{1}{q^{1+\varepsilon}}$$

has infinitely many solutions $\mathbf{p} \in \mathbb{Z}^n$, $q \in \mathbb{N}$.

Likewise, $\mathbf{a} \in \mathbb{R}^n$ is *badly approximable* ($\mathbf{a} \in BA$), if there exists $0 < c < 1$ such that the inequality

$$\|q\mathbf{a} - \mathbf{p}\|^n < \frac{c}{q}$$

has no integer solutions.

An example of a badly approximable vector can be constructed using the following result (see [**Sch80**]). If $p(t)$ is a polynomial of degree n with coefficients in $\mathbb{Z}$ that is irreducible over $\mathbb{Q}$, then $\mathbf{a} = (a, a^2, \dots, a^{n-1}) \in BA$ for every real root a (this generalizes the property that $a \in BA$ for any quadratic irrational $a \in \mathbb{R}$).

LEMMA 29.12. $l(VWA) = 0$.

Proof. As in Lemma 29.8, let

$$A_\varepsilon^q = \{\mathbf{a} \in [0,1]^n : \min_{\|\mathbf{p}\| \le q} \|q\mathbf{a} - \mathbf{p}\|^n < 1/q^{1+\varepsilon}\}.$$

Then $l(A_\varepsilon^q) \le 2^n/q^{1+\varepsilon}$, and hence $\sum_{q=1}^\infty l(A_\varepsilon^q) < \infty$. The rest is evident. □

Later we will prove the equality $l(BA) = 0$ from ergodic considerations. At the same time, the family $BA \subset \mathbb{R}^n$ is of Hausdorff dimension n (Schmidt [**Sch69**])[2].

Theorem 29.9 is generalized in the multi-dimensional case as follows (see also [**Dod**]).

THEOREM 29.13 (Khintchine–Groshev, [**Gro**]). *Let $\psi(q)$ be a positive function nonincreasing in $q \in \mathbb{N}$. Then the set*

$$A_\psi = \{\mathbf{a} \in \mathbb{R}^n : \textit{the inequality } \|q\mathbf{a} - \mathbf{p}\| < \psi(q) \textit{ has infinitely many solutions } q \in \mathbb{N},\ \mathbf{p} \in \mathbb{Z}^n\}$$

is of measure $0 \iff \sum_q \psi(q)^n < \infty$, *and is of full measure* $\iff \sum_q \psi(q)^n = \infty$.

We observe that the property $\mathbf{a} \in VWA$ does not imply, in general, that $a_i \in VWA$, $i = 1, \dots, n$. For instance, $(b, b) \in VWA$ regardless of Diophantine properties of the number $b \in \mathbb{R}$. More generally, $\mathbf{a} = (a_1, a_2) \in VWA$ if the numbers $1, a_1, a_2$ are linearly dependent over $\mathbb{Q}$. Thus, the subset VWA of the plane $\mathbb{R}^2$ contains any line $k + la_1 + ma_2 = 0$, where $k, l, m \in \mathbb{Z}$.

Now choose a smooth curve $K \subset \mathbb{R}^n$. It is of interest to find how many well-approximable points it contains (with respect to the Lebesgue measure on the curve). As the case $n = 2$ shows, all points on a line can be well approximated. But this is not so for other curves. It was conjectured in the thirties by Mahler that almost all points of the curve $K_0 = \{(t, t^2, \dots, t^n),\ t \in \mathbb{R}\}$, are not VWA. For $n = 2$ this was proved in the 50s by Kubilius. In the general case, Mahler's conjecture was settled in the sixties by Sprindzhuk (see [**Sp77**]). At the same time, almost

---

[2] Actually, Schmidt proved a stronger result, saying that the set BA is (α, β)-winning (see [**Sch69**] for more).

all points on a line may not belong to the class VWA. For instance, according to Schmidt (see [**Sch80**]), if $a \notin VWA$, then $(a,t) \notin VWA$ and $(t, at) \notin VWA$ for almost all $t \in \mathbb{R}$.

Let $f : U \to \mathbb{R}^n$ be an analytic map from an open region $U \subset \mathbb{R}^k$ to $\mathbb{R}^n$, where the functions $1, f_1, \ldots, f_n$ are linearly independent over $\mathbb{R}$ (in other words, the manifold $M = f(U)$ is not contained in a proper affine subspace of $\mathbb{R}^k$). Such an analytic manifold $M \subset \mathbb{R}^n$ will be referred to as a *curved* manifold. It was conjectured in the survey [**Sp80**] that, for a curved manifold M, the set $M \cap VWA$ is null with respect to the Lebesgue measure l_M on the manifold M. Schmidt (see [**Sch80**]) proved the conjecture for $n = 2$, and recently Beresnevich and Bernik [**BeB**] did it for $n = 3$.

A class larger than $VWA \subset \mathbb{R}^n$ arises as follows. Given $\mathbf{a} \in \mathbb{R}^n$, we introduce the notation

$$\Pi(\mathbf{a}) = \prod_1^n |a_i|, \qquad \Pi_+(\mathbf{a}) = \prod_1^n |a_i|_+, \qquad \text{where } |x|_+ = \max(|x|, 1).$$

We say that $\mathbf{a} \in \mathbb{R}^n$ is *very well multiplicatively approximable* ($\mathbf{a} \in VWMA$) if for some $\varepsilon > 0$ there are infinitely many points $q \in \mathbb{Z} - \{0\}$, $\mathbf{p} \in \mathbb{Z}^n$ such that

$$\Pi(q\mathbf{a} + \mathbf{p}) \leq |q|^{-1-\varepsilon}.$$

Clearly, $VWA \subset VWMA$.

An example of numbers not in the class $VWMA$ is given by the following deep generalization of Roth's theorem.

THEOREM 29.14 (Schmidt, [**Sch80**]). *Let $a_1, \ldots, a_n \in \mathbb{R}$ be algebraic numbers such that $1, a_1, \ldots, a_n$ are linearly independent over $\mathbb{Q}$. Then $\mathbf{a} = (a_1, \ldots, a_n) \notin VWMA$.*

The equality $l(VWMA) = 0$ was established by Spencer [**Spen**]. Moreover, he showed that

$$\lim_{q \to \infty} \min_{\mathbf{p} \in \mathbb{Z}^n} q (\ln q)^{n+\delta} \Pi(q\mathbf{a} + \mathbf{p}) = \infty$$

for almost all $\mathbf{a} \in \mathbb{R}^n$ and arbitrarily small $\delta > 0$.

Of course, proving the property $l_M(M \cap VWMA) = 0$ is even more difficult than proving the property $l_M(M \cap VWA) = 0$. For instance, the multiplicative analogue of Mahler's conjecture for the curve $K_0 \subset \mathbb{R}^n$ (called Baker's conjecture) was open until recently. Sprindzhuk [**Sp80**] formulated a general conjecture that almost all points of a curved manifold $M \subset \mathbb{R}^n$ are not $VWMA$; he himself proved the conjecture for $n = 2$. In full scope it was proved by Kleinbock and Margulis (see §30).

4. Diophantine approximation of real matrices. Let L be a real $n \times m$ matrix treated as a linear operator from $\mathbb{R}^m$ to $\mathbb{R}^n$. Then the transpose matrix $L^\intercal \in \mathrm{Mat}(m,n)$ is an operator from $\mathbb{R}^n$ to $\mathbb{R}^m$. The matrix L will be also considered as a collection of n homogeneous linear forms $L_j : \mathbb{R}^m \to \mathbb{R}$.

One of the basic tools in the approximation of matrices is as follows.

THEOREM 29.15 (Khintchine's transfer principle, see [**Ca57**]). *Let* $0 < c < 1 \le X$, *and let there exist* $x \in \mathbb{Z}^m - \{0\}$ *and* $y \in \mathbb{Z}^n$ *such that*

$$\|L(x) + y\| \le c \quad \textit{and} \quad \|x\| \le X.$$

Then there exist $u \in \mathbb{Z}^n - \{0\}$ *and* $v \in \mathbb{Z}^m$ *such that*

$$\|L^{\intercal}(u) + v\| \le d \quad \textit{and} \quad \|u\| \le U.$$

Explicitly,

$$d = (l-1)X^{\frac{1-n}{l-1}} c^{\frac{n}{l-1}}, \qquad U = (l-1)X^{\frac{m}{l-1}} c^{\frac{1-m}{l-1}},$$

where $l = n + m$.

The proof makes use of the following result.

LEMMA 29.16 [**Ca57**]. *Let* $\{f_i(z),\ i = 1, \dots, l\}$ *be a collection of* l *homogeneous linear forms in* $z \in \mathbb{R}^l$. *Similarly, let* $\{g_i(w),\ i = 1, \dots, l\}$ *be a collection of* l *homogeneous linear forms in* $w \in \mathbb{R}^l$, *where the matrix* G *given by the forms* g_i *has determinant* $\det(G) = \Delta \ne 0$. *Assume that the bilinear form*

$$\Phi(z, w) = \sum_{1}^{l} f_i(z) g_i(w)$$

has integer coefficients. Then if there exists $z \in \mathbb{Z}^l - \{0\}$ *such that* $|f_i(z)| \le 1$, $i = 1, \dots, l$, *then there exists* $w \in \mathbb{Z}^l - \{0\}$ *such that*

$$|g_i(w)| \le (l-1)|\Delta|^{\frac{1}{l-1}}, \qquad i = 1, \dots, l.$$

Proof. Fix $z \in \mathbb{Z}^l - \{0\}$ from the hypothesis of the lemma. With no loss of generality, $0 < \max_i |f_i(z)| = |f_l(z)| = \lambda \le 1$. The determinant of the system composed from l linear forms $\Phi(z, w), g_1(w), \dots, g_{l-1}(w)$, is $f_l(z) = \lambda\Delta$. By Minkowski's lemma, there exists $w \in \mathbb{Z}^l - \{0\}$ such that

$$|\Phi(z, w)| < 1 \quad \text{and} \quad |g_i(w)| \le |\Delta|^{\frac{1}{l-1}}.$$

But then $\Phi(z, w) = 0$, and hence $|g_l(w)| \le (l-1)|\Delta|^{\frac{1}{l-1}}$. ☐

Proof of Theorem 29.15. We introduce variables $y \in \mathbb{R}^n$ and $v \in \mathbb{R}^m$. Let $f_k(x, y)$ be a linear form on $\mathbb{R}^l$, given as

$$f_k(x, y) = c^{-1}(L_k(x) + y_k),\ 1 \le k \le n, \ \text{ and } \ f_k(x, y) = X^{-1} x_{k-n},\ n < k \le l.$$

In a similar fashion, let

$$g_k(u, v) = c u_k,\ 1 \le k \le n, \ \text{ and } \ g_k(u, v) = X(-L^{\intercal}_{k-n}(u) + v_{k-n}),\ n < k \le l.$$

The determinant of the system composed from g_k is $\Delta = c^n X^m$. Note that

$$\sum_{k=1}^{l} f_k g_k = \sum_{j \le n} u_j y_j + \sum_{i \le m} v_i x_i.$$

By the hypothesis of the theorem, there exist $(x, y) \in \mathbb{Z}^l - \{0\}$ such that $|f_k(x, y)| \le 1$, $k = 1, \dots, l$. Applying Lemma 29.16, we find a pair $(u, v) \in \mathbb{Z}^l - \{0\}$ such that

$$c\|u\| \le (l-1)\Delta^{\frac{1}{l-1}} \quad \text{and} \quad X\| - L^{\intercal}(u) + v\| \le (l-1)\Delta^{\frac{1}{l-1}}.$$

The rest is left to the reader. ☐

We say that a matrix L is *very well approximable*, that is, $L \in VWA$, if for some $\varepsilon > 0$ the inequality

$$\|L(q) - p\|^n < \frac{1}{\|q\|^{m+\varepsilon}}$$

has infinitely many solutions $p \in \mathbb{Z}^n$, $q \in \mathbb{Z}^m - \{0\}$.

Likewise, a matrix L is *badly approximable*, that is, $L \in BA$, if there exists $0 < c < 1$ such that

$$\|L(q) - p\|^n > \frac{c}{\|q\|^m} \quad \text{for all } p \in \mathbb{Z}^n,\ q \in \mathbb{Z}^m - \{0\}.$$

It follows from Theorem 29.15 that

COROLLARY 29.17. *$L \in VWA \iff L^{\mathsf{T}} \in VWA$; $L \in BA \iff L^{\mathsf{T}} \in BA$.*

In particular ($n = 1$), the linear form $L(x) = \mathbf{a}\mathbf{x} = a_1x_1 + \cdots + a_mx_m$ on $\mathbb{R}^m$ has the same Diophantine properties as the vector $\mathbf{a} = (a_1, \ldots, a_m) \in \mathbb{R}^m$. In other words, if $\mathbf{a} \in \mathbb{R}^m$, then $\mathbf{a} \in VWA$ whenever the following two equivalent conditions hold:

1) For some $\varepsilon > 0$, there are infinitely many $\mathbf{q} \in \mathbb{Z}^m - \{0\}$ and $p \in \mathbb{Z}$ such that $|\mathbf{q}\mathbf{a} + p| \cdot \|\mathbf{q}\|^m \leq \|\mathbf{q}\|^{-m\varepsilon}$.

1′) For some $\varepsilon > 0$, there are infinitely many $q \in \mathbb{Z} - \{0\}$ and $\mathbf{p} \in \mathbb{Z}^m$ such that $\|q\mathbf{a} + \mathbf{p}\|^m \cdot |q| \leq |q|^{-\varepsilon}$.

Similarly, $\mathbf{a} \in VWMA$ if the following two equivalent conditions hold:

2) For some $\varepsilon > 0$, there are infinitely many $\mathbf{q} \in \mathbb{Z}^m - \{0\}$ and $p \in \mathbb{Z}$ such that $|\mathbf{q}\mathbf{a} + p| \cdot \Pi_+(\mathbf{q}) \leq \Pi_+(\mathbf{q})^{-\varepsilon}$,

2′) For some $\varepsilon > 0$, there are infinitely many $q \in \mathbb{Z} - \{0\}$ and $\mathbf{p} \in \mathbb{Z}^m$ such that $\Pi(q\mathbf{a} + \mathbf{p}) \cdot |q| \leq |q|^{-\varepsilon}$.

It follows, for instance, that $\mathbf{a} = (a_1, a_2) \in VWMA$ if either of the numbers a_1, a_2 has order of approximation > 2 (that is, belongs to the class VWA). On the other hand, $\mathbf{a} = (a_1, a_2) \in VWA$ if a_1 or a_2 has order of approximation > 3.

The equality $l(BA) = 0$ was proved by Groshev [**Gro**] (see also [**Sch80**]). We will prove it in §30.

A matrix $L \in \mathrm{Mat}(n, m)$ is said to be *singular* (in Khintchine's sense) if, for any $\varepsilon > 0$ and large enough N, the system of inequalities

$$\|L(q) - p\|^n < \frac{\varepsilon}{N^m}, \qquad 0 < \|q\| < N,$$

has a solution[3] $p \in \mathbb{Z}^n$, $q \in \mathbb{Z}^m$.

Clearly, a matrix L is singular (and, in addition, VWA), if the kernel of the operator $(p, q) \to L(q) - p$ contains at least one vector $(p, q) \in \mathbb{Z}^{n+m} - \{0\}$. In particular ($n = m = 1$), every number $a \in \mathbb{Q}$ is singular. On the other hand, the theory of continued fractions tells us that, given an irrational number $a \in \mathbb{R}$ and the sequence p_i/q_i of best approximations, the following two inequalities hold:

$$\begin{cases} |qa - p| \geq |q_ia - p_i| & \text{for all } p, q \in \mathbb{Z},\ 0 \leq q < q_{i+1}, \\ q_{i+1}|q_ia - p_i| > 1/2. \end{cases}$$

[3] Till the end of §29 all vectors are again denoted by italic letters.

It follows that $|qa - p| > \frac{1}{2q_{i+1}}$ for all $p, q \in \mathbb{Z}$, $0 \le q < q_{i+1}$, and hence no irrational number can be singular.

On the contrary, if $n > 1$ or $m > 1$, then, given a singular matrix $L \in \mathrm{Mat}(n, m)$, it can happen that no pair $p \in \mathbb{Z}^n$, $q \in \mathbb{Z}^m - \{0\}$ satisfies $L(q) - p = 0$.

Consider, for instance, the case $n = 1$, $m = 2$. Let $L = (a_1, a_2)$. Then L is singular if $q_1 a_1 + q_2 a_2 - p = 0$ for some q_1, q_2, $p \in \mathbb{Z}$. Hence the set of singular matrices $L \in \mathrm{Mat}(1, 2)$ contains the countable family of lines given by equations with integer coefficients. But this family does not exhaust all singular matrices. This is seen from the following result.

LEMMA 29.18 [**Ca57**]. *For any function $w : \mathbb{N} \to \mathbb{R}^+$ there exists a pair (a_1, a_2) such that*

1) *the inequality $|q_1 a_1 + q_2 a_2 - p| < w(n)$ has an integer solution $q = (q_1, q_2) \in \mathbb{Z}^2$, $p \in \mathbb{Z}$, where $0 < \|q\| \le n$, and*

2) *the numbers 1, a_1, a_2 are linearly independent over $\mathbb{Q}$.*

In particular, if $n^2 w(n) \to 0$ as $n \to +\infty$, we obtain a singular pair $(a_1, a_2) \in \mathbb{R}^2$ which does not belong to a line with integer coefficients.

Certainly, a matrix $L \in \mathrm{Mat}(n, m)$ is singular $\iff L^\intercal \in \mathrm{Mat}(m, n)$ is singular.

LEMMA 29.19. *The set of singular matrices has measure* 0 *in* $\mathrm{Mat}(n, m)$.

Proof. It suffices to examine the unit cube in the space $\mathrm{Mat}(n, m)$ consisting of matrices $L = \{a_{i,j}\}$ such that $0 \le a_{i,j} \le 1$, $i = 1, \dots, n$, $j = 1, \dots, m$. We fix $q \in \mathbb{Z}^m - \{0\}$ and assume that $q_1 \ne 0$. Given elements $a_{i,2}, \dots, a_{i,m}$, the measure of numbers $0 \le a_{i,1} \le 1$ satisfying the inequality $|L(q)_i - p_i| < \varepsilon R^{-m/n}$ for some $p_i \in \mathbb{Z}$ does not exceed $2\varepsilon R^{-m/n}$. Hence, given $q \in \mathbb{Z}^m - \{0\}$, the set of matrices L in the unit cube satisfying the inequality $\|L(q) - p\| \le \varepsilon R^{-m/n}$ for some $p \in \mathbb{Z}^n$ is of measure at most $(2\varepsilon)^n R^{-m}$. But the number of points $q \in \mathbb{Z}^m$ with $\|q\| \le R$ does not exceed $(2R + 1)^m < (3R)^m$. Hence, given $R > 1$, the measure of the set

$$A_R = \{L : \ \|L(q) - p\| \le \varepsilon R^{-m/n} \text{ for some } p \in \mathbb{Z}^n, \ q \in \mathbb{Z}^m, \ \|q\| \le R\}$$

is at most $\varepsilon_1 = 3^m 2^n \varepsilon^n$. Since the set of singular matrices is $\bigcup_R \bigcap_{r \ge R} A_r$, its measure is also bounded by an arbitrarily small number ε_1. □

The singularity property plays an important role while approximating Diophantine solutions of the equation $L(y) = x$, where $x \in \mathbb{R}^n$, $y \in \mathbb{R}^m$. More specifically, for any $x \in \mathbb{R}^n$ the system of equations

$$\|L(q) - p - x\|^n < \frac{C}{\|q\|^m}$$

has infinitely many solutions $p \in \mathbb{Z}^n$, $q \in \mathbb{Z}^m$ for some $C > 0$ if and only if the matrix $L \in \mathrm{Mat}(n, m)$ is not singular (Khintchine, [**Kh48**]). For $n = m = 1$ the result goes back to Chebyshev.

§30. Diophantine approximation from the ergodic viewpoint

Here we interpret Diophantine properties of matrices and vectors in terms of ergodic theory and discuss Littlewood's conjecture. Then we set forth some important results on Diophantine approximation on manifolds due to Kleinbock and Margulis.

1. Diophantine properties of matrices. We will work with $k \times l$ matrices. In what follows,

$$G = \mathrm{SL}(k+l, \mathbb{R}), \qquad \Gamma = \mathrm{SL}(k+l, \mathbb{Z}), \qquad D_t = \begin{pmatrix} d_k(t) & 0 \\ 0 & d_l(-tk/l) \end{pmatrix},$$

where $d_n(t) = \mathrm{diag}(e^t, \dots, e^t)$ stands for the diagonal $n \times n$ matrix with elements e^t along the main diagonal. Our study will be concerned with positive semiorbits of the flow $(G/\Gamma, D_{\mathbb{R}})$. The expanding horospherical subgroup for $D_{\mathbb{R}}$ is the kl-dimensional Abelian matrix group of elements

$$h_L = \begin{pmatrix} I_k & L \\ 0 & I_l \end{pmatrix}, \qquad L \in \mathrm{Mat}(k, l),$$

where I_n is the identity $n \times n$ matrix. The neutral subgroup consists of matrices $\begin{pmatrix} A & 0 \\ 0 & B \end{pmatrix}$, where $A \in \mathrm{GL}(k, \mathbb{R})$, $B \in \mathrm{GL}(l, \mathbb{R})$, and the contracting subgroup consists of matrices

$$\begin{pmatrix} I_k & 0 \\ C & I_l \end{pmatrix}, \qquad C \in \mathrm{Mat}(l, k).$$

Each element $g \in G$ decomposes into the product

$$g = \begin{pmatrix} A & 0 \\ C & B \end{pmatrix} \begin{pmatrix} I_k & L \\ 0 & I_l \end{pmatrix} \gamma, \qquad \gamma \in \Gamma.$$

It is clear that the behavior of a trajectory $D_t g \Gamma$ as $t \to +\infty$ depends only on the properties of the matrix L in the decomposition of the element g. Hence the classes of divergent and bounded positive semiorbits of our flow correspond to families of matrices $L \in \mathrm{Mat}(k, l)$.

THEOREM 30.1 (Dani [**Da85**]). *A semiorbit $\{D_t h_L \Gamma,\ t \geq 0\}$ is divergent $\Longleftrightarrow$ the matrix L is singular.*

Proof. 2) $\Rightarrow$ 1). Let $B_\delta = \{x \in \mathbb{R}^{k+l} : \|x\| < \delta\}$, where $0 < \delta < 1$, and let $\Lambda = h_L(\mathbb{Z}^{k+l})$. According to Mahler's criterion, it suffices to find $T = T(\delta)$ such that $D_t(\Lambda) \cap B_\delta \neq \{0\}$ for all $t \geq T$. Fix $0 < \varepsilon < \delta^{1+l/k}$. Since the matrix is singular, there exists N_0 such that for all $N \geq N_0$ there exist $q \in \mathbb{Z}^l - \{0\}$ and $p \in \mathbb{Z}^k$ for which

$$\|L(q) + p\| < \frac{\varepsilon}{N^{l/k}} \quad \text{and} \quad \|q\| < N.$$

Choose $y = (L(q) + p, q) \in \Lambda$. Then $D_t(y) = (e^t(L(q) + p), e^{-tk/l} q)$. Hence

$$D_t(y) \in B_\delta \quad \text{whenever} \quad e^{-tk/l} N < \delta \text{ and } \frac{\varepsilon e^t}{N^{l/k}} < \delta.$$

Let $I_N = (\frac{l}{k} \log \frac{N}{\delta}, \log \frac{\delta N^{l/k}}{\varepsilon})$. Then $D_t(y) \cap B_\delta \neq \{0\}$ for all $t \in I_N$. Since $\varepsilon < \delta^{k+l}$, it follows that I_N is a nonempty interval, and, for large enough N, the intervals I_N and I_{N+1} overlap. Hence $\bigcup I_N$ contains an interval of the form $(T, +\infty)$.

1) $\Rightarrow$ 2). Assume that L is nonsingular. Then there exist $0 < \varepsilon < 1$ and a sequence $N_i \to +\infty$ for which

$$\|L(q) + p\| > \varepsilon N_i^{-l/k} \quad \text{for all } p \in \mathbb{Z}^k \text{ and } q \in \mathbb{Z}^l \text{ such that } 0 < \|q\| < N_i.$$

Let t_i satisfy the inequality $N_i < e^{t_i k/l} < \frac{N_i}{\varepsilon}$. Then $D_{t_i}(\Lambda) \cap B_\varepsilon = \{0\}$. Indeed, let $y = (L(q) + p, q) \in \Lambda$, where $(p, q) \in \mathbb{Z}^{k+l}$. If $\|q\| \geq N_i$, then $\|D_{t_i}(y)\| \geq$

$e^{-t_i k/l} N_i > \varepsilon$. But if $0 < \|q\| < N_i$, then $\|D_{t_i}(y)\| > \varepsilon e^{t_i} N_i^{-l/k} > \varepsilon$. Since $t_i \to +\infty$, the semiorbit $\{D_t h_L \Gamma,\ t \geq 0\}$ does not exit to infinity by Mahler's criterion. □

Let $k = 1$, $l = 2$. Then a matrix L determines the linear form $L(x) = a_1 x_1 + a_2 x_2$ on $\mathbb{R}^2$. In this case

$$D_t = \begin{pmatrix} e^t & 0 & 0 \\ 0 & e^{-t/2} & 0 \\ 0 & 0 & e^{-t/2} \end{pmatrix}, \qquad h_L = \begin{pmatrix} 1 & a_1 & a_2 \\ 0 & 1 & 0 \\ 0 & 0 & 1 \end{pmatrix}.$$

If the form L is singular, then the semiorbit $\{D_t h_L \operatorname{SL}(3, \mathbb{Z}),\ t > 0\}$ is divergent. We recall (see §25) that the semiorbit is degenerate if the lattice $\mathbb{Z}^3 \subset \mathbb{R}^3$ contains a subgroup Σ such that $\det(D_t h_L(\Sigma)) \to 0$, $t \to +\infty$. In our case this can happen only if $h_L(\mathbb{Z}^3) \cap W^- \neq 0$, where $W^- = \operatorname{Span}(e_2, e_3)$ stands for the contracting subspace in $\mathbb{R}^3$ relative to $D_{\mathbb{R}}$. In other words, the semiorbit is degenerate if and only if the numbers $1, a_1, a_2$ are linearly dependent over $\mathbb{Q}$. But the form $L(x) = a_1 x_1 + a_2 x_2$ can be singular even if the numbers are independent over $\mathbb{Q}$ (see Lemma 29.18). This gives us an example of a nondegenerate divergent semiorbit.

We recall that a matrix L is *badly approximable* if there exists $c > 0$ such that

$$\|L(q) + p\| > c\|q\|^{-l/k} \quad \text{for all } q \in \mathbb{Z}^l - \{0\},\ p \in \mathbb{Z}^k.$$

THEOREM 30.2 (DANI [**Da85**]). *A semiorbit $\{D_t h_L \Gamma,\ t \geq 0\}$ is bounded if and only if the matrix L is badly approximable.*

Proof. 2) $\Rightarrow$ 1). Let a constant $0 < c < 1$ be chosen from the bad approximation condition and let $\delta^{1+l/k} < c$. Suppose that $y = (L(q) + p, q) \in \Lambda$ and $D_t(y) \in B_\delta$. Then $e^{-tk/l}\|q\| < \delta$ and $e^t \|L(q) + p\| < \delta$. It follows that

$$\|L(q) + p\| < \delta e^{-t} < \delta^{1+l/k} \|q\|^{-l/k} < c\|q\|^{-l/k},$$

which is possible only if $q = 0$. But now from $D_t(y) \in B_\delta$, where $t \geq 0$, it follows that $y = 0$ and hence $D_t(\Lambda) \cap B_\delta = \{0\}$.

1) $\Rightarrow$ 2). Let $\delta > 0$ be such that $D_t(\Lambda) \cap B_\delta = \{0\}$ for all $t \geq 0$. Choose $q \in \mathbb{Z}^l - \{0\}$, $p \in \mathbb{Z}^k$ and $y = (L(q) + p, q) \in \Lambda$. Let $e^{-tk/l}\|q\| = \delta/2$. Then from $\|D_t(y)\| \geq \delta$ it follows that $e^t \|L(q) + p\| \geq \delta$ and hence

$$\|L(q) + p\| \geq c\|q\|^{-l/k}, \qquad \text{where } c = \frac{\delta^{1+l/k}}{2^{l/k}}.$$

□

As a corollary of Theorems 30.1 and 30.2, we deduce that the sets of singular and badly approximable matrices have measure 0.

The reader is referred to [**Kl98b**] for more on relations between Diophantine properties of matrices and ergodic theory.

2. Ergodic interpretation of Diophantine properties of vectors. Now we consider the property of a vector $\mathbf{y} \in \mathbb{R}^k$ being very well multiplicatively approximable. Let

$$h_{\mathbf{y}} = \begin{pmatrix} 1 & \mathbf{y}^{\mathsf{T}} \\ 0 & I_k \end{pmatrix} \mathbb{Z}^{k+1}.$$

In $\mathrm{SL}(k+1,\mathbb{Z})$ we distinguish the k-dimensional Cartan subgroup of diagonal matrices

$$g_{\mathbf{t}} = \mathrm{diag}(e^t, e^{-t_1}, \dots, e^{-t_k}), \qquad \text{where } \mathbf{t} = (t_1, \dots, t_k) \in \mathbb{R}^k \text{ and } t = \sum_1^k t_i.$$

We will be concerned with the action of the semigroup $\{g_{\mathbf{t}},\ \mathbf{t} \in \mathbb{R}^k_+\}$ on $\mathrm{SL}(k+1,\mathbb{R})/\mathrm{SL}(k+1,\mathbb{Z})$.

Let us equip the space of unimodular lattices in $\mathbb{R}^{k+1}$ with a function

$$\delta(\Lambda) = \inf_{\mathbf{v}\in\Lambda-\{0\}} \|\mathbf{v}\|.$$

LEMMA 30.3 [**KM98**]. *Let $\mathbf{y} \in \mathbb{R}^k$. Then $\mathbf{y} \in VWMA$ if and only if there exists $\alpha > 0$ such that the inequality $\delta(g_{\mathbf{t}} h_{\mathbf{y}} \mathbb{Z}^{k+1}) \le e^{-\alpha t}$ has infinitely many solutions $\mathbf{t} \in \mathbb{Z}^k_+$.*

Proof. 1) $\Rightarrow$ 2). Let $\mathbf{q} \in \mathbb{Z}^k - \{0\}$, $p \in \mathbb{Z}$, be such that $|\mathbf{q}\mathbf{y} + p| \le \Pi_+(\mathbf{q})^{-1-\varepsilon}$. If $q_j \ne 0$ then let $e^{s_j} = q_j$, and if $q_j = 0$ then let $s_j = 0$. We have $\Pi_+(\mathbf{q}) = e^s$. Note that

$$\begin{pmatrix} 1 & \mathbf{y}^{\mathsf{T}} \\ 0 & I_k \end{pmatrix} \begin{pmatrix} p \\ \mathbf{q} \end{pmatrix} = \begin{pmatrix} \mathbf{q}\mathbf{y} + p \\ \mathbf{q} \end{pmatrix}.$$

Define $t_j = s_j + \frac{s\varepsilon}{k+1}$; then $t = s(1 + \frac{k\varepsilon}{k+1})$. Now

$$\delta(g_{\mathbf{t}} h_{\mathbf{y}} \mathbb{Z}^{k+1}) \le \max(e^{s(1+\frac{k\varepsilon}{k+1})}|\mathbf{q}\mathbf{y} + p|, e^{-\frac{s\varepsilon}{k+1}}) = e^{-\frac{s\varepsilon}{k+1}} = e^{-\alpha t},$$

where $\alpha = \frac{\varepsilon}{k+1+\varepsilon k}$. Taking a sequence $\mathbf{q}, p \to \infty$, we obtain a sequence $\mathbf{t} \to \infty$. Replacing every vector $\mathbf{t} \in \mathbb{R}^k_+$ by the nearest vector in $\mathbb{Z}^k_+$, we get at our statement (with a slightly smaller coefficient α).

The converse statement is straightforward. □

Note that the function $\delta(g\mathbb{Z}^{k+1})$, $g \in G = \mathrm{SL}(k+1,\mathbb{R})$, describes the distance between the marked point $\Gamma = \mathrm{SL}(k+1,\mathbb{Z})$ and the point $g\Gamma \in G/\Gamma$. More specifically, the ratio

$$\frac{1 + \log(1/\delta(g\mathbb{Z}^{k+1}))}{1 + d(\Gamma, g\Gamma)}$$

is uniformly bounded between two positive constants [**KM98**]. Hence the lemma can be formulated as follows: a vector $\mathbf{y} \in \mathbb{R}^k$ does not belong to the class $VWMA$ if and only if $d(\Gamma, g_{\mathbf{t}} h_{\mathbf{y}} \Gamma)$ is a sublinear function in t as $\mathbf{t} \in \mathbb{R}^k_+$.

It is even easier to find a criterion for the condition $\mathbf{y} \in VWA$. It suffices to consider the one-parameter diagonal subgroup

$$\{a_t = \mathrm{diag}(e^{kt}, e^{-t}, \dots, e^{-t})\} \subset \mathrm{SL}(k+1,\mathbb{R}).$$

Note that $a_{\mathbb{R}}$ belongs to the Cartan subgroup $\{g_{\mathbf{t}},\ \mathbf{t} \in \mathbb{R}^k\}$.

LEMMA 30.4. *Let $\mathbf{y} \in \mathbb{R}^k$. Then $\mathbf{y} \in VWA$ if and only if there exists $\alpha > 0$ such that the inequality $\delta(a_t h_{\mathbf{y}} \mathbb{Z}^{k+1}) \le e^{-\alpha t}$ has infinitely many solutions $t \in \mathbb{Z}_+$.*

In other words, $\mathbf{y} \in \mathbb{R}^k$ does not belong to the class VWA if and only if $d(\Gamma, a_t h_{\mathbf{y}} \Gamma)$ is a sublinear function in $t > 0$.

It would be of interest to understand whether for $k > 1$ there exists a singular vector $\mathbf{y} \in \mathbb{R}^k$ such that $\mathbf{y} \notin VWA$. If so, we obtain a semiorbit

$$\{a_t h_{\mathbf{y}} \operatorname{SL}(k+1,\mathbb{Z}),\ t > 0\} \subset \operatorname{SL}(k+1,\mathbb{R})/\operatorname{SL}(k+1,\mathbb{Z}),$$

which very slowly (with sublinear speed) exits to infinity.

3. Littlewood's conjecture. Let $\mathbf{y} = (y_1, y_2) \in \mathbb{R}^2$. A conjecture due to Littlewood (see [**Ca59**], [**Sch80**]) says that

$$\liminf_{n\to+\infty} n\langle ny_1\rangle\langle ny_2\rangle = 0, \qquad \text{where } \langle x\rangle = \operatorname{dist}(x,\mathbb{Z}) = \min(\{x\}, 1-\{x\}).$$

In other words,

$$\text{for any } \varepsilon > 0 \text{ there exist } q \in \mathbb{Z} - \{0\} \text{ and } \mathbf{p} \in \mathbb{Z}^2 \text{ such that}$$
$$|q| \cdot \Pi(q\mathbf{y} + \mathbf{p}) < \varepsilon.$$

Using Khintchine's transfer principle (more precisely, its multiplicative analogue), it is not hard to see that the conjecture is equivalent to the following statement:

$$\text{for any } \varepsilon > 0 \text{ there exist } \mathbf{q} \in \mathbb{Z}^2 - \{0\} \text{ and } p \in \mathbb{Z} \text{ such that}$$
$$\Pi_+(\mathbf{q}) \cdot |\mathbf{q}\mathbf{y} + p| < \varepsilon.$$

Minkowski's lemma ensures, obviously, that if $\varepsilon = 1$, then both statements hold for every $\mathbf{y} \in \mathbb{R}^2$. Certainly, Littlewood's conjecture is trivial for $\mathbf{y} \in VWMA$. Also there are no problems if one of the numbers y_1, y_2 is not badly approximable. It follows that Littlewood's conjecture is true for almost all vectors $\mathbf{y} \in \mathbb{R}^2$.

It was noticed by Margulis (and implicitly in [**CS**]) that Littlewood's conjecture follows from a statement on orbit structures of an Anosov $\mathbb{R}^2$-flow.

More specifically, let

$$G = \operatorname{SL}(3,\mathbb{R}), \qquad \Gamma = \operatorname{SL}(3,\mathbb{Z}),$$

$$g_{\mathbf{t}} = \begin{pmatrix} e^t & 0 & 0 \\ 0 & e^{-t_1} & 0 \\ 0 & 0 & e^{-t_2} \end{pmatrix}, \qquad \text{where } \mathbf{t} = (t_1, t_2) \text{ and } t = t_1 + t_2.$$

The subgroup $D = \{g_{\mathbf{t}},\ \mathbf{t} \in \mathbb{R}^2\}$, is a two-dimensional Cartan subgroup of G. A general conjecture (formulated by Margulis in [**Ma91b**]) says that all orbit closures and all D-ergodic measures of the D-action on G/Γ are homogeneous. Indeed, as was shown in §2 of [**Ma97**], Littlewood's conjecture is true if all bounded D-orbits in G/Γ are compact.[4]

We will show a connection between Littlewood's conjecture and the dynamics of $(G/\Gamma, D)$ in a more elementary way. To prove Littlewood's conjecture, it suffices to show that the "semiorbit" $\{g_{\mathbf{t}} h_{\mathbf{y}}\Gamma \subset G/\Gamma,\ \mathbf{t} \in \mathbb{R}^2_+\}$ is not bounded for all $\mathbf{y} \in \mathbb{R}^2$ (note that the opposite "semiorbit" $\{g_{\mathbf{t}} h_{\mathbf{y}}\Gamma \subset G/\Gamma,\ \mathbf{t} \in \mathbb{R}^2_-\}$ is closed and exits to infinity in all directions).

In other words,

$$\text{for any } \varepsilon > 0 \text{ there exists a vector } \mathbf{t} \in \mathbb{R}^2_+ \text{ such that } \delta(g_{\mathbf{t}} h_{\mathbf{y}} \mathbb{Z}^3) < \varepsilon.$$

[4]Probably the reader has already noticed a remarkable similarity with the proof of the Oppenheim–Davenport conjecture.

This means that there exist $p \in \mathbb{Z}$ and $\mathbf{q} \in \mathbb{Z}^2$ such that

$$e^{t_1+t_2}|\mathbf{q}\mathbf{y}+p| < \varepsilon, \qquad |e^{-t_1}q_1| < \varepsilon, \qquad |e^{-t_2}q_2| < \varepsilon.$$

Multiplying the three inequalities (and keeping in mind that one of the coordinates of the vector $\mathbf{q}$ can be zero), we obtain $\Pi_+(\mathbf{q}) \cdot |\mathbf{q}\mathbf{y}+p| < \varepsilon^2$, as desired.

4. Diophantine approximation on manifolds. We recall Sprindzhuk's conjecture from §29. Let U be an open subset in $\mathbb{R}^d$ and let $f : U \to \mathbb{R}^n$ be an analytic map. Assume that the functions $1, f_1, \dots, f_n$ are linearly independent over $\mathbb{R}$ (that is, the manifold $M = f(U)$ is *curved*). The conjecture is that almost all points of the manifold M (with respect to the Lebesgue measure on M) are not *VWMA*. We present the idea of its proof from [**KM98**].

As we already know, it suffices to show that for almost every $\mathbf{x} \in U$ and any $\gamma > 0$, the inequality $\delta(g_\mathbf{t}\Lambda_{f(\mathbf{x})}) \le e^{-\gamma t}$ has finitely many solutions $\mathbf{t} \in \mathbb{Z}^n_+$, where

$$\Lambda_{f(\mathbf{x})} = \begin{pmatrix} 1 & f(\mathbf{x})^\mathsf{T} \\ 0 & I_n \end{pmatrix} \mathbb{Z}^{n+1}.$$

We choose $\mathbf{t} \in \mathbb{Z}^n_+$, $\gamma > 0$, a point $\mathbf{x}_0 \in U$, and its neighborhood $B = B(\mathbf{x}_0)$. Look at the set

$$E_\mathbf{t} = \{\mathbf{x} \in B : \ \delta(g_\mathbf{t}\Lambda_{f(\mathbf{x})}) \le e^{-\gamma t}\}.$$

If we manage to show that $\sum_{\mathbf{t}\in\mathbb{Z}^n_+} |E_\mathbf{t}| < \infty$ (where $|\cdot|$ stands for the Lebesgue measure), then Sprindzhuk's conjecture will follow immediately from the Borel–Cantelli lemma.

It turns out that the following result holds.

THEOREM 30.5 [**KM98**]. *Under the above conditions there exist a ball $B \subset U$ centered at $\mathbf{x}_0 \in U$ and constants $D, \rho > 0$, $l \in \mathbb{N}$ such that for any $\mathbf{t} \in \mathbb{R}^n_+$ and $0 < \varepsilon \le \rho$, we have*

$$(*) \qquad |\{\mathbf{x} \in B : \ \delta(g_\mathbf{t}\Lambda_{f(\mathbf{x})}) < \varepsilon\}| \le D(\varepsilon/\rho)^{1/dl}|B|.$$

In particular, $|E_\mathbf{t}| < D_1 e^{-\gamma\beta t}$ for large enough $\|\mathbf{t}\|$. This proves Sprindzhuk's conjecture (and also Baker's).

The inequality (*) can be commented on as follows. Let

$$G = \mathrm{SL}(n+1,\mathbb{R}), \quad \Gamma = \mathrm{SL}(n+1,\mathbb{Z}), \quad A_\varepsilon = \{g\Gamma \in G/\Gamma : \ \delta(g\mathbb{Z}^{n+1}) < \varepsilon\}.$$

We know that near 0 the function $\delta(g\mathbb{Z}^{n+1})$ behaves as $c'e^{-\alpha' \operatorname{dist}(\Gamma, g\Gamma)}$ for some $c', \alpha' > 0$. On the other hand, it follows from the description of the Siegel fundamental domain (see [**Rag**]) for G/Γ that

$$\mathrm{vol}(g\Gamma : \ \operatorname{dist}(\Gamma, g\Gamma) > R) < c''e^{-\alpha'' R} \quad \text{for some } c'', \alpha'' > 0.$$

Hence $\mathrm{vol}(A_\varepsilon) < c\varepsilon^\alpha$ with universal constants $c, \alpha > 0$ depending only on n. The action of $g_\mathbf{t}$, $\mathbf{t} \in \mathbb{Z}^n_+$, on the commutative matrix subgroup

$$H = \left\{ h_\mathbf{x} = \begin{pmatrix} 1 & \mathbf{x}^\mathsf{T} \\ 0 & I_n \end{pmatrix}, \ \mathbf{x} \in \mathbb{R}^n \right\}$$

consists in expansion of the subgroup in all directions. Hence one can treat the inequality (*) as a variant of the "uniform distribution" of the family of expanding translates of the area $h_{f(B)} \subset H$: the fraction of times we hit A_ε, as well as the measure of the set A_ε, is bounded from above by a power of ε.

In addition, in [**KM98**] the results of Margulis and Dani on the nondivergence of unipotent trajectories in the space of lattices were substantially refined. More specifically, according to Theorem 20.5, for any $c > 0$ and any unimodular lattice $\Lambda \subset \mathbb{R}^k$ there exists $\varepsilon > 0$ such that, for all unipotent subgroups $u_{\mathbb{R}} \subset \mathrm{SL}(k, \mathbb{R})$ and all $T > 0$,

$$|\{x \in [-T, T] : \delta(u_x \Lambda) < \varepsilon\}| \leq cT.$$

Sometimes one is interested in the explicit form of the dependence between c and ε. The answer is given by the following result.

THEOREM 30.6 [**KM98**]. *For any unimodular lattice $\Lambda \subset \mathbb{R}^k$ there exist constants $\rho = \rho(\Lambda) > 0$ and $D = D(k) > 0$ such that for all unipotent subgroups $u_{\mathbb{R}} \subset \mathrm{SL}(k, \mathbb{R})$ and all $T > 0$ and $\varepsilon \leq \rho$ we have*

$$(**) \qquad |\{x \in [-T, T] : \delta(u_x \Lambda) < \varepsilon\}| \leq D \left(\frac{\varepsilon}{\rho}\right)^{1/k^2} T.$$

Clearly, every unipotent subgroup $u_{\mathbb{R}} \subset \mathrm{SL}(k, \mathbb{R})$ is determined by a collection of polynomials of degree at most k^2. In the course of the proof of all results on the nondivergence of unipotent trajectories, the following property of polynomials is crucial.

Let V be an open subset in $\mathbb{R}^d$ and ϕ a continuous function on V. Given $B \subset V$, we define $\|\phi\|_B = \sup_{\mathbf{x} \in B} |\phi(\mathbf{x})|$. A function ϕ will be called (C, α)-*good on* V for some $C, \alpha > 0$, if for any open ball $B \subset V$ and any $\varepsilon > 0$

$$|\{\mathbf{x} \in B : |\phi(\mathbf{x})| < \varepsilon\}| \leq C \left(\frac{\varepsilon}{\|\phi\|_B}\right)^{\alpha} |B|.$$

The proof of Lemma 12.3 shows that the family of polynomials of degree at most k consists of $(C, 1/k)$-good functions on $\mathbb{R}$ (for some $C = C(k)$). On the other hand, given $\mathbf{t}$, entries of the matrix $g_{\mathbf{t}} h_{f(\mathbf{x})}$ are linear combinations of $1, f_1, \ldots, f_n$. Surprisingly, the linear hull of analytic functions $1, f_1, \ldots, f_n$ consists of $(C, 1/dl)$-good functions on a neighborhood of the point $\mathbf{x} \in U$ (for some $C > 0$ and $l \in \mathbb{N}$). This allows one to use the same strategy while proving $(*)$ and $(**)$. Note that the estimate (*) is uniform in all $\mathbf{t} \in \mathbb{R}^n_+$, whereas (**) is uniform in all unipotent subgroups $u_{\mathbb{R}} \subset \mathrm{SL}(k, \mathbb{R})$.

Actually, Kleinbock and Margulis proved Sprindžuk's conjecture not only for analytic but also for smooth manifolds. More specifically, let $f : U \to \mathbb{R}^n$ be a C^m-smooth map. We say that a point $\mathbf{y} = f(\mathbf{x}) \in M$ is *nondegenerate* if for some $l \leq m$ the partial derivatives of f at $\mathbf{x}$ generate $\mathbb{R}^n$. This is an infinitesimal version of the manifold being curved (if the map f is analytic and the manifold $M = f(U)$ is curved, then all points $\mathbf{y} \in M$ are nondegenerate). The following theorem holds.

THEOREM 30.7 [**KM98**]. *Let U be an open subset in $\mathbb{R}^d$, let $f : U \to \mathbb{R}^n$ be a C^m-smooth map, and let almost all points of $M = f(U) \subset \mathbb{R}^n$ be nondegenerate. Then almost all points of M are not VWMA.*

In particular ($n = 2$), we obtain a result of Schmidt (see [**Sch80**]): almost all points of a smooth curve $M \subset \mathbb{R}^2$ are not *VWA* if M has nonzero curvature almost everywhere. The fact that almost all points of $M \subset \mathbb{R}^2$ are not *VWMA* was proved by Sprindzhuk (see [**Sp80**]).

§31. Counting lattice points in $\mathbb{R}^n$

First we study the asymptotics of the number of lattice points inside polyhedra, and then we proceed to count lattice points in homogeneous $\mathbb{Q}$-manifolds in $\mathbb{R}^n$.

1. The distribution of lattice points in polyhedra. Here we discuss Skriganov's papers [**Skr89,93,98**] on the number of lattice points in polyhedra. In the two-dimensional case these results go back to Hardy and Littlewood [**HL**] and Khintchine [**Kh23**].

Let P be a compact region in $\mathbb{R}^n$ with piecewise smooth boundary, $\Lambda \subset \mathbb{R}^n$ a unimodular lattice, and $N(P, \Lambda)$ the number of points of the lattice Λ inside P. We introduce the error term $R(P, \Lambda) = \text{vol}(P) - N(P, \Lambda)$. By $tP \subset \mathbb{R}^n$ we denote the region obtained from P by the uniform dilation with coefficient $t > 0$. We will be concerned with the asymptotics $R(tP, \Lambda)$ as $t \to \infty$. (Note that recently a significant advance in estimating $R(tP, \Lambda)$, where P is an ellipsoid, was achieved by Bentkus and Gotze [**BG**], [**BG97**]. In particular, they improved the classical estimate obtained by Landau in 1915.)

We will examine the case where P is a polyhedron. Then $R(tP, \Lambda) = O(t^{n-1})$, and the estimate is best possible, say, for $\Lambda = \mathbb{Z}^n$ and the cube $P = [-1, 1]^n$. But under certain conditions one can obtain anomalously small (for instance, logarithmic) asymptotics for $R(tP, \Lambda)$.

In the two-dimensional case one can reduce counting the asymptotics $R(tP, \mathbb{Z}^2)$ to the estimate of the remainder term in the ergodic theorem applied to an irrational rotation of the circle. We start by considering a triangle $P = ABC$ in the upper half-plane. Let us draw vertical segments from the points A, B, C to points Z_1, Z_2, Z_3 on the horizontal axis. Assume that $Z_1 < Z_3 < Z_2$ and that the point C sits inside the trapezoid with vertices A, B, Z_1, Z_2. Then the triangle ABC is obtained from $T_1 = ABZ_2Z_1$ by removing the trapezoids $T_2 = CBZ_2Z_3$ and $T_3 = ACZ_3Z_1$. So we have to estimate $R(T_1, \mathbb{Z}^2)$.

Let T_1 be given by the system of inequalities

$$Z_1 < x \leq Z_2, \qquad 0 < y \leq a_1 x + b_1,$$

where $y = a_1 x + b_1$ is the equation of the line through A and B. We have

$$R(T_1, \mathbb{Z}^2) = \text{vol}(T_1) - N(T_1, \mathbb{Z}^2) = \sum_{Z_1 < k \leq Z_2} \{a_1 k + b_1\} + r_{AZ_1} + r_{BZ_2},$$

where r_{AZ_1} and r_{BZ_2} stand for the "contribution" to $R(T_1, \mathbb{Z}^2)$ of vertical sides of the trapezoid. Since $R(P, \mathbb{Z}^2) = R(T_1, \mathbb{Z}^2) - R(T_2, \mathbb{Z}^2) - R(T_3, \mathbb{Z}^2)$, we do not need the explicit form of r_{AZ_1} and r_{BZ_2}, because they will disappear under the subtraction. Doing the same with T_2 and T_3, we obtain the estimate

$$R(P, \mathbb{Z}^2) \sim \sum_{Z_1 < k \leq Z_2} \{a_1 k + b_1\} - \sum_{Z_1 < k \leq Z_3} \{a_3 k + b_3\} - \sum_{Z_3 < k \leq Z_2} \{a_2 k + b_2\},$$

where $y = a_2 x + b_2$ and $y = a_3 x + b_3$ are equations of the lines containing BC and AC (we cannot write the exact equality because of the edge effect near the vertices A, B, C. But the error does not exceed 3 for any triangle ABC).

In the notation of §29.2, we have

$$\sum_{Z_1 < k \leq Z_2} \{ak + b\} = \frac{[Z_2] - [Z_1]}{2} + r_a([Z_2] - [Z_1]).$$

We already know the asymptotics of the remainder term $r_a(m)$ for badly approximable numbers a and for almost all numbers a. Moreover, Roth's Theorem 29.6 enables one to estimate $r_a(m)$ for algebraic irrational numbers a as well (see [**Skr93**]). Now the asymptotics for $R(tP, \mathbb{Z}^2)$ is obtained by adding together the asymptotics r_{a_i}, $i = 1, 2, 3$. In a similar fashion one can handle an arbitrary polyhedron $P \subset \mathbb{R}^2$. We summarize our results as follows.

THEOREM 31.1 [**HL**], [**Kh23**], [**Skr89,93**]. *Suppose that a polyhedron $P \subset \mathbb{R}^2$ with m sides is such that the i-th side is parallel to a vector $(1, a_i)$, $i = 1, \dots, m$. If all the numbers a_i, $1 \le i \le m$, are badly approximable, then $R(tP, \mathbb{Z}^2) = O(\ln t)$. If all the numbers are algebraic irrationals, then $R(tP, \mathbb{Z}^2) = O(t^\varepsilon)$ with arbitrarily small $\varepsilon > 0$. Finally, for almost all collections $(a_1, \dots, a_m)$, $R(tP, \mathbb{Z}^2) = O(\ln^{1+\varepsilon} t)$ with arbitrarily small $\varepsilon > 0$.*

Remark. The first assertion was implicitly contained in [**HL**], where the authors considered the right triangle P with vertices $(0,0), (0,a), (1,0)$ and counted the "contribution" of the hypotenuse to $R(tP, \mathbb{Z}^2)$. The second assertion was proved in [**Skr93**], and the third in [**Kh23**] and [**Skr98**].

To prove similar results in $\mathbb{R}^n$, we will use ergodic theory on the space $\mathcal{L}_n = \mathrm{SL}(n, \mathbb{R}) / \mathrm{SL}(n, \mathbb{Z})$ of unimodular lattices in $\mathbb{R}^n$. The Haar measure on $\mathcal{L}_n$ will be denoted by μ. We endow $\mathbb{R}^n$ with the norm

$$\|v\| = \max_{1,\dots,n} |v_i|$$

(relative to the natural basis in $\mathbb{R}^n$ that generates the lattice $\mathbb{Z}^n$). By $(\cdot, \cdot)$ we denote the scalar product. In addition, we define

$$\mathrm{Nm}(v) = \Pi_1^n v_i.$$

Given $\Lambda \in \mathcal{L}_n$, let $\Lambda^\perp$ stand for the *dual* unimodular lattice determined by the condition $(\Lambda, \Lambda^\perp) \subset \mathbb{Z}$. Clearly, if $\Lambda = g\mathbb{Z}^n$ for some $g \in \mathrm{SL}(n, \mathbb{R})$, then $\Lambda^\perp = (g^{-1})^\intercal \mathbb{Z}^n$.

Recall that $\delta(\Lambda) = \min_{v \in \Lambda - \{0\}} \|v\|$. Now we introduce the new parameter

$$\nu(\Lambda, \rho) = \min\{|\mathrm{Nm}(v)|, \ v \in \Lambda, \ 0 < \|v\| < \rho\}.$$

Then $\nu(\Lambda, \rho) = 0$ for all $\rho < \delta(\Lambda)$. Let

$$\mathrm{Nm}(\Lambda) = \inf\{\nu(\Lambda, \rho), \ \rho \ge \delta(\Lambda)\} = \inf\{|\mathrm{Nm}(v)|, \ v \in \Lambda - \{0\}\}.$$

One says that a lattice Λ is *admissible* if $\mathrm{Nm}(\Lambda) > 0$. An example in the two-dimensional case is given by $\Lambda = \mathbb{Z}(1,1) + \mathbb{Z}(a, a')$, where $a \ne a'$ is a pair of badly approximable numbers. In the multi-dimensional case one can take an "algebraic" lattice

$$\Lambda = \mathbb{Z}(1, 1, \dots, 1) + \mathbb{Z}(a_1, a_2, \dots, a_n) + \dots + \mathbb{Z}(a_1^{n-1}, a_2^{n-1}, \dots, a_n^{n-1}),$$

where $a_1, a_2, \dots, a_n$ are distinct real roots of a polynomial of degree n with integer coefficients.

A lattice Λ is said to be *weakly admissible* if $\nu(\Lambda, \rho) > 0$ for all $\rho > \delta(\Lambda)$. Clearly, Λ is weakly admissible $\iff$ Λ has no nonzero points on the coordinate hyperplanes.

Now the $(n-1)$-dimensional subgroup $D \subset \mathrm{SL}(n, \mathbb{R})$ of diagonal matrices comes into play. It is important that the subgroup preserves Nm:

$$\mathrm{Nm}(v) = \mathrm{Nm}(dv) \quad \text{for all } d \in D,\ v \in \mathbb{R}^n.$$

We will also use the discrete subgroup

$$\Delta = \{\mathrm{diag}(2^{m_1}, \dots, 2^{m_n}),\ m_i \in \mathbb{Z},\ m_1 + \cdots + m_n = 0\}.$$

Clearly, $\Delta \simeq \mathbb{Z}^{n-1}$ is a lattice in D, and Δ acts ergodically on $\mathcal{L}_n$. In Δ we take the "ball"

$$\Delta_r = \{\mathrm{diag}(2^{m_1}, \dots, 2^{m_n}) \in \Delta,\ |m_i| \le r\}.$$

The family Δ_r, $r \to \infty$, is averaging, and the Birkhoff–Khintchine ergodic theorem (more precisely, its multi-parameter analogue) holds: for each $f \in L^1(\mathcal{L}_n, \mu)$,

$$\frac{\sum_{d \in \Delta_r} f(d\Lambda)}{\mathrm{card}(\Delta_r)} \to \int f\, d\mu \quad \text{as } r \to \infty \text{ for almost all } \Lambda \in \mathcal{L}_n.$$

Note that $\mathrm{card}(\Delta_r) \sim c_n r^{n-1}$.

One can verify that the function $\delta(\Lambda)^{-n}$ is not summable on the space $\mathcal{L}_n$. However, as was shown in [**Skr98**], for any $\varepsilon > 0$, the function $\delta(\Lambda)^{\varepsilon - n}$ is already summable (see Proposition 7.1 in [**KM99**] for a simpler proof). Using the ergodic theorem, it is now easy to derive that

$$S(\Lambda, r) = \sum_{d \in \Delta_r} \delta(d\Lambda)^{-n} = O(r^{n-1+\varepsilon}), \qquad r \to \infty, \tag{1}$$

for any $\varepsilon > 0$ and almost all lattices $\Lambda \in \mathcal{L}_n$.

Note that $\delta(d\Lambda) \ge \mathrm{Nm}(\Lambda)^{1/n}$ and, by Mahler's criterion, *a lattice Λ is admissible $\Longleftrightarrow$ the orbit $\Delta\Lambda \subset \mathcal{L}_n$ is bounded $\Longleftrightarrow$ the dual lattice $\Lambda^\perp$ is admissible.* Also it is easy to see that, for such a lattice, $S(\Lambda, r) = O(r^{n-1})$.

Moreover, using the Borel–Cantelli lemma, the following estimate is proved in [**Skr98**]:

$$\nu(\Lambda, \rho) > c_\varepsilon(\Lambda)(\log \rho)^{1-n-\varepsilon}, \qquad \rho \to \infty, \tag{2}$$

for any $\varepsilon > 0$ and almost all lattices $\Lambda \in \mathcal{L}_n$.

We proceed to study the error term $R(P, \Lambda)$. Clearly, for any bounded region $\mathcal{O} \subset \mathbb{R}^n$,

$$N(\mathcal{O}, \Lambda) = \sum_{v \in \Lambda} \chi_{\mathcal{O}}(v),$$

where $\chi_{\mathcal{O}}$ denotes the characteristic function of the region $\mathcal{O}$. On the other hand, $\mathrm{vol}(\mathcal{O}) = \int_{\mathcal{O}} d^n v$ and, by the Poisson summation formula,

$$\sum_{v \in \Lambda} \chi_{\mathcal{O}}(v) = \sum_{v \in \Lambda^\perp} \widetilde{\chi}_{\mathcal{O}}(v),$$

where

$$\widetilde{\chi}_{\mathcal{O}}(v) = \int_{\mathbb{R}^n} e^{2\pi i (v,u)} \chi_{\mathcal{O}}(u) d^n u$$

is the Fourier transform of the function $\chi_{\mathcal{O}}$. It follows that

$$R(\mathcal{O}, \Lambda) = \sum_{v \in \Lambda^\perp - \{0\}} \widetilde{\chi}_{\mathcal{O}}(v). \tag{3}$$

We need the explicit form of the function $\widetilde{\chi}_P$ for a polyhedron P. By the Stokes formula,

$$\int_P \Big(\sum_{j=1}^{n} \frac{\partial^2}{\partial v_j^2} e^{2\pi i(u,v)}\Big) d^n v = \int_{\partial P} \Big(\frac{\partial}{\partial \nu_v} e^{2\pi i(u,v)}\Big) d^{n-1} v,$$

where ν_v is the normal vector to ∂P at $v \in \partial P$. By induction, the following formula is established in [**Skr98**]:

$$\widetilde{\chi}_P(v) = \Big(\frac{1}{2\pi i}\Big)^n \sum_f \lambda(g_f v) e^{2\pi i (P_f^0, v)}, \tag{4}$$

where the summation is taken over all *flags* f of the polyhedron P. Recall that a flag is a sequence of faces

$$f = \{P = P_f^n \supset P_f^{n-1} \supset \cdots \supset P_f^0,\ \dim P_f^j = j\},$$

where P_f^j is a face of P_f^{j+1}, $0 \le j \le n-1$. With each flag f one associates a matrix $g_f \in \mathrm{SO}(n, \mathbb{R})$ whose j-th row represents the unit vector parallel to P_f^{n-j+1} and orthogonal to P_f^{n-j}. Moreover, the function λ in (4) is of the form

$$\lambda(v) = \frac{\Phi(v)}{\mathrm{Nm}(v)},$$

where $\Phi(v)$ depends only on the angular coordinates of the vector $v \in \mathbb{R}^n$ and is infinitely smooth on the unit sphere in $\mathbb{R}^n$.

This information will be applied in presenting the idea of the proof of the following result, which is crucial for counting the asymptotics $R(tP, \Lambda)$.

THEOREM 31.2 [**Skr98**]. *Let P be a convex polyhedron in $\mathbb{R}^n$, and Λ a unimodular lattice in $\mathbb{R}^n$. Then*

$$R(tP, \Lambda) < c\Big(t^{n-1}\rho^{-1/2} + \sum_f S(g_f\Lambda^\perp, r_f)\Big), \qquad \text{where } r_f = \log\Big(\frac{\rho^n}{\nu(g_f\Lambda^\perp, \rho)}\Big);$$

here the parameter $\rho > 1$ is arbitrarily large, the summation is taken over all flags f of the polyhedron P, and $c = c(P, \Lambda)$.

This theorem easily implies the following result.

THEOREM 31.3 [**Skr98**]. *Let Π be a parallelepiped with sides parallel to the coordinate axes, and Λ an admissible unimodular lattice in $\mathbb{R}^n$. Then $R(t\Pi, \Lambda) = O((\log t)^{n-1})$.*

Indeed, the matrices g_f in this case form the group of permutations and reorientations of coordinate axes. Hence $S(g_f\Lambda^\perp, r) = S(\Lambda^\perp, r)$ and $\nu(g_f\Lambda^\perp, r) = \nu(\Lambda^\perp, r)$. If the lattice Λ is admissible, so is $\Lambda^\perp$. Hence $\nu(\Lambda^\perp, r) > \mathrm{Nm}(\Lambda^\perp) > 0$ and $r = O(\log \rho)$. It follows that

$$S(\Lambda^\perp, r) = O(r^{n-1}) = O((\log \rho)^{n-1})$$

and

$$R(t\Pi, \Lambda) = O(t^{n-1}\rho^{-1/2} + (\log \rho)^{n-1}).$$

It suffices to take $\rho = t^{2n-2}$. □

Note that for $n > 2$, no examples are known of convex polyhedra other than parallelepipeds with $R(tP, \Lambda) = O((\log t)^{n-1})$.

THEOREM 31.4 [**Skr98**]. *Let P be a compact polyhedron. Then for almost all unimodular lattices $\Lambda \subset \mathbb{R}^n$ we have $R(tP, \Lambda) = O((\log t)^{n-1+\varepsilon})$, with arbitrarily small $\varepsilon > 0$.*

Proof. Using partitions, the problem is reduced to the case of a convex polyhedron. Since the polyhedron P has finitely many flags, we derive that $S(g_f \Lambda^\perp, r) = O(r^{n-1+\varepsilon})$ for every flag f and almost all lattices $\Lambda \in \mathcal{L}_n$. It follows from (2) that $r_f = O(\log \rho)$ for almost all lattices $\Lambda \in \mathcal{L}_n$. Hence

$$R(tP, \Lambda) = O(t^{n-1}\rho^{-1/2} + (\log \rho)^{n-1+\varepsilon}).$$

It now suffices to take $\rho = t^{2n-2}$. $\square$

To demonstrate the relation between our Diophantine problem and the D-action on $\mathcal{L}_n$, we give a "pseudo-proof" of Theorem 31.2. More specifically, we will obtain a weaker estimate for $R(tP, \Lambda)$.

It follows from (3) and (4) that

$$R(tP, \Lambda) = \frac{1}{(2\pi i)^n} \sum_f \sum_{v \in \Lambda^\perp - \{0\}} \lambda(g_f v) e^{2\pi i (tP_f^0, v)}.$$

This series is difficult to estimate because it is not absolutely convergent. Hence one has to introduce an auxiliary parameter ρ and apply a "smoothing" procedure. It is proved in [**Skr98**] that a properly smoothed series $R(tP, \Lambda)$ breaks into the sum of two series $R_1(t, P, \Lambda) + R_2(t, P, \Lambda)$, where the first series is finite (all its terms vanish for $|v| \geq \rho$) and the second is absolutely convergent. Moreover, the sum of the second series is bounded from above by $ct^{n-1}\rho^{-1/2}$, and the first consists of terms of the form $\lambda(g_f v) w_\rho(v) e^{2\pi i (t_\rho^\pm P_f^0, v)}$. Here $w_\rho \in C^\infty(\mathbb{R}^n)$, $w_\rho(v) = 0$ for $\|v\| \geq \rho$, and $|w_\rho(v)| < A$ for all $\rho > 0$, $v \in \mathbb{R}^n$, where A is a universal constant (we do not need the explicit form of $t_\rho^\pm$).

We will use the rough estimate

$$R_1(t, P, \Lambda) \leq c(P) A \sum_f \sum_{v \in B_\rho \cap g_f \Lambda^\perp} \frac{1}{|\operatorname{Nm}(v)|}, \tag{5}$$

where $B_\rho = \{v \in \mathbb{R}^n : 0 < \|v\| < \rho\}$. If one takes into account that $R_1(t, P, \Lambda)$ contains terms with opposite signs, then Theorem 31.2 can be proved in its full scope (see [**Skr98**]). But even this estimate enables one to demonstrate the relation with the D-action on $\mathcal{L}_n$ (and to obtain a slightly weaker asymptotics).

LEMMA 31.5. *Let a lattice $\Lambda \in \mathcal{L}_n$ be weakly admissible. Then for large enough ρ,*

$$\sum_{v \in B_\rho \cap \Lambda} \frac{1}{|\operatorname{Nm}(v)|} \leq c(\Lambda) \sum_{d \in \Delta_r} \frac{\log(\rho/\delta(d\Lambda))}{\delta(d\Lambda)^n}, \qquad \textit{where } r = \log\left(\frac{\rho^n}{\nu(\Lambda, \rho)}\right).$$

Suppose that the lemma is proved. Then it is easy to see that under the conditions of Theorem 31.3 we obtain the asymptotics $O((\log t)^n)$, and under the conditions of Theorem 31.4 the asymptotics $O((\log t)^{n+\varepsilon})$. Thus, power exponents at logarithms increased by one (because of $\log \rho$). As we already mentioned, for the "proper" proof of the theorems one needs a more delicate estimate of $R_1(t, P, \Lambda)$ than that given in (5).

We proceed to the proof of the lemma. The action of the discrete group Δ breaks each coordinate orthant of the space $\mathbb{R}^n$ into sectors, and elements $d \in \Delta$ permute the latter. We restrict our attention to the positive orthant $R^n_+ = \{x_i > 0,\ i = 1, \ldots, n\}$ and select the sector S containing the ray $x_1 = \cdots = x_n$. Then there exist $a > 1,\ b > 0$ such that

$$\frac{\|v\|^n}{a} < \mathrm{Nm}(v) < a\|v\|^n, \qquad v \in S,$$

and

$$\|dv\| \geq b\|v\|, \qquad v \in S,\ d \in \Delta.$$

For every v, it is not difficult to estimate the norm of an element $d \in \Delta$ such that $dv \in S$. Indeed, let $d = \mathrm{diag}(2^{t_1}, \ldots, 2^{t_n}) \in D$ be an element such that $(dv)_1 = \cdots = (dv)_n = (\mathrm{Nm}(v))^{1/n}$. If $t_i > 0$, then

$$t_i \leq \log \frac{(\mathrm{Nm}(v))^{1/n}}{\min_i v_i} \leq \log \frac{\|v\|^{n-1}}{(\mathrm{Nm}(v))^{1-1/n}} < \log \frac{\|v\|^n}{\mathrm{Nm}(v)}. \tag{6}$$

If we are looking for $d \in \Delta$ such that $dv \in S$, then a universal constant τ_n must be added to the right side of (6) (see [**Skr98**]). Of course, we can disregard it for large enough $\|v\|$. In a similar way we consider the case $t_i < 0$ and obtain the same estimate.

Let $B^+_\rho = B_\rho \cap \mathbb{R}^n_+$. Then

$$B^+_\rho \cap \Lambda \subset \bigcup_{d \in \Delta_r} d^{-1}S,$$

where $r = r(\Lambda, \rho)$ is as in the lemma. It now suffices to note that

$$\sum_{v \in B^+_\rho \cap \Lambda} \frac{1}{\mathrm{Nm}(v)} < \sum_{d \in \Delta_r} \sum_{v \in S \cap d(B^+_\rho \cap \Lambda)} \frac{1}{\mathrm{Nm}(v)}$$
$$< a \sum_{d \in \Delta_r} \sum_{v \in B^+_{\rho/b} \cap d\Lambda} \|v\|^{-n} < c \sum_{d \in \Delta_r} \frac{\log(\rho / b\delta(d\Lambda))}{\delta(d\Lambda)^n},$$

since

$$\sum_{v \in B^+_\rho \cap \Lambda} \|v\|^{-n} < \sum_{v \in B^+_\rho \cap \delta(\Lambda)\mathbb{Z}^n} \|v\|^{-n} \quad \text{and} \quad \sum_{v \in B^+_R \cap \mathbb{Z}^n} \|v\|^{-n} < c' \log R.$$

□

Remark. It follows from Theorem 31.4 that for almost all $g \in \mathrm{SL}(n, \mathbb{R})$, one has the asymptotics $R(tP, g\mathbb{Z}^n) = O((\log t)^{n-1+\varepsilon})$ with arbitrarily small $\varepsilon > 0$. It makes sense to find the typical asymptotics in the class of orthogonal rotations as well. It was proved in [**Skr98**] that $R(tP, g\mathbb{Z}^n) = O((\log t)^{2(n-1)+\varepsilon})$ for almost all $g \in \mathrm{SO}(n, \mathbb{R})$. However, one can prove (see the paper [**SS**] of Skriganov and the author) that in the class of orthogonal rotations the typical asymptotics is the same as in Theorem 31.4. This estimate is derived from Theorem 31.4 and, as is easily seen, sharpens the latter. Note that for $n = 2$ the result was already proved by Khintchine [**Kh23**].

2. Counting lattice points on manifolds. Now we study the asymptotics of the number of lattice points on algebraic manifolds in $\mathbb{R}^n$. We make use of the

$\mathbb{Q}$-structure on the space given by the lattice $\mathbb{Z}^n \subset \mathbb{R}^n$. Suppose that $V \subset \mathbb{R}^n$ is an affine manifold over $\mathbb{Q}$ (that is, the set of zeros in $\mathbb{R}^n$ of a polynomial with rational coefficients), and let $V_{\mathbb{Z}} = V \cap \mathbb{Z}^n$ be the set of of lattice points in V. One of the basic problems in Diophantine analysis is to find the asymptotics of the number $N(T, V)$ of points in the intersection $V_{\mathbb{Z}} \cap D_T$ of the discrete set $V_{\mathbb{Z}}$ with the Euclidean ball D_T of radius T as $T \to \infty$.

Usually, this problem is solved by the Hardy–Littlewood circle method (see [**Bi**], [**Sch85**], [**BR**]). But the method is not universal. If the manifold V admits a transitive group action, one can apply ergodic theory (see [**Li**], [**Sa91**]). Recently new results have been obtained in this direction.

We assume that V is an orbit of a reductive algebraic $\mathbb{Q}$-group $G \subset \mathrm{SL}(n, \mathbb{R})$; that is, $V = Gv_0$ for some element $v_0 \in V_{\mathbb{Z}}$. Then the stabilizer $H = \{g \in G : \ gv_0 = v_0\}$ is a reductive algebraic $\mathbb{Q}$-group as well, and $V \simeq G/H$. In addition, we assume that G and H admit no nontrivial $\mathbb{Q}$-characters. Then by the Borel–Harish-Chandra theorem, $G_{\mathbb{Z}}$ and $H_{\mathbb{Z}}$ are lattices, and $V_{\mathbb{Z}}$ is a finite union of $G_{\mathbb{Z}}$-orbits. Therefore it suffices to estimate the number $N(T, V, \mathcal{O})$ of lattice points with norm at most T in the orbit $\mathcal{O} = G_{\mathbb{Z}}v_0$.

We normalize Haar measures on G and H by the condition $\mathrm{vol}(G/G_{\mathbb{Z}}) = \mathrm{vol}(H/H_{\mathbb{Z}}) = 1$, and denote by λ the induced G-invariant measure on G/H. Let

$$R_T = \{gH : \ gv_0 \in D_T\} \subset G/H$$

be the image of $V \cap D_T$ under the identification $V \simeq G/H$. The following theorem holds.

THEOREM 31.6 [**DRS**], [**EM**]. *Let G be a semisimple group and let the lattice $G_{\mathbb{Z}} \subset G$ be irreducible. Assume that $H \subset G$ is the subgroup of fixed points of an involution of G, and that $H_{\mathbb{Z}}$ is a lattice in H. Then, asymptotically, $N(T, V, \mathcal{O}) \sim \lambda(R_T)$; that is, $N(T, V, \mathcal{O})/\lambda(R_T) \to 1$ as $T \to \infty$.*

The condition imposed on H is equivalent to saying that $V = G/H$ is an affine symmetric manifold. We mention some applications of the theorem found by Duke, Rudnick, and Sarnak in [**DRS**].

Example 31.7. Let

$$F(x_1, \dots, x_n) = \sum_{i,j=1}^{n} f_{i,j} x_i x_j$$

be a nondegenerate quadratic form with integer coefficients and signature (p, q), where $p + q = n, \ p, q \geq 1$. Now choose $k \in \mathbb{N}$ and consider the hyperboloid

$$V = \{x \in \mathbb{R}^n : \ F(x) = k\}.$$

The group $G = \mathrm{SO}(F) \simeq \mathrm{SO}(p, q)$ acts transitively on V, and the stabilizer H of the point $v_0 \in V_{\mathbb{Z}}$ is isomorphic to $\mathrm{SO}(p-1, q)$. If $n \geq 4$, then the subgroup H is semisimple and $H_{\mathbb{Z}}$ is a lattice in H. It is not difficult to check that the manifold $\mathrm{SO}(p, q)/\mathrm{SO}(p-1, q)$ is symmetric, and hence all the assumptions of the theorem are satisfied. The computation of the volume $\lambda(R_T)$ made in [**DRS**] enables one to derive the asymptotics $N(T, V) \sim c_{k,F} T^{n-2}$.

The case $n = 3$ is the most interesting one. Let

$$F(x_1, x_2, x_3) = x_1^2 + x_2^2 - dx_3^2, \quad \text{where } d \in \mathbb{N}.$$

We consider the one-sheeted hyperboloid $V = \{x \in \mathbb{R}^3 : F(x) = 1\}$. The stabilizer H of the point $v_0 = (1,0,0) \in V_{\mathbb{Z}}$ is isomorphic to $\mathrm{SO}(1,1)$; that is, in this case the group H is reductive and one-dimensional. One can show that the group $H_{\mathbb{Z}}$ is nontrivial $\iff d$ is not a square integer. Elementary divisor theory (see, for instance, [**Sco**]) says that $N(T,V) \sim c_d T$ if $\sqrt{d} \notin \mathbb{Z}$ (which is consistent with Theorem 31.6), and $N(T,V) \sim c_d T \log T$ if $\sqrt{d} \in \mathbb{Z}$.

Example 31.8. Take the manifold

$$V = \{x \in \mathrm{Mat}(n,n) : \det x = k\}.$$

We supply the vector space $\mathrm{Mat}(n,n)$ with the Hilbert–Schmidt norm $\|(x_{i,j})\| = \sqrt{\sum_{i,j} x_{i,j}^2}$. The group $G = \mathrm{SL}(n,\mathbb{R}) \times \mathrm{SL}(n,\mathbb{R})$ acts transitively on V by the formula $(g_1, g_2)(x) = g_1^{-1} x g_2$. Here $V \simeq G/H$, where $H \simeq \mathrm{SL}(n,\mathbb{R})$ is the diagonal subgroup of G.[5] Applying Theorem 31.6, one proves (see [**DRS**]) the asymptotics

$$N(T,V) \sim c_{n,k} T^{n(n-1)}.$$

Example 31.9. Let S be the manifold of symmetric $n \times n$ matrices with determinant $k \in \mathbb{N}$. The group $G = \mathrm{SL}(n,\mathbb{R})$ acts on S by the formula $g(x) = g^{\mathsf{T}} x g$. Here S is a union of homogeneous spaces of the form $\mathrm{SL}(n,\mathbb{R})/\mathrm{SO}(p,q)$, where $p + q = n$ and the number q is even. In [**DRS**] the following asymptotics was proved:

$$N(T,S) \sim d_{n,k} T^{n(n-1)/2}.$$

Note that in [**EK**] (by different methods) the asymptotics

$$N(T,S') \sim c_n T^{n(n-1)/2} \log T$$

was obtained for the manifold S' of degenerate symmetric $n \times n$ matrices of rank $n-1$.

The main idea in [**DRS**] is to prove that

$$g\mu_H \to \mu_G \quad \text{as } gH \to \infty \text{ in } G/H,$$

where μ_H and μ_G stand for invariant probability measures on $H/H_{\mathbb{Z}}$ and $G/G_{\mathbb{Z}}$ (actually, it suffices to prove the relation "on the average" in R_T as $T \to \infty$). This was done in [**DRS**] by methods of spectral analysis on $G/G_{\mathbb{Z}}$. Eskin and McMullen [**EM**] suggested a more simple and universal approach based on the mixing property of the G-action on $G/G_{\mathbb{Z}}$.

Subsequently, Eskin, Mozes and Shah in [**EMS96**] were able to avoid the symmetry condition imposed on V. Recall that, according to Theorem 20.8 (proved by them), all limit probability measures for the family $\{g\mu_H,\ g \in G\}$ are algebraic. Moreover, if the identity component H_0 is a maximal connected algebraic $\mathbb{Q}$-subgroup of G (and this is so under the assumptions of Theorem 31.6), then the limit measure coincides with μ_G by virtue of Corollary 20.9. Hence the same asymptotic estimate as in Theorem 31.6 holds true.

There exist two obstacles in the general case. First, the family of measures $\{g\mu_H,\ g \in G\}$ must be relatively compact (otherwise one can obtain a limit measure

[5] Actually, we treat G as a group of linear transformations of the vector space $\mathrm{Mat}(n,n)$; that is, we embed G into $\mathrm{SL}(n^2,\mathbb{R})$.

supported at infinity). By Theorem 20.7, this happens if and only if the space $\pi(Z_G(H)) \subset G/G_{\mathbb{Z}}$ is compact, where $\pi : G \to G/G_{\mathbb{Z}}$ is the natural projection (or, equivalently, H is not contained in a parabolic $\mathbb{Q}$-subgroup of G).

Second, the sets $R_T \subset G/H$ must "uniformly cover" the space G/H as $T \to \infty$. Following [**EMS96**], we say that a sequence $R_{T_n} \subset G/H$ *focuses to* L/H, where $L \supset H$ is a reductive $\mathbb{Q}$-subgroup of G, if there exists a compact set $C \subset G$ such that

$$\lim_{n\to\infty} \frac{\lambda(q(CL(Z_G(H_0) \cap G_{\mathbb{Z}})) \cap R_{T_n})}{\lambda(R_{T_n})} > 0,$$

where $q : G \to G/H$ is the natural projection. Note that $q(CL(Z_G(H_0) \cap G_{\mathbb{Z}}))$ is closed in G/H. Indeed, the product $LG_{\mathbb{Z}}$ is closed in G. Hence $L(Z_G(H_0) \cap G_{\mathbb{Z}})$ is closed in G and has its image in G/H closed.

We note that the focusing condition, though unusual, is not impossible: an example given in [**EMS96**] shows that for $G = \mathrm{SL}(2, \mathbb{C})$, $L \simeq \mathrm{SL}(2, \mathbb{R})$ and the diagonal subgroup $H \subset L$, the sequence $R_n \subset G/H$ focuses to L/H as $n \to \infty$.

Now we can formulate our main result.

THEOREM 31.10 [**EMS96**]. *Let $H \subset G$ be two reductive $\mathbb{Q}$-subgroups of the group $\mathrm{SL}(n, \mathbb{R})$, where $H_{\mathbb{Z}}$ and $G_{\mathbb{Z}}$ are lattices, and let $V = Gv_0 \simeq G/H$ for some $v_0 \in \mathbb{Z}^n$. Assume that H_0 is not contained in a parabolic $\mathbb{Q}$-subgroup of G and that no sequence $R_{T_n} = V \cap D_{T_n} \subset G/H$ focuses. Then $N(T, V, \mathcal{O}) \sim \lambda(R_T)$.*

The paper [**EMS96**] provides two examples of application of Theorem 31.10.

Example 31.11. Take a polynomial $p(t) = t^n + a_1 t^{n-1} + \cdots + a_n$ with integer coefficients irreducible over $\mathbb{Q}$. Note that

$$V = \{A \in \mathrm{Mat}(n, n) : \det(tE - A) = p(t)\}$$

is a closed G-orbit of the matrix

$$v_0 = \begin{pmatrix} 0 & \dots & 0 & -a_n \\ 1 & \dots & 0 & -a_{n-1} \\ \dots & \dots & \dots & \dots \\ 0 & \dots & 1 & -a_1 \end{pmatrix}$$

under the action by conjugations of the group G of $n \times n$ matrices with determinant ± 1. The centralizer H of the element v_0 is a maximal $\mathbb{Q}$-torus of dimension $n-1$ with no nontrivial $\mathbb{Q}$-characters. Since any maximal $\mathbb{Q}$-torus in a proper parabolic $\mathbb{Q}$-subgroup admits a nontrivial $\mathbb{Q}$-character, H_0 is not contained in such a subgroup. After checking the nonfocusing condition and applying Theorem 31.10, the following estimate is proved in [**EMS96**]

$$N(T, V) \sim c_p T^{n(n-1)/2},$$

where c_p is an explicitly computable constant. It is worth noting that H is not the subgroup of fixed points of an involution, and hence Theorem 31.6 cannot be applied in this case.

Example 31.12. Let A be a nondegenerate indefinite quadratic form with integer coefficients in $n \geq 3$ variables of signature (p, q), where $p \geq q \geq 1$ and $n = p + q$. In addition, let B be a definite quadratic form with integer coefficients in $m \leq p$ variables. We consider the $\mathbb{Q}$-manifold

$$V = \{X \in \mathrm{Mat}(m, n) : XAX^t = B\}.$$

Assume that the set $V_{\mathbb{Z}}$ is nonempty. The group $G = \mathrm{SO}(A)$ acts on $\mathrm{Mat}(m,n)$ by multiplication on the right, and the restriction of the action onto $V_{A,B}$ is transitive. The stabilizer H of a point $\xi \in V_{\mathbb{Z}}$ is isomorphic to $\mathrm{SO}(J_0)$, where J_0 is a diagonal matrix with $n-m$ elements ± 1, and the centralizer $Z_G(H_0)$ is isomorphic to $\mathrm{SO}(m)$ and hence compact. After checking the nonfocusing condition and computing the volume involved, the following asymptotics is derived in [**EMS96**]:

$$N(T,V) \sim cT^{r(n-r-1)}, \qquad \text{where } r = \min(m,q) \geq 1.$$

Clearly, in the case $r = m = 1$ we obtain precisely the asymptotics from Example 31.7.

References

[AB68] L. Auslander and J. Brezin, *Almost algebraic Lie algebras*, J. Algebra **8** (1968), 295–313.

[AB71] ———, *Uniform distribution in solvmanifolds*, Adv. Math. **7** (1971), 111–144.

[AG] L. Auslander and L. Green, *G-induced flows*, Amer. J. Math. **88** (1966), 43–60.

[AGH] L. Auslander, L. Green, and F. Hahn, *Flows on homogeneous spaces*, Princeton Univ. Press, Princeton, NJ, 1963.

[Al] V. M. Alekseev, *Symbolic dynamics*, Eleventh Math. Summer School (Kolomyya, 1973), Izdanie Inst. Mat. Akad. Nauk Ukrain. SSR, Kiev, 1976, pp. 5–210. (Russian)

[An67] D. V. Anosov, *Geodesic flows on closed Riemannian manifolds of negative curvature*, Trudy Mat. Inst. Steklov. **90** (1967); English transl., Proc. Steklov Inst. Math. **90** (1969).

[An73] ———, *The additive functional homology equation that is connected with an ergodic rotation of the circle*, Izv. Akad. Nauk SSSR Ser. Mat. **37** (1973), 1259–1274; English transl., Math. USSR–Izv. **7** (1973), 1257–1271.

[Ar] V. I. Arnold, *Small denominators.* I: *Mapping the circle onto itself*, Izv. Akad. Nauk SSSR Ser. Mat. **25** (1961), 21–86; English transl., Amer. Math. Soc. Transl. (2) **46** (1965), 213–284.

[AS] D. V. Anosov and V. V. Solodov, *Hyperbolic manifolds*, Itogi Nauki i Tekhn.: Sovremennye Problemy Matematiki: Fundamental′nye Napravleniya, vol. 66, VINITI, Moscow, 1991, pp. 12–99; English transl. in Encyclopaedia of Math. Sci., vol. 66 (Dynamical Systems, IX), Springer–Verlag, Berlin, 1995.

[Au] L. Auslander, *An exposition of the structure of solvmanifolds*, Bull. Amer. Math. Soc. **79** (1973), 227–285.

[B] A. Borel, *Values of indefinite quadratic forms at integral points and flows on spaces of lattices*, Bull. (N. S.) Amer. Math. Soc. **32** (1995), 184–204.

[BB] F. Bien and A. Borel, *Sous-groupes epimorphiques de groupes algébriques linéaires. I*, C. R. Acad. Sci. Paris Sér. I Math. **315** (1992), 649–655.

[Be] D. Berend, *Minimal sets on tori*, Ergodic Theory Dynam. Systems **4** (1984), 499–507.

[BeB] V. Beresnevich and V. Bernik, *On a metrical theorem of W. Schmidt*, Acta Arith. **75** (1996), 219–233.

[Ben] D. Benardete, *Topological equivalence of flows on homogeneous spaces, and divergence of one-parameter subgroups of Lie groups*, Trans. Amer. Math. Soc. **306** (1988), 499–527.

[BG] V. Bentkus and F. Gotze, *Lattice point problems and distribution of values of quadratic forms*, Ann. of Math. (to appear).

[BG97] ———, *On the lattice point problem for ellipsoids*, Acta Arith. **80** (1997), 101–125.

[Bi] B. J. Birch, *Forms in many variables*, Proc. Roy. Soc. London Ser. A **265** (1962), 245–263.

[BM] J. Brezin and C. C. Moore, *Flows on homogeneous spaces: a new look*, Amer. J. Math. **103** (1981), 571–613.

[Bo71] R. Bowen, *Entropy for group endomorphisms and homogeneous spaces*, Trans. Amer. Math. Soc. **153** (1971), 401–414.

[Bo75] ———, *Equilibrium states and the ergodic theory of Anosov diffeomorphisms*, Springer–Verlag, Berlin, 1975.

[Bo76] ———, *Weak mixing and unique ergodicity on homogeneous spaces*, Israel J. Math. **23** (1976), 267–273.

[Bo79] ———, *Methods of symbolic dynamics*, "Mir", Moscow, 1979 (Russian); see also *Symbolic dynamics for hyperbolic flows*, Amer. J. Math. **95** (1973), 429–460.

[BP] A. Borel and G. Prasad, *Values of isotropic quadratic forms at S-integral points*, Compositio Math. **83** (1992), 347–372.

[BR] M. Borovoi and Z. Rudnick, *Hardy-Littlewood varieties and semisimple groups*, Invent. Math. **119** (1995), 37–66.

[Bre] T. Brennan, *Distribution of values of diagonal quadratic at integer points*, Undergraduate Thesis, Princeton Univ., 1994.

[BS] M. Burger and P. Sarnak, *Ramanujan duals.* II, Invent. Math. **106** (1991), 1–11.

[BSh] Z. I. Borevich and I. R. Shafarevich, *Number theory*, "Nauka", Moscow, 1964; English transl., Academic Press, New York, 1966.

[Bu] M. Burger, *Horocycle flow on geometrically finite surfaces*, Duke Math. J. **61** (1990), 779–803.

[Ca57] J. W. S. Cassels, *An introduction to Diophantine approximation*, Cambridge Univ. Press, Cambridge, 1957.

[Ca59] ———, *An introduction to the geometry of numbers*, Springer–Verlag, Berlin, 1959.

[Ca78] ———, *Rational quadratic forms*, Academic Press, New York, 1978.

[Cow] M. Cowling, *Sur les coefficients des représentations unitaries des groupes de Lie simples*, Analyse Harmonique sur les Groupes de Lie (Sém., Nancy/Strasbourg, 1976–1978), Lecture Notes in Math., vol. 739, Springer–Verlag, Berlin, 1979, pp. 132–178.

[Cr] C. Croke, *Rigidity for surfaces of nonpositive curvature*, Comment. Math. Helv. **65** (1990), 150–169.

[CS] J. W. S. Cassels and H. P. F. Swinnerton-Dyer, *On the product of three homogeneous forms and indefinite ternary quadratic forms*, Philos. Trans. Roy. Soc. London Ser. A **248** (1955), 73–96.

[Da76a] S. G. Dani, *Bernoullian translations and minimal horospheres on homogeneous spaces*, J. Indian Math. Soc. **40** (1976), 245–284.

[Da76b] ———, *Kolmogorov automorphisms on homogeneous spaces*, Amer. J. Math. **98** (1976), 119–163.

[Da77] ———, *Spectrum of an affine transformations*, Duke Math. J. **44** (1977), 125–156.

[Da80a] ———, *A simple proof of Borel's density theorem*, Math. Z. **174** (1980), 81–94.

[Da80b] ———, *Strictly nonergodic actions on homogeneous spaces*, Duke Math. J. **47** (1980), 633–639.

[Da81] ———, *Invariant measures and minimal sets of horospherical flows*, Invent. Math. **64** (1981), 357–385.

[Da82] ———, *On uniformly distributed orbits of certain horocycle flows*, Ergodic Theory Dynam. Systems **2** (1982), 139–158.

[Da84] ———, *On orbits of unipotent flows on homogeneous spaces*, Ergodic Theory Dynam. Systems **4** (1984), 25–34.

[Da85] ———, *Divergent trajectories of flows on homogeneous spaces*, J. Reine Angew Math. **359** (1985), 55–89.

[Da86a] ———, *Orbits of horospherical flows*, Duke Math. J. **53** (1986), 177–188.

[Da86b] ———, *On orbits of unipotent flows on homogeneous spaces.* II, Ergodic Theory Dynam. Systems **6** (1986), 167–182.

[Da86c] ———, *Bounded orbits of flows on homogeneous spaces*, Comment. Math. Helv. **61** (1986), 636–660.

[Da88] ———, *On orbits of endomorphisms of tori and the Schmidt game*, Ergodic Theory Dynam. Systems **8** (1988), 523–529.

[Da89] ———, *Dense orbits of horospherical flows*, Dynamical Syatems and Ergodic Theory (Warsaw, 1986), Banach Center Publ., vol. 23, PWN, Warsaw, 1989, pp. 173–195.

[Da96] ———, *Flows on homogeneous spaces: a review*, Ergodic Theory of $\mathbb{Z}_d$-actions (Warwick, 1993–1994), London Math. Soc. Lecture Note Ser., vol. 228, Cambridge Univ. Press, Cambridge, 1996, pp. 63–112.

[Da99] ———, *Dynamical systems on homogeneous spaces*, Encyclopaedia of Math. Sci., vol. 100, Springer–Verlag, Berlin, 1999, pp. 63–112.

[DM89] S. G. Dani and G. A. Margulis, *Values of quadratic forms at primitive integral points*, Invent. Math. **98** (1989), 405–424.

[DM90a] ———, *Orbit closures of generic unipotent flows on homogeneous spaces of* $\mathrm{SL}(3,\mathbb{R})$, Math. Ann. **286** (1990), 101–128.

[DM90b] ———, *Values of quadratic forms at integral points: an elementary approach*, Enseignement Math. **36** (1990), 143–174.

[DM93] ———, *Limit distributions of orbits of unipotent flows and values of quadratic forms*, I. M. Gelfand Seminar, Advances in Soviet Math., vol. 16, part 1, Amer. Math. Soc., Providence, RI, 1993, pp. 91–137.

[Dod] M. M. Dodson, *Geometric and probabilistic ideas in metric Diophantine approximation*, Uspekhi Mat. Nauk **48** (1993), no. 5, 77–106; English transl., Russian Math. Surveys **48** (1993), no. 5, 73–102.

[DR] H. Davenport and H. Ridout, *Indefinite quadratic forms*, Proc. London Math. Soc. (3) **9** (1959), 544–555.

[DRS] W. Duke, Z. Rudnick, and P. Sarnak, *Density of integer points on affine homogeneous varieties*, Duke Math. J. **71** (1993), 143–180.

[DS] S. G. Dani and J. Smillie, *Uniform distribution of horocycle flows for Fuchsian groups*, Duke Math. J. **51** (1984), 185–194.

[EK] A. Eskin and J. Katznelson, *Singular symmetric matrices*, Duke Math. J. **79** (1995), 515–547.

[EM] A. Eskin and C. McMullen, *Mixing, counting and equidistribution in Lie groups*, Duke Math. J. **71** (1993), 181–209.

[EMM] A. Eskin, G. A. Margulis, and S. Mozes, *Upper bounds and asymptotics in a quantitative version of the Oppenheim conjecture*, Ann. of Math. (2) **147** (1998), 93–141.

[EMS96] A. Eskin, S. Mozes, and N. Shah, *Unipotent flows and counting lattice points on homogeneous varieties*, Ann. of Math. (2) **143** (1996), 253–299.

[EMS97] ———, *Non-divergence of translates of certain algebraic measures*, Geom. Funct. Anal. **7** (1997), 48–80.

[EP] R. Ellis and W. Perrizo, *Unique ergodicity of flows on homogeneous spaces*, Israel J. Math. **29** (1978), 276–284.

[F] L. Flaminio, *An extension of Ratner's rigidity theorem to n-dimensional hyperbolic space*, Ergodic Theory Dynam. Systems **7** (1987), 73–92.

[Fed] H. Federer, *Geometric measure theory*, Springer–Verlag, Berlin, 1969.

[FO] J. Feldman and D. Ornstein, *Semi-rigidity of horocycle flows over compact surfaces of variable negative curvature*, Ergodic Theory Dynam. Systems **7** (1987), 49–72.

[FS] L. Flaminio and R. Spatzier, *Geometrically finite groups, Patterson-Sullivan measures and Ratner's rigidity theorem*, Invent. Math. **99** (1990), 601–626.

[Fu61] H. Furstenberg, *Strict ergodicity and transformations of the torus*, Amer. J. Math. **83** (1961), 573–601.

[Fu63] ———, *The structure of distal flows*, Amer. J. Math. **85** (1963), 477–515.

[Fu72] ———, *The unique ergodicity of the horocycle flow*, Recent Advances in Topological Dynamics (A. Besk, ed.), Springer–Verlag, Berlin, 1972, pp. 95–115.

[G] L. Green, *Spectra of nilflows*, Bull. Amer. Math. Soc. **67** (1961), 414–415.

[GF] I. M. Gelfand and S. V. Fomin, *Geodesic flows on manifolds of constant negative curvature*, Uspekhi Mat. Nauk **7** (1952), no. 1, 118–137; English transl., Amer. Math. Soc. Transl. (2) **1** (1955), 49–65.

[GGP] I. M. Gelfand, M. I. Graev, and I. I. Piatetskii-Shapiro, *Generalized functions.* Vol. 6: *Representation theory and automorphic functions*, "Nauka", Moscow, 1966; English transl., W. B. Saunders, Philadelphia, PA, 1969.

[GH] W. Gottschalk and G. Hedlund, *Topological dynamics*, Colloquium Publ., vol. 38, Amer. Math. Soc., Providence, RI, 1955.

[Gh] E. Ghys, *Dynamique des flots unipotents sur les espaces homogènes*, Astérisque **206** (1992), 93–136.

[GNS] A. I. Galochkin, Yu. V. Nesterenko, and A. B. Shidlovskii, *Introduction to number theory*, Izdat. Moskov. Gos. Univ., Moscow, 1984. (Russian)

[Go] V. V. Gorbatsevich, *Splitting of Lie groups and its applications to the study of homogeneous spaces*, Izv. Akad. Nauk SSSR Ser. Mat. **43** (1979), 1127–1158; English transl., Math. USSR-Izv. **15** (1979), 441–467.

[GR] H. Garland and M. S. Raghunathan, *Fundamental domains for lattices in* $\mathbb{R}$*-rank 1 semisimple Lie groups*, Ann. of Math. (2) **92** (1970), 279–326.

[Gro] A. V. Groshev, *Une théorème sur les systèmes des formes linéaires*, C. R. (Dokl.) Acad. Sci. URSS **19** (1938), 151–152.

[Gu] B. M. Gurevich, *The entropy of a horocycle flow*, Dokl. Akad. Nauk SSSR **136** (1961), 768–770; English transl., Soviet Math. Dokl. **2** (1961), 124–126.

[GW] ———, *Minimal transformations with no common factors need not be disjoint*, Israel J. Math. **45** (1983), 1–8.

[H] M. Herman, *Construction of a minimal diffeomorphism with nonzero topological entropy*, Ergodic Theory Dynam. Systems **1** (1981), 65–76.

[He36] G. A. Hedlund, *Fuchsian groups and transitive horocycles*, Duke Math. J. **2** (1936), 530–542.

[He39] ———, *The dynamics of geodesic flows*, Bull. Amer. Math. Soc. **45** (1939), 241–260.

[HL] G. H. Hardy and J. E. Littlewood, *Some problems of Diophantine approximation: the lattice points of a right-angled triangle*, Proc. London Math. Soc. (2) **20** (1922), 15–36.

[HM] G. Hochschild and G. D. Mostow, *Representations and representative functions of Lie groups*, Ann. of Math. (2) **66** (1957), 495–542.

[Ho36] E. Hopf, *Fuchsian groups and ergodic theory*, Trans. Amer. Math. Soc. **39** (1936), 299–314.

[How] R. Howe, *A notion of rank for unitary representations of the classical groups*, Harmonic Analysis and Group Representations, Liguori, Naples, 1982, pp. 223–331.

[Hu] J. E. Humphreys, *Linear algebraic groups*, Springer–Verlag, Berlin, 1975.

[Ja] N. Jacobson, *Lie algebras*, Interscience, New York, 1962.

[Jo] R. Johnson, *On almost periodic linear differential systems of Millionshchikov and Vinograd*, J. Math. Anal. **85** (1982), 452–461.

[K75] A. B. Katok, *Time change, monotone equivalence and standard dynamical systems*, Dokl. Akad. Nauk SSSR **223** (1975), 789–792; English transl., Soviet Math. Dokl. **16** (1975), 986–990.

[K82] ———, *Entropy and closed geodesics*, Ergodic Theory Dynam. Systems **2** (1982), 339–365.

[Ka] S. A. Kalikow, *Twofold mixing implies threefold mixing for rank one transformations*, Ergodic Theory Dynam. Systems **4** (1984), 237–259.

[Kat] Y. Katznelson, *Ergodic automorphisms of T^n are Bernoulli shifts*, Israel J. Math. **10** (1971), 186–195.

[Kh23] A. Khintchine, *Ein Satz über Kettenbruche, mit arithmetischen Anwendungen*, Math. Z. **18** (1923), 289–306.

[Kh48] ———, *Regular systems of linear equations*, Izv. Akad. Nauk SSSR Ser. Mat. **12** (1948), 249–258. (Russian)

[Kh78] ———, *Continued fractions*, University of Chicago Press, Chicago, IL, 1964.

[Kir] A. A. Kirillov, *Elements of the theory of representations*, "Nauka", Moscow, 1978; English transl., Springer–Verlag, Berlin, 1976.

[KK] V. V. Kozlov and N. N. Kolesnikov, *Integrability of Hamiltonian systems*, Vestnik Moskov Univ. Ser. I Mat. Mekh. **1979**, no. 6, 88–91; English transl., Moscow Univ. Math. Bull. **34** (1979), no. 5/6, 48–51.

[Kl98a] D. Y. Kleinbock, *Nondense orbits of homogeneous flows*, Ergodic Theory Dynam. Systems **18** (1998), 373–396.

[Kl98b] ———, *Flows on homogeneous spaces and Diophantine properties of matrices*, Duke Math. J. **95** (1998), 107–124.

[KM96] D. Y. Kleinbock and G. A. Margulis, *Bounded orbits of non-quasiunipotent flows on homogeneous spaces*, Sinai's Moscow Seminar on Dynamical Systems, Amer. Math. Soc. Transl. (2) **171** (1996), 141–171.

[KM98] ———, *Flows on homogeneous spaces and Diophantine approximation on manifolds*, Ann. of Math. (2) **148** (1998), 339–360.

[KM99] ———, *Logarithm laws for flows on homogeneous spaces*, Invent. Math. (to appear).

[Ko] A. N. Kolmogorov, *On dynamical systems with integral invariant on a torus*, Dokl. Akad. Nauk SSSR **93** (1953), 763–766; English transl. in his *Selected Works*. Vol. I: *Mathematics and Mechanics*, Kluwer, Dordrecht, 1991, pp. 344–348.

[KS94] A. Katok and R. J. Spatzier, *First cohomology of Anosov actions of higher rank abelian groups and applications to rigidity*, Inst. Hautes Études Sci. Publ. Math. **79** (1994), 131–156.

[KS96] ______, *Invariant measures for higher rank hyperbolic abelian actions*, Ergodic Theory Dynam. Systems **16** (1996), 751–778.

[KSF] I. P. Kornfel′d, Ya. G. Sinai, and S. V. Fomin, *Ergodic theory*, "Nauka", Moscow, 1980; English transl. (by Cornfeld, Fomin, and Sinai), Springer–Verlag, Berlin, 1982.

[Ku] A. G. Kushnirenko, *An estimate from above of the entropy of a classical dynamical systems*, Dokl. Akad. Nauk SSSR **161** (1965), 37–38; English transl., Soviet Math. Dokl. **6** (1965), 360–362.

[KW] Y. Katznelson and B. Weiss, *When all points are recurrent/generic*, Ergodic Theory and Dynamical Systems. I (College Park, MD, 1979–80) (A. Katok, ed.), Birkhäuser, Boston, 1981, pp. 195–210.

[L] F. Ledrappier, *Un champ markovien peut être d'entropie nulle et mélangeant*, C. R. Acad. Sci. Paris Ser. A **287** (1978), 561–563.

[Le] E. Lesigne, *Théorèmes ergodiques pour une translation sur un nilvariété*, Ergodic Theory Dynam. Systems **9** (1989), 115–126.

[Li] Yu. V. Linnik, *Ergodic properties of algebraic fields*, Izdat. Leningrad. Univ., Leningrad, 1967; English transl., Springer–Verlag, Berlin, 1968.

[Lin] D. A. Lind, *Dynamical properties of quasihyperbolic toral automorphisms*, Ergodic Theory Dynam. Systems **2** (1982), 49–68.

[Liv] A. N. Livshitz, *Cohomology of dynamical systems*, Izv. Akad. Nauk SSSR Ser. Mat. **36** (1972), 1296–1320; English transl., Math. USSR–Izv. **6** (1972), 1278–1301.

[LW] E. Lindenstrauss and B. Weiss, *On sets invariant under the action of the diagonal group*, Preprint (1999).

[LY] F. Ledrappier and L.-S. Young, *The metric theory of diffeomorphisms*, Ann. of Math. (2) **122** (1985), 509–539.

[M] A. I. Mal′tsev, *On solvable Lie algebras*, Izv. Akad. Nauk SSSR Ser. Mat. **9** (1945), 329–352; English transl., Amer. Math. Soc. Transl. (1) **9** (1962), 228–262.

[Ma70] G. A. Margulis, *On some measures related to U-flows on compact manifolds*, Funktsional. Anal. i Prilozhen. **4** (1970), no. 1, 62–76; English transl. in Functional Anal. Appl. **4** (1970).

[Ma71] ______, *On the action of unipotent subgroups in a space of lattices*, Mat. Sb. **86** (1971), 552–556; English transl., Math. USSR–Sb. **15** (1972), 549–554.

[Ma87] ______, *Formes quadratiques indéfinies et flots unipotents sur les espaces homogènes*, C. R. Acad. Sci. Paris Sér. I Math. **304** (1987), 249–253.

[Ma88] ______, *Lie groups and ergodic theory*, Algebra–Some Current Trends (Varna, 1986), Springer–Verlag, Berlin, 1988, pp. 130–146.

[Ma89] ______, *Discrete subgroups and ergodic theory*, Proc. Conf. in Honour of A. Selberg (Oslo, 1987), Academic Press, New York, 1989, pp. 377–398.

[Ma90] ______, *Orbits of group actions and values of quadratic forms at integral points*, Festschrift in Honour of I. I. Piatetski-Shapiro (Ramat Aviv, 1989), Israel Math. Conf. Proc., vol. 3, Weizmann Science Press, Jerusalem, 1990, pp. 127–150.

[Ma91a] ______, *Compactness of minimal closed invariant sets of actions of unipotent groups*, Geom. Dedicata **37** (1991), 1–7.

[Ma91b] ______, *Dynamical and ergodic properties of subgroups actions on homogeneous spaces with applications to number theory*, Proc. Internat. Congress Math. (Kyoto, 1990), Math. Soc. Japan, Kyoto, 1991, pp. 193–215.

[Ma91c] ______, *Discrete subgroups of semisimple Lie groups*, Springer–Verlag, Berlin, 1991.

[Ma97] ______, *Oppenheim conjecture*, Fields Medalists Lectures, World Sci. Publ., River Edge, NJ, 1997, pp. 272–327.

[Ma99] ______, *Problems and conjectures in rigidity theory*, Mathematics: Frontiers and Perspectives, Amer. Math. Soc., Providence, RI, 2000, pp. 161–174.

[Mar75] B. Marcus, *Unique ergodicity of the horocycle flow: the variable curvature case*, Israel J. Math **21** (1975), 131–144.

[Mar78] ______, *The horocycle flow is mixing of all degrees*, Invent. Math. **46** (1978), 201–209.

[Mar83] ______, *Topological conjugacy of horocycle flows*, Amer. J. Math. **105** (1983), 623–632.

[Mau] F. Mautner, *Geodesic flows on symmetric Riemann spaces*, Ann. of Math. (2) **65** (1957), 416–431.

[MF] A. S. Mishchenko and A. T. Fomenko, *A generalized Liouville method for the integration of Hamiltonian systems*, Funktsional. Anal. i Prilozhen. **12** (1978), no. 2, 45–56; English transl., Funct. Anal. Appl. **12** (1978), 113–121.

[Mi68] V. M. Millionshchikov, *Proof of the existence of irregular systems of linear differential equations with almost periodic coefficients*, Differentsial′nye Uravneniya **4** (1968), 391–396; English transl. in Differential Equations **4** (1968).

[Mi76] ______, *A formula for the entropy of a smooth dynamical system*, Differentsial′nye Uravneniya **12** (1976), 2188–2192; English transl. in Differential Equations **12** (1977).

[Mo66] C. C. Moore, *Ergodicity of flows on homogeneous spaces*, Amer. J. Math. **88** (1966), 154–178.

[Mo80] ______, *The Mautner phenomenon for general unitary representations*, Pacific J. Math. **86** (1980), 154–169.

[Mo87] ______, *Exponential decay of correlation coefficients for geodesic flows*, Group Representations, Ergodic Theory, Operator Algebras, and Mathematical Physics (Berkeley, Calif, 1984), Math. Sci. Res. Inst. Publ., vol. 6, Springer–Verlag, Berlin, 1987, pp. 163–181.

[Mor] H. M. Morse, *Recurrent geodesics on surface of negative curvature*, Trans. Amer. Math. Soc. **22** (1921), 84–100.

[Moz92] S. Mozes, *Mixing of all orders of Lie group actions*, Invent. Math. **107** (1992), 235–241.

[Moz95a] ______, *Epimorphic subgroups and invariant measures*, Ergodic Theory Dynam. Systems **15** (1995), 1207–1210.

[Moz95b] ______, *Correction to the paper "Mixing of all orders of Lie group actions"*, Invent. Math. **119** (1995), 399.

[MS] S. Mozes and N. Shah, *On the space of ergodic invariant measures of unipotent flows*, Ergodic Theory Dynam. Systems **15** (1995), 149–159.

[MT94] G. A. Margulis and G. Tomanov, *Invariant measures for actions of unipotent groups over local fields on homogeneous spaces*, Invent. Math. **116** (1994), 347–392.

[MT96] ______, *Measure rigidity for almost linear groups and its applications*, J. Analyse Math. **69** (1996), 25–54.

[MW] S. Mozes and B. Weiss, *Minimality and unique ergodicity for subgroup actions*, Ann. Inst. Fourier (Grenoble) **48** (1998), 1533–1541.

[N] N. N. Nekhoroshev, *Action–angle variables and their generalizations*, Trudy Moskov. Mat. Obshch. **26** (1972), 181–198; English transl., Trans. Moscow Math Soc. **26** (1974), 180–198.

[Ne] V. V. Nemytskii, *Topological questions of the theory of dynamical systems*, Uspekhi Mat. Nauk **4** (1949), no. 6, 91–153; English transl, Amer. Math. Soc. Transl. (1) **5** (1962), 414–497.

[Ni] P. J. Nicholls, *The ergodic theory of discrete groups*, London Math. Soc. Lecture Note Ser., vol. 148, Cambridge Univ. Press, Cambridge, 1989.

[O] D. S. Ornstein, *Ergodic theory, randomness, and dynamical systems*, Yale Univ. Press, New Haven, CT, 1994.

[Op] A. Oppenheim, *Values of quadratic forms.* I,II, Quart. J. Math. Oxford Ser. (2) **4** (1953), 54-59, 60–66.

[Ot] J.-P. Otal, *Le spectre marqué des longueurs des surfaces à courbure negative*, Ann. of Math. (2) **131** (1990), 151–162.

[OW] D. Ornstein and B. Weiss, *Geodesic flows are Bernoullian*, Israel J. Math. **14** (1973), 184–198.

[P69] W. Parry, *Ergodic properties of affine transformations and flows on nilmanifolds*, Amer. J. Math. **91** (1969), 751–777.

[P70] ______, *Spectral analysis of G-extensions of dynamical systems*, Topology **9** (1970), 217–224.

[P71] ______, *Metric classification of ergodic nilflows and unipotent affines*, Amer. J. Math. **93** (1971), 819–828.

[P72] ______, *Cocycles and velocity changes*, J. London Math. Soc. **5** (1972), 511–516.

[P73] ______, *Dynamical representation in nilmanifolds*, Compositio Math. **26** (1973), 150–174.

[Pa] O. S. Parasjuk, *Horocycle flow on surfaces of constant negative curvature*, Uspekhi Mat. Nauk **8** (1953), no. 3, 124–126. (Russian)

[Pe] Ya. B. Pesin, *Characteristic Lyapunov exponents and smooth ergodic theory*, Uspekhi Mat. Nauk **32** (1977), no. 4, 55–111; English transl., Russian Math. Surveys **32** (1977), no. 4, 55–114.

[Pi] M. S. Pinsker, *Dynamical systems with completely positive or zero entropy*, Dokl. Akad. Nauk SSSR **133** (1960), 1025–1026; English transl., Soviet Math. Dokl. **1** (1960), 937–938.

[PR] G. Prasad and M. S. Raghunathan, *Cartan subgroups and lattices in semisimple Lie groups*, Ann. of Math. (2) **96** (1972), 296–317.

[PS] C. Pugh and M. Shub, *Ergodic elements of ergodic actions*, Compositio Math. **23** (1971), 115–122.

[R78] M. Ratner, *Horocycle flows are loosely Bernoullian*, Israel J. Math. **31** (1978), 122–132.

[R79] ———, *The Cartesian square of the horocycle flow is not loosely Bernoullian*, Israel J. Math. **34** (1978), 72–96.

[R82a] ———, *Rigidity of horocycle flows*, Ann. of Math. (2) **115** (1982), 597–614.

[R82b] ———, *Factors of horocycle flows*, Ergodic Theory Dynam. Systems **2** (1982), 465–489.

[R83] ———, *Horocycle flows, joinings and rigidity of products*, Ann. of Math. (2) **118** (1983), 277–313.

[R84] ———, *Ergodic theory in hyperbolic space*, Conf. Modern Anal. Probab., Contemp. Math., vol. 26, Amer. Math. Soc., Providence, RI, 1984, pp. 309–334.

[R86] ———, *Rigidity of time changes for horocycle flows*, Acta Math. **156** (1986), 1–32.

[R87a] ———, *The rate of mixing for geodesic and horocycle flows*, Ergodic Theory Dynam. Systems **7** (1987), 267–288.

[R87b] ———, *Rigid reparametrizations and cohomology for horocycle flows*, Invent. Math. **88** (1987), 341–374.

[R90a] ———, *Invariant measures for unipotent translations on homogeneous spaces*, Proc. Nat. Acad. Sci. USA **87** (1990), 4309–4311.

[R90b] ———, *Strict measure rigidity for unipotent subgroups of solvable groups*, Invent. Math. **101** (1990), 449–482.

[R90c] ———, *On measure rigidity of unipotent subgroups of semisimple groups*, Acta Math. **165** (1990), 229–309.

[R91a] ———, *On Raghunathan's measure conjecture*, Ann. of Math. (2) **134** (1991), 545–607.

[R91b] ———, *Raghunathan's topological conjecture and distributions of unipotent flows*, Duke Math. J. **63** (1991), 235–280.

[R92] ———, *Raghunathan's conjecture for* $\mathrm{SL}(2,\mathbb{R})$, Israel J. Math. **80** (1992), 1–31.

[R93] ———, *Raghunathan's conjectures for p-adic Lie groups*, Internat. Math. Res. Notices **1993**, 141–146.

[R94] ———, *Invariant measures and orbit closures for unipotent actions on homogeneous spaces*, Geom. and Funct. Anal. **4** (1994), 236–256.

[R95a] ———, *Raghunathan's conjectures for cartesian products of real and p-adic Lie groups*, Duke Math. J. **77** (1995), 275–382.

[R95b] ———, *Interactions between ergodic theory, Lie groups, and number theory*, Proc. Internat. Congress Math. (Zürich, 1994), Birkhäuser, Basel, 1995, pp. 157–182.

[R98] ———, *On the p-adic and S-arithmetic generalizations of Raghunathan's conjectures*, Proc. Internat. Colloq. Lie Groups and Ergodic Theory (Mumbai 1996), Narosa, New Delhi, 1998, pp. 167–202.

[Rag] M. Raghunathan, *Discrete subgroups of Lie groups*, Springer–Verlag, Berlin, 1972.

[Re] R. Remage, *On minimal sets in the plane*, Proc. Amer. Math. Soc. **13** (1962), 41–47.

[Ree] M. Rees, *Some* $\mathbb{R}^2$*-Anosov flows*, Univ. of Minnesota Math. Report 82–123.

[Ro] V. A. Rokhlin, *On a decomposition of dynamical systems into transitive components*, Mat. Sb. **25** (1949), 235–249. (Eussian)

[Ry91] V. V. Ryzhikov, *On a connection between the mixing properties of a flow with an isomorphism entering into its translations*, Mat. Zametki **49** (1991), no. 6, 98–106; English transl. in Math. Notes **49** (1991).

[Ry93] ———, *Joinings and multiple mixing of actions of finite rank*, Funktsional. Anal. i Prilozhen. **27** (1993), no. 2, 63–78; English transl. in Funct. Anal. Appl. **27** (1993).

[Ry97] ———, *Polymorphisms, joinings and tensor simplicity of dynamical systems*, Funktsional. Anal. i Prilozhen. **31** (1997), no. 2, 45–57; English transl., Funct. Anal. Appl. **31** (1977), 42–57.

[Sa81] P. Sarnak, *Asymptotic behavior of periodic orbits of the horocycle flow and Eisenstein series*, Comm. Pure Appl. Math. **34** (1981), 719–739.

[Sa91] ______, *Diophantine problems and linear groups*, Proc. Internat. Congress Math. (Kyoto, 1990), Math. Soc. Japan, Kyoto, 1991, pp. 459–471.

[Saf] A. V. Safonov, *On the spectral type of ergodic G-induced flows*, Izv. Vyssh. Uchebn. Zaved. Mat. **1982**, no. 6, 42–47; English transl., Soviet Math. (Izv. VUZ) **26** (1982), no. 6, 47–54.

[Sc] K. Schmidt, *Dynamical systems of algebraic origin*, Birkhäuser, Basel, 1995.

[Sch69] W. M. Schmidt, *Badly approximable systems of linear forms*, J. Number Theory **1** (1969), 139–154.

[Sch80] ______, *Diophantine approximation*, Lecture Notes in Math., vol. 785, Springer–Verlag, Berlin, 1980.

[Sch85] ______, *The density of integer points on homogeneous varieties*, Acta Math. **154** (1985), 243–296.

[Sco] E. J. Scourfield, *The divisors of a quadratic polynomial*, Proc. Glasgow Math. Assoc. **5** (1961), 8–20.

[Se73] J.-P. Serre, *A course in arithmetic*, Springer–Verlag, Berlin, 1973.

[Sh91] N. Shah, *Uniformly distributed orbits of certain flows on homogeneous spaces*, Math. Ann. **289** (1991), 315–334.

[Sh94] ______, *Limit distributions of polynomial trajectories on homogeneous spaces*, Duke Math. J. **75** (1994), 711–732.

[Sh96] ______, *Limit distributions of expanding translates of certain orbits on homogeneous spaces*, Proc. Indian Acad. Sci. Math. Sci. **106** (1996), 105–125.

[Sh98] ______, *Invariant measures and orbit closures on homogeneous spaces for actions of subgroups generated by unipotent elements*, Proc. Internat. Colloq. Lie Groups and Ergodic Theory (Mumbai, 1996), Narosa, New Delhi, 1998, pp. 229–271.

[Si60] Ya. G. Sinai, *Geodesic flows on manifolds of negative curvature*, Dokl. Akad. Nauk SSSR **131** (1960), 752–755; English transl., Soviet Math. Dokl. **1** (1960), 335–339.

[Si66] ______, *Dynamical system with Lebesgue spectrum of countable multiplicity*. II, Izv. Akad. Nauk SSSR Ser. Mat. **30** (1966), 15-68; English transl., Amer. Math. Soc. Transl. (2) **68** (1968), 54–88.

[Si68] ______, *Construction of Markov partitions*, Funktsional. Anal. i Prilozhen. **2** (1968), no. 3, 70–80; English transl. in Functional Anal. Appl. **2** (1968).

[Si72] ______, *Gibbs measures in ergodic theory*, Uspekhi Mat. Nauk **27** (1972), no. 4, 21–63; English transl., Russian Math. Surveys **27** (1972), no. 4, 21–69.

[Si96] ______, *Lectures in ergodic theory*, Phasis, Moscow, 1996. (Russian)

[Skl] M. D. Šklover, *Classical dynamical systems on the torus with continuous spectrum*, Izv. Vyss. Ucebn. Zaved., Matematika **1967**, no. 10, 113–124. (Russian)

[Skr89] M. M. Skriganov, *Lattices in algebraic number fields and uniform distribution modulo 1*, Algebra i Analiz **1** (1989), no. 2, 207–228; English transl., Leningrad Math. J. **1** (1990), 535–558.

[Skr93] ______, *On integer points in polygons*, Ann. Inst. Fourier (Grenoble) **43** (1993), no. 2, 313–323.

[Skr98] ______, *Ergodic theory on* SL(n)*, Diophantine approximations and anomalies in the lattice point problem*, Invent. Math. **132** (1998), 1–72.

[Sp77] V .G. Sprindzhuk, *Metric theory of Diophantine approximation*, "Nauka", Moscow, 1977; English transl., Wiley, New York, 1979.

[Sp80] ______, *Achievements and problems in Diophantine approximation theory*, Uspekhi Mat. Nauk **35** (1980), no. 4, 3–68; English transl., Russian Math. Surveys **35** (1980), no. 4, 1–80.

[Spen] D. Spencer, *The lattice points of tetrahedra*, J. Math. and Phys. Mass. Inst. Tech. **21** (1942), 189–197.

[SS] M. M. Skriganov and A. N. Starkov, *A remark on logarithmically small errors in the lattice point problem*, Preprint, POMI, St. Petersburg, 1998.

[SSS] A. V. Safonov, A. N. Starkov, and A. M. Stepin, *Dynamical systems with transitive symmetry group*, Itogi Nauki i Tekhn.: Sovremennye Problemy Mat.: Fundamental′nye

Napravleniya, vol. 66, VINITI, Moscow, 1991, pp. 187–243; English transl. in Encyclopaedia of Math. Sci., vol. 66 (Dynamical Systems, IX), Springer–Verlag, Berlin, 1995.

[St83] A. N. Starkov, *Ergodic behavior of flows on homogeneous spaces*, Dokl. Akad. Nauk SSSR **273** (1983), 538–540; English transl., Soviet Math. Dokl. **28** (1983), 675–676.

[St84a] ______, *Flows on compact solvmanifolds*, Mat. Sb. **123** (1984), 549–556; English transl., Math. USSR-Sb. **51** (1985), 549–556.

[St84b] ______, *A counterexample to a theorem on lattices in Lie groups*, Vestnik Moskov. Univ. Ser. I Mat. Mekh. **1984**, no. 5, 68-69; English transl., Moscow Univ. Math. Bull. **39** (1984), no. 5, 65–67.

[St86a] ______, *On spaces of finite volume*, Vestnik Moskov. Univ. Ser. I Mat. Mekh. **1986**, no. 5, 64–66; English transl., Moscow Univ. Math. Bull. **41** (1986), no. 5, 56–58.

[St86b] ______, *Nonergodic homogeneous flow*, Dokl. Akad. Nauk SSSR **288** (1986), 560–562; English transl., Soviet Math. Dokl. **33** (1986), 729–731.

[St87a] ______, *On a criterion for the ergodicity of G-induced flows*, Uspekhi Mat. Nauk **42** (1987), no. 3, 197–198; English transl., Russian Math. Surveys **42** (1987), no. 3, 233–234.

[St87b] ______, *Solvable homogeneous flows*, Mat. Sb. **134** (1987), 242–259; English transl., Math. USSR–Sb. **62** (1989), 243–260.

[St88] ______, *Reduction of the theory of homogeneous flows to the case of a discrete isotropy subgroup*, Dokl. Akad. Nauk SSSR **301** (1988), 1328–1331; English transl., Soviet Math. Dokl. **38** (1989), 232–236.

[St89] ______, *Ergodic decomposition of flows on homogeneous spaces of finite volume*, Mat. Sb. **180** (1989), 1614–1633; English transl., Math. USSR–Sb. **68** (1991), 483–502.

[St90] ______, *Structure of orbits of homogeneous flows and the Raghunathan conjecture*, Uspekhi Mat. Nauk **45** (1990), no. 2, 219–220; English transl., Russian Math. Surveys **45** (1990), no. 2, 227–228.

[St91] ______, *Horospherical flows on homogeneous finite-volume spaces*, Mat. Sb. **182** (1991), 774–784; English transl., Math. USSR-Sb. **73** (1992), 161–170.

[St93] ______, *Multiple mixing of homogeneous flows*, Dokl. Ross. Akad. Nauk **333** (1993), 442–445; English transl., Russian Acad. Sci. Dokl. Math. **48** (1994), 573–578.

[St94] ______, *Rigidity of lattices in solvable Lie groups*, Proc. Indian Acad. Sci. Math. Sci. **104** (1994), 495–514.

[St95a] ______, *Minimal sets of homogeneous flows*, Ergodic Theory Dynam. Systems **15** (1995), 361–377.

[St95b] ______, *Fuchsian groups from the dynamical viewpoint*, J. Dynam. Contr. Syst. **3** (1995), 427–445.

[St95c] ______, *On mutual isomorphisms of translations of a homogeneous flow*, Mat. Zametki **58** (1995), 98–110; English transl., Math. Notes **58** (1995), 744–751.

[St97] ______, *New progress in the theory of homogeneous flows*, Uspekhi Mat. Nauk **52** (1997), no. 4, 87–192; English transl., Russian Math. Surveys **52** (1997), 721–818.

[St98] ______, *The first cohomology group, mixing, and minimal sets of the commutative group of algebraic actions on a torus*, J. Math. Sci. (New York) **95** (1999), 2576–2582.

[St99a] ______, *On Anosov infra-homogeneous actions*, Preprint, 1999.

[St99b] ______, *Minimality and unique ergodicity of homogeneous actions*, Matem. Zametki **66** (1999), 293–301; English transl. in Math. Notes **66** (1999).

[Ste69] A.M. Stepin, *Flows on solvable manifolds*, Uspekhi Mat. Nauk **24** (1969), no. 2, 241–242. (Russian)

[Ste73] ______, *Dynamical systems on homogeneous spaces of semisimple Lie groups*, Izv. Akad. Nauk SSSR Ser. Mat. **37** (1973), 1091–1107; English transl., Math. USSR–Izv. **7** (1975), 1089–1104.

[Ste82] ______, *On a homology equation in the theory of dynamical systems*, Researches in the Theory of Functions of Several Real Variables, Izdat. Yaroslav. Gos. Univ., Yaroslavl′, 1982, pp. 106–117. (Russian)

[SW] N. Shah and B. Weiss, *On actions of epimorphic subgroups on homogeneous spaces*, Ergodic Theory Dynam. Systems (to appear).

[T] G. Tomanov, *On unique ergodicity for subgroup actions on homogeneous spaces*, Preprint (1998).

[Th] J.-P. Thouvenot, *Some properties and applications of joinings in ergodic theory*, Ergodic Theory and Its Connections with Harmonic Analysis (Alexandria, 1993), London Math. Soc. Lecture Note Ser., vol. 205, Cambridge Univ. Press, Cambridge, 1995, pp. 207–235.

[Tom] P. Tomter, *On the classification of Anosov flows*, Topology **14** (1975), 179–189.

[V75] W. A. Veech, *Minimality of horospherical flows*, Israel J. Math. **21** (1975), 233–239.

[V77] ———, *Unique ergodicity of horospherical flows*, Amer. J. Math. **99** (1977), 827–859.

[V86] ———, *Periodic points and invariant pseudomeasures for toral endomorphisms*, Ergodic Theory Dynam. Systems **6** (1986), 449–473.

[Ve] A. Verjovsky, *Arithmetic, geometry and dynamics in the unit tangent bundle of the modular orbifold*, Dynamical Systems (Santiago, 1990), Pitman Res Notes Math., vol. 285, Longman Sci. Tech., Harlow, 1993, pp. 263–298.

[Ven] B. A. Venkov, *On Markov's extremal problem for ternary quadratic forms*, Izv. Akad. Nauk SSSR, Ser. Mat. **9** (1945), 429–434. (Russian)

[Ver] A. M. Veršik, *A measurable realization of continuous groups of automorphisms of a unitary ring*, Izv. Akad. Nauk SSSR Ser. Mat. **29** (1965), 127–136; English transl., Amer. Math. Soc. Transl. (2) **84** (1969), 69–81.

[VGO] E. B. Vinberg, V. V. Gorbatsevich, and A. L. Onishchik, *Structure of Lie groups and Lie algebras*, Itogi Nauki i Tekhn.: Sovremennye Problemy Mat.: Fundamental′nye Napravleniya, vol. 41, VINITI, Moscow, 1990; English transl., Encyclopaedia of Math. Sci., vol. 41 (Lie Groups and Lie Algebras, III), vol. 41, Springer–Verlag, Berlin, 1994.

[VGS] E. B. Vinberg, V. V. Gorbatsevich, and O. V. Shvartsman, *Discrete subgroups of Lie groups*, Itogi Nauki i Tekhn.: Sovremennye Problemy Mat.: Fundamental′nye Napravleniya, vol. 21, VINITI, Moscow, 1988, pp. 5–120; English transl. in Encyclopaedia of Math. Sci., vol. 21 (Lie Groups and Lie Algebras, I), Springer–Verlag, Berlin, 1993.

[Vi] R. E. Vinograd, *On a problem of N. P. Erugin*, Differentsial′nye Uravneniya **11** (1975), 632–638; English transl. in Differential Equations **11** (1975).

[Vin] I. M. Vinogradov, *Elements of number theory*, "Nauka", Moscow, 1972; English transl. of earlier ed., Dover, New York, 1954.

[VO] E. B. Vinberg and A. L. Onishchik, *Seminar on Lie groups and algebraic groups*, "Nauka", Moscow, 1988; English transl., *Lie groups and algebraic groups*, Springer–Verlag, Berlin, 1990.

[VS] E. B. Vinberg and O. V. Shvartsman, *Discrete groups of motions of spaces of constant curvature*, Itogi Nauki i Tekhn.: Sovremennye Problemy Mat.: Fundamental′nye Napravleniya, vol. 29, VINITI, Moscow, 1988, pp. 147–259; English transl. in Encyclopaedia of Math. Sci., vol. 29 (Geometry, II), Springer–Verlag, Berlin, 1993.

[W] B. Weiss, *Finite dimensional representations and subgroup actions on homogeneous spaces*, Israel J. Math. **106** (1998), 189–207.

[We] A. Weil, *L'intégration dans les groupes topologiques et ses applications*, Hermann, Paris, 1940.

[Wi85] D. Witte, *Rigidity of some translations on homogeneous spaces*, Invent. Math. **81** (1985), 1–27.

[Wi87] ———, *Zero-entropy affine maps on homogeneous spaces*, Amer. J. Math. **109** (1987), 927–961.

[Wi90] ———, *Topological equivalence of foliations of homogeneous spaces*, Trans. Amer. Math. Soc. **317** (1990), 143–166.

[Wi94] ———, *Measurable quotients of unipotent translations on homogeneous spaces*, Trans. Amer. Math. Soc. **345** (1994), 577–594.

[Wi96] ———, *Zero-entropy affine maps on homogeneous spaces: corrections*, Amer. J. Math. **118** (1996), 1137–1140.

[Win] A. Wintner, *The linear difference equation of first order for angular variables*, Duke Math. J. **12** (1945), 445–449.

[Wu] T. S. Wu, *Products of subgroups in Lie groups*, Illinois J. Math. **29** (1985), 687–695.

[Z] R. Zimmer, *Ergodic theory and semisimple groups*, Birkhäuser, Boston, 1984.

[Z87] ———, *Amenable actions and dense subgroups of Lie groups*, J. Funct. Anal. **72** (1987), 58–64.

[Za] D. Zagier, *Eisenstein series and the Riemann zeta-function*, Automorphic Forms, Representation Theory and Arithmetic (Internat. Colloq., Bombay, 1979), Tata Inst. Fund. Res. Stud. Math., vol. 10, Springer–Verlag, Berlin, 1981, pp. 275–301.

[Ze] A. Zeghib, *Ensembles invariants des flots géodésiques des variétés localement symétriques*, Ergodic Theory Dynam. Systems **15** (1995), 379–412.

Index

Selected Titles in This Series

(Continued from the front of this publication)

154 **V. V. Prasolov and A. B. Sossinsky,** Knots, links, braids and 3-manifolds. An introduction to the new invariants in low-dimensional topology, 1997

153 **S. Kh. Aranson, G. R. Belitsky, and E. V. Zhuzhoma,** Introduction to the qualitative theory of dynamical systems on surfaces, 1996

152 **R. S. Ismagilov,** Representations of infinite-dimensional groups, 1996

151 **S. Yu. Slavyanov,** Asymptotic solutions of the one-dimensional Schrödinger equation, 1996

150 **B. Ya. Levin,** Lectures on entire functions, 1996

149 **Takashi Sakai,** Riemannian geometry, 1996

148 **Vladimir I. Piterbarg,** Asymptotic methods in the theory of Gaussian processes and fields, 1996

147 **S. G. Gindikin and L. R. Volevich,** Mixed problem for partial differential equations with quasihomogeneous principal part, 1996

146 **L. Ya. Adrianova,** Introduction to linear systems of differential equations, 1995

145 **A. N. Andrianov and V. G. Zhuravlev,** Modular forms and Hecke operators, 1995

144 **O. V. Troshkin,** Nontraditional methods in mathematical hydrodynamics, 1995

143 **V. A. Malyshev and R. A. Minlos,** Linear infinite-particle operators, 1995

142 **N. V. Krylov,** Introduction to the theory of diffusion processes, 1995

141 **A. A. Davydov,** Qualitative theory of control systems, 1994

140 **Aizik I. Volpert, Vitaly A. Volpert, and Vladimir A. Volpert,** Traveling wave solutions of parabolic systems, 1994

139 **I. V. Skrypnik,** Methods for analysis of nonlinear elliptic boundary value problems, 1994

138 **Yu. P. Razmyslov,** Identities of algebras and their representations, 1994

137 **F. I. Karpelevich and A. Ya. Kreinin,** Heavy traffic limits for multiphase queues, 1994

136 **Masayoshi Miyanishi,** Algebraic geometry, 1994

135 **Masaru Takeuchi,** Modern spherical functions, 1994

134 **V. V. Prasolov,** Problems and theorems in linear algebra, 1994

133 **P. I. Naumkin and I. A. Shishmarev,** Nonlinear nonlocal equations in the theory of waves, 1994

132 **Hajime Urakawa,** Calculus of variations and harmonic maps, 1993

131 **V. V. Sharko,** Functions on manifolds: Algebraic and topological aspects, 1993

130 **V. V. Vershinin,** Cobordisms and spectral sequences, 1993

129 **Mitsuo Morimoto,** An introduction to Sato's hyperfunctions, 1993

128 **V. P. Orevkov,** Complexity of proofs and their transformations in axiomatic theories, 1993

127 **F. L. Zak,** Tangents and secants of algebraic varieties, 1993

126 **M. L. Agranovskiĭ,** Invariant function spaces on homogeneous manifolds of Lie groups and applications, 1993

125 **Masayoshi Nagata,** Theory of commutative fields, 1993

124 **Masahisa Adachi,** Embeddings and immersions, 1993

123 **M. A. Akivis and B. A. Rosenfeld,** Élie Cartan (1869–1951), 1993

122 **Zhang Guan-Hou,** Theory of entire and meromorphic functions: deficient and asymptotic values and singular directions, 1993

121 **I. B. Fesenko and S. V. Vostokov,** Local fields and their extensions: A constructive approach, 1993